Thyme

The genus *Thymus*

Medicinal and Aromatic Plants – Industrial Profiles

Individual volumes in this series provide both industry and academia with in-depth coverage of one major medicinal or aromatic plant of industrial importance.

Edited by Dr Roland Hardman

(Continued)

Thyme
The genus *Thymus*

Edited by

Elisabeth Stahl-Biskup

Institut für Pharmazie, Abteilung Pharmazeutische Biologie,
Universität Hamburg, Germany

and

Francisco Sáez

Facultad de Biología, Departamento de Biología Vegetal (Botánica),
Universidad de Murcia, Spain

CRC Press
Taylor & Francis Group
Boca Raton London New York

CRC Press is an imprint of the
Taylor & Francis Group, an **informa** business
A TAYLOR & FRANCIS BOOK

CRC Press
Taylor & Francis Group
6000 Broken Sound Parkway NW, Suite 300
Boca Raton, FL 33487-2742

First issued in paperback 2019

© 2002 by Taylor & Francis Group, LLC
CRC Press is an imprint of Taylor & Francis Group, an Informa business

Typeset in Garamond by
Integra Software Services Pvt. Ltd, Pondicherry, India

ISBN-13: 978-0-415-28488-2 (hbk)
ISBN-13: 978-0-367-39584-1 (pbk)

British Library Cataloguing in Publication Data
A catalogue record for this book is available from the British Library
Library of Congress Cataloging in Publication Data
A catalog record for this book has been requested

Visit the Taylor & Francis Web site at
http://www.taylorandfrancis.com

and the CRC Press Web site at
http://www.crcpress.com

For Rainer, Inma, Natalia, Angel and Rubén

Contents

Preface to the series

There is increasing interest in industry, academia and the health sciences in medicinal and aromatic plants. In passing from plant production to the eventual product used by the public, many sciences are involved. This series brings together information which is currently scattered through an ever increasing number of journals. Each volume gives an in-depth look at one plant genus, about which an area specialist has assembled information ranging from the production of the plant to market trends and quality control.

Many industries are involved such as forestry, agriculture, chemical food, flavour, beverage, pharmaceutical, cosmetic and fragrance. The plant raw materials are roots, rhizomes, bulbs, leaves, stems, barks, wood, flowers, fruits and seeds. These yield gums, resins, essential (volatile) oils, fixed oils, waxes, juices, extracts and spices for medicinal and aromatic purposes. All these commodities are traded worldwide. A dealer's market report for an item m1ay say 'Drought in the country of origin has forced up prices'.

Natural products do not mean safe products and account of this has to be taken by the above industries; which are subject to regulation. For example, a number of plants which are approved for use in medicine must not be used in cosmetic products.

The assessment of safe to use starts with the harvested plant material which has to comply with an official monograph. This may require absence of, or prescribed limits of, radioactive material, heavy metals, aflatoxic, pesticide residue, as well as the required level of active principle. This analytical control is costly and tends to exclude small batches of plant material. Large scale contracted mechanised cultivation with designated seed or plantlets is now preferable.

Today, plant selection is not only for the yield of active principle, but for the plant's ability to overcome disease, climatic stress and the hazards caused by mankind. Such methods as *in vitro* fertilization, meristem cultures and somatic embryogenesis are used. The transfer of sections of DNA is giving rise to controversy in the case of some end-uses of the plant material.

Some suppliers of plant raw material are now able to certify that they are supplying organically-farmed medicinal plants, herbs and spices. The European Union directive (CVO/EU No. 2092/91) details the specifications for the *obligatory* quality controls to be carried out at all stages of production and processing of organic products.

Fascinating plant folklore and ethnopharmacology leads to medicinal potential. Examples are the muscle relaxants based on the arrow poison, curare, from species of Chondrodendron, and the anti-malarials derived from species of *Cinchona* and *Artemisia*. The methods of detection of pharmacological activity have become increasingly reliable and specific, frequently involving enzymes in bioassays and avoiding the use of laboratory animals. By using bioassay linked fractionation of crude plant juices or extracts, compounds can be specifically targeted which, for example, inhibit blood platelet aggregation, or have anti-tumour, or anti-viral, or any other required activity. With the assistance of robone devices all the members of a genus may be readily screened. However, the plant material must be fully authenticated by a specialist.

The medicinal traditions of ancient civilisations such as those of China and India have a large armamenoaria of plants in their pharmacopoeias which are used throughout South-East Asia. A similar situation exists in Africa and South America. Thus, a very high percentage of the World's population relies on medicinal and aromatic plants for their medicine. Western medicine is also responding. Already in Germany all medical practitioners have to pass an examination in phytotherapy before being allowed to practise. It is noticeable that throughout Europe and the USA, medical, pharmacy and health related schools are increasingly offering training in phytotherapy.

Multinational pharmaceutical companies have become less enamoured of the single compound magic bullet cure. The high costs of such ventures and the endless competition from 'me too' compounds from rival companies often discourage the attempt Independent phytomedicine companies have been very strong in Germany. However, by the end of 1995, eleven (almost all) had been acquired by the multinational pharmaceutical firms, acknowledging the lay public's growing demand for phytomedicines in the Western World.

The business of dictary supplements in the Western World has expanded from the health store to the pharmacy. Alternative medicine includes plant-based products. Appropriate measures to ensure the quality, safety and efficacy of these either already exist or are being answered by greater legislative control by such bodies as the Food and Drug Administration of the USA and the recently created European Agency for the Evaluation of Medicinal Products, based in London.

In the USA, the Dietary Supplement and Health Education Act of 1994 recognised the class of phytotherapeutic agents derived from medicinal and aromatic plants. Furthermore, under public pressure, the US Congress set up an Office of Alternative Medicine and this office in 1994 assisted the filing of several Investigational New Drug (IND) applications, required for clinical trials of some Chinese herbal preparations. The significance of these applications was that each Chinese preparation involved several plants and yet was handled as a *single* IND. A demonstration of the contribution to efficacy, of *each* ingredient of *each* plans, was not required. This was a major step forward towards more sensible regulations in regard to phytomedicines.

My thanks are due to the staffs of Harwood Academic Publishers and Taylor & Francis who have made this series possible and especially to the volume editors and their chapter contributors for the authoritative information.

Roland Hardman

Preface

When Roland Hardman asked us to edit the book "The Genus *Thymus*" within his remarkable series "Medicinal and Aromatic Plants – Industrial Profiles" it was a big challenge for us because the amount of material concerning the genus *Thymus* has increased continuously and immensely up to the present. We knew that it would not be an easy job to compile all the results revealed by more than 2000 scientific publications. Nevertheless, from the beginning we were enthusiastically dedicated to this task – as editors and authors – always convinced that we were doing a very valuable job discussing *Thymus* under different aspects. We both spent several years with scientific studies on the botany, chemistry and systematology of *Thymus* always being aware of the urgent need of a compilatory work as a fundamental basis for further research projects. Therefore with the publication of this book not only the wish of the series editor but also our own wish for a comprehensive report on the status quo of the genus *Thymus* came true.

Everyone who has ever dealt with the genus *Thymus* knows that *Thymus vulgaris* L. deserves close attention. Its pleasant aroma and flavour as well as its potent pharmacological activities give him the predicate of one of the most popular plants widely used as flavouring agent, culinary herb, herbal medicine, and it is used in perfumes as well as being a commercial source of the monoterpene thymol. Therefore *Thymus vulgaris* L. became the central species in quite a few contributions to the book, especially in those dealing with thyme as a source of commercial products and as an herbal drug on the one hand and on the other hand in those articles discussing field culture, harvesting, post-harvest handling and processing of thyme. Nevertheless for us the presentation of the material covering the whole genus *Thymus* has been an aim of prime importance. Therefore we made every endeavour to find authors who were willing to elucidate the scientific achievements of the whole genus and discuss the problems of this species-rich plant genus, of a taxonomically complex group present in all temperate regions of the northern hemisphere. The result is a review of the history, the botany, and the taxonomy of the genus *Thymus* including several aspects of the population structure in thyme as well as the complete essential oil and flavonoid chemistry of the genus. Following our own scientific inclinations a separate chapter was dedicated to the problem of essential oil polymorphism in the genus *Thymus*.

We hope we have succeeded in presenting an informative overview of the information presently available on the genus *Thymus*: botany, taxonomy, chemistry, and pharmacology which also takes into account several applied aspects such as field culture, harvesting and processing of thyme. The readers may consider this book to be a reliable basis and feel stimulated to invest further efforts into the research of this promising genus which still shows many gaps which are worth filling during the coming decades.

Elisabeth Stahl-Biskup
Francisco Sáez

Contributors

Dr Esperanza Crespo
Departamento de Farmacología
Facultad de Farmacia
Campus de La Cartuja
Universidad de Granada
18071-Granada, Spain

Dr Brian M. Lawrence
c/o R.J. Reynolds Tobacco Company
Research and Development
Bowman Gray Technical Center
950 Reynolds Boulevard
Winston-Salem, NC 27105, USA

Dr Ramón Morales
Real Jardín Botánico de Madrid, CSIC
Plaza de Murillo, 2
28014-Madrid, Spain

Charles Rey
Station fédérale de recherches en
    production végétale de Changins
Centre des Fougères
CH-1964 Conthey, Switzerland

Dr Francisco Sáez
Departamento de Biología Vegetal
    (Botánica)
Facultad de Biología
Universidad de Murcia
30100-Espinardo, Murcia, Spain

Prof. Dr Elisabeth Stahl-Biskup
Institut für Pharmazie
Universität Hamburg
Abteilung Pharmazeutische
    Biologie and Mikrobiologie
Bundesstrasse 45
D-20146 Hamburg, Germany

Dr John D. Thompson
Centre d'Ecologie Fonctionelle
    et Evolutive
CNRS
1919 Route de Mende
34293 Montpellier Cedex 5
France

Dr Arthur O. Tucker
Department of Agriculture and Natural
    Resources
Delaware State University
Dover DE 19901-2277, USA

Dr Petras R. Venskutonis
Department of Food Technology
Kaunas University of Technology
Radvilénu pl. 19
Kaunas
LT-3028, Lithuania

Dr Roser Vila
Unitat de Farmacologia i
    Farmacognòsia
Facultat de Farmàcia
Avda. Diagonal, 643
08028 Barcelona, Spain

Dr Antonio Zarzuelo
Departamento de Farmacología
Facultad de Farmacia
Campus de La Cartuja
Universidad de Granada
18071-Granada, Spain

1 The history, botany and taxonomy of the genus *Thymus*

Ramón Morales

INTRODUCTION

Within the Labiate family, with about 220 genera, the genus *Thymus* is one of the eight most important genera with regard to the number of species included, although this number varies depending on the taxonomical point of view. If we choose criteria to minimise variability, available data report 215 species for the genus, a number only exceeded by the genera *Salvia, Hyptis, Scutellaria, Stachys, Teucrium, Nepeta,* and *Plectranthus*.

The common English word 'thyme' has traditionally been used to name both the genus and its most commercially used species, *Thymus vulgaris*, sometimes leading to misunderstandings. Generally speaking, thyme is an aromatic plant used for medicinal and spice purposes almost everywhere in the world. The genus *Thymus* is very frequent in the Mediterranean region, where some species form a special type of bushy vegetation not more than 50 cm high, well adapted to hot and dry summer weather. The Spanish name for these vegetation communities, 'tomillares', include other Labiate species such as *Sideritis, Satureja, Salvia* or *Lavandula*, with similar climatic and edaphic patterns.

A common feature of these and many other aromatic plants is the presence of countless glandular hairs of different forms which contain volatile essential oils that evaporate when the glandular hairs are damaged. This way they produce an intensive fragrance that embraces the plant. It is probably due to the strong scent that humans have always been attracted to these plants and have exploited their essential oils for popular and industrial purposes.

HISTORICAL BACKGROUND

The history of *Thymus* before Linné

Several explanations exist concerning the origin of the name 'Thymus'. Some authors assume that the Latin name *Thymus* comes from the Greek word *thyo* (perfume). Another interpretation of its etymology considers the Greek word *thymos* (courage, strength). Originally 'thymus' described a group of aromatic plants with similar aspects which were used as stimulants of vital functions. Many popular names in the Romance languages are derived from the Latin name. The same occurs with the English name.

In his work about medicinal plants and poisons, Dioscorides (First century, translation of Laguna, 1555) writes about '*Thymo*'. Laguna however did not find there any *Thymus* species, but a plant corresponding to the genus *Satureja*. On page 294 Laguna describes the *Serpol*, presenting two varieties, a cultivated and a wild one. The latter, *Zygis*, resembles a *Thymus* species. It is presented as an erect plant, whereas the former shows a creeping habit.

In his Natural History, Book 21, Chapter 10 (translation of Huerta, 1629), Plinio (First century) reports on *T. vulgaris* as follows: 'in the Narbonne province, the stony fields are full of thyme, and thousands of sheep come from very far provinces to feed on it'. Later (page 289) he speaks about two different varieties of thyme, a white and a black one, and he comments on their therapeutic attributes. In Chapter 62 of his first book, Clusius (1576) refers to *T. vulgaris* with his *Thymum durius sive Plinii*. The subsequent chapter 'De Serpyllo silvestri Zygi' includes a description of *T. zygis*, which is one of the most common species in Spain; and in Chapter 64 entitled 'De Tragorigano' he writes: 'Multis Hispaniae locis provenit solo arido petroso cum Stoechade permista', refering to *T. mastichina*, whose Spanish name is 'sarilla'. Some years later, in the book of icons of Lobelius (1581) five drawings of thyme are presented all being very difficult to identify.

In the beginning of the seventeenth century, like preceding authors, Dodonaeus (1616) also described two varieties saying: 'Thymo: unum cephaloton dictum, alterum durius'. Today we can be sure that with the first he refers to *Thymbra capitata* and with the second to *Thymus vulgaris*. His *Serpyllo vulgari* seems to be a *Thymus* species of the section *Serpyllum*. Furthermore, in his chapter 'De Serpyllo ex Dioscoride, Theophrasto et aliis', he comments on the different ideas about *Serpyllum* expressed by several authors. He describes *T. mastichina*, the first plant which he treats in his Chapter 18 on *Tragoriganum*, applying the criteria of Dioscorides. We can find in the work of Bauhin (1623) a few years later, that he divides *Thymus* into four parts: the first (*T. vulgaris folio tenuiore*), as well as the second (*T. vulgaris folio latiore*) seems to be *T. vulgaris*; the third is called *Thymus capitatus* (today *Thymbra capitata*), and the fourth is *Thymum inodorum*. Within his *Serpyllum* nine different varieties are considered; the last one, 'Serpyllum folio Thymi', has turned out to be identical with the *Zygis* of Dioscorides.

In the eighteenth century Barrelier (1714) presents a book of icons. Icon number 788 represents *T. moroderi* (Martínez, 1936) from 'the kingdom of Valencia'; icon number 780 shows *T. hyemalis* (Figure 1.1) and number 694, entitled *Marum hispanicum*, contains a drawing of *T. piperella*. In his list of names with short explanations Tournefort (1719) described six varieties of *T. lusitanicus*, four of them are *T. lotocephalus* and another is *T. moroderi* (Figure 1.2). Within *Thymbra* he considers '*Thymbra hispanica*', with *T. mastichina* and *T. zygis*.

The Linnean *Thymus*

It is very interesting to observe the changes made by Linné in his different works about the *Thymus* species. Most of his knowledge is based on experiences of former authors. In Hortus Cliffortianus (1737, pp. 305–306) he describes six species. Nowadays we know that two of them, the latter ones, do not refer to *Thymus* but to *Satureja* and *Acinos*. His No. 1, *T. erectus* turned out to be *T. vulgaris* (Figure 1.3), No. 2 *T. repens* is a species within the section *Serpyllum*, No. 3 is *Thymbra capitata*, and No. 4 *T. mastichina* (Figure 1.4). In his *Hortus Upsaliensis* (1748, pp. 160–161) only *T. vulgaris* and *T. mastichina* are mentioned. The reference work for the binomial

Figure 1.1 Drawings from Barrelier (1714) work, number 780 corresponds to *T. hyemalis*.

system of nomenclature in Botany '*Species Plantarum* 1st edition' (1753) includes the following eight species in *Thymus*: 1. *T. serpyllum*, 2. *T. vulgaris*, 3. *T. zygis*, 4. *T. acinos* (today *Acinos arvensis*), 5. *T. alpinus* (today *Acinos alpinus*), 6. *T. cephalotos* (today *T. lotocephalus*), 7. *T. villosus*, and 8. *T. pulegioides* (Figure 1.5). Within *Satureja* we find 4. *Satureja mastichina* (today *T. mastichina*).

In Genera Plantarum (1754, p. 257) Linnaeus lists in 646. *Thymus*: *Serpyllum*, *Acinos*, and *Mastichina*. *Species Plantarum* 2nd edition (1762, pp. 825–827) transferred *T. mastichina*, former *Satureja mastichina*, as number 8 into *Thymus*. In the 1st edition this number was

Figure 1.2 Drawings from Barrelier (1714) work, number 788 corresponds to *T. moroderi*.

established for *T. pulegioides*. This transfer Linnaeus commented on literally: 'Ambigit media inter Saturejam et Thymum, sed cum stamina delitescant in fundo corollae, et stylus corolla longior ad Thymum refero'. In *Systema Naturae* 2 (12th edition, 1767, P. 400) for the first time *T. piperella* appears.

Figure 1.3 Typus of *T. vulgaris* in Linné (1737) Hortus Cliffortianus.

After Linnaeus

Brotero (1804) described a new species, *T. caespititius*. Also Hoffmannsegg and Link (1809), in their magnificent and big work about the Flora of Portugal, described some new species: *T. albicans*, *T. capitellatus*, *T. camphoratus*, and *T. sylvestris*. It was Bentham (1834) who, for the first time, divided the genus *Thymus* into sections: *Mastichina*, with *T. mastichina* and *T. tomentosus*; *Serpyllum* with *T. vulgaris*, *T. piperella*,

Figure 1.4 Typus of *T. mastichina* in *Linné (1737)* Hortus Cliffortianus.

T. villosus, *T. capitellatus*, and *T. capitatus*; and *Pseudothymbra*, with *T. cephalotos* (today *T. lotocephalus*).

Edmund Boissier (1839–1845), the famous Swiss botanist from Geneva, studied and described new *Thymus* species from the Iberian Peninsula, a result of years of research travelling through Spain. He also left valuable descriptions of *Thymus* species from the north of Africa (Figure 1.6) as well as from Greece and Turkey written down

Thymus repens, foliis planis, floribus verticillato-fpicatis. *Hort. cliff.* 306. *Roy. lugdb.* 325.
Serpyllum vulgare minus. *Bauh. pin.* 220.
Serpyllum vulgare. *Dod. pempt.* 277.
β. Serpyllum vulgare majus. *Bauh. pin.* 220.
γ. Serpyllum vulgare minus, capitulis lanuginofis. *Tour-
nef. inft.* 197. *It. gotl.* 219.
δ. Serpyllum angustifolium hirsutum. *Bauh. pin.* 220.
ε. Serpyllum foliis citri odore. *Bauh. pin.* 220.
Habitat in Europæ *aridis apricis.* ♄

2. THYMUS erectus, foliis revolutis ovatis, floribus *vulgaris.*
verticillato-fpicatis. *Hort. cliff.* 305. *Hort. upf.* 160.
Mat. med. 281. *Roy. lugdb.* 325. *Sauv. monfp* 148.
Thymus vulgaris, folio tenuiore. *Bauh. pin.* 219.
β. Thymus vulgaris, folio latiore. *Bauh. pin.* 219.
Thymum durius. *Dod. pempt.* 276.
Habitat in G. Narbonenfis, Hifpaniæ *montofis faxo-
fis.* ♄

3. THYMUS floribus verticillato-fpicatis, caule fuffru- *Zygis.*
ticofo, foliis linearibus bafi ciliatis. *Læft.*
Thymo vulgatiori rigidiori fimile. *Bauh. hift.* 2. *p.* 271.
Thymum angusto longiorique folio. *Barr. ic.* 777.
Serpyllum fylveftre Zygis diofcoridis. *Cluf. hift.* 358.
Serpyllum folio thymi. *Bauh. pin.* 220?
Habitat in Hifpania.
Facies T. vulgaris, at Folia bafi ciliata.

4. THYMUS floribus verticillatis, pedunculis uniflor's, *Acinos.*
caulibus erectis fubramofis, foliis acutis ferratis.
Fl. fuec. 478.
Thymus caulibus vix ramofis, foliis ovatis acutis, pe-
dunculis plurimis unifloris. *Hort. cliff.* 306. *Roy.*
lugdb. 325.
Clinopodium arvenfe, ocymi facie. *Bauh. pin.* 225.
Clinopodium vulgare. *Lob. ic.* 506.
Habitat in Europæ *glareofis, cretaceis, ficcis.* ☉

5. THYMUS verticillis fexfloris, foliis obtufiufculis *alpinus.*
concavis fubferratis.
Clinopodium verticillis paucifloris in fpicam congeftis.
Hall. helv. 653.
Clinopodium montanum. *Bauh. pin.* 225. *Bocc. muf.*
2. *p.* 50. *t.* 45.
Acini pulchra fpecies. *Bauh. hift.* 3. *p.* 620.

Ha-

Figure 1.5 Page from Linné (1753) Species Plantarum, where *T. vulgaris* and *T. zygis* are described.

Tab. 141.

Figure 1.6 T. broussonetii (Boissier, 1839–1845, tab. 141).

in his *Flora Orientalis* (1867–1884). Fortunately various beautiful illustrations are available. In his *Elenchus* (1838) he describes *T. willdenowii*, *T. granatensis*, *T. longiflorus*, and *T. membranaceus*. In 1845, he created the section *Pseudothymbra* and later he described *T. carnosus*, *T. lusitanicus*, and *T. baeticus* to be a variety of *T. hirtus*.

Willkomm (1868), a German botanist and author of the *Prodromus Florae Hispanicae*, together with his Danish colleague Lange, stated that the genus comprises five sections: *Mastichina*, *Zygis*, *Piperella*, *Serpyllum*, and *Pseudothymbra*. The section *Serpyllum* includes two groups: the first one with *T. chamaedrys*, *T. serpyllum*, and *T. herba-barona* and the second one with *T. bracteatus*, *T. serpylloides*, and *T. granatensis*.

Briquet (1897) edited the Labiatae in Engler's monumental work, and considers two sections, *Pseudothymbra* and *Serpyllum*, the latter with five subsections: *Bracteatae* (*T. capitellatus*, *T. villosus*, *T. algarbiensis*, *T. albicans*); *Serpylla*; *Piperellae* (*T. piperella*, *T. caespititius*, *T. origanoides*, *T. bovei*); *Vulgares* (*T. vulgaris*, *T. sabulicola*, *T. hyemalis*, *T. zygis*, *T. carnosus*, *T. hirtus*); and *Mastichinae* (*T. mastichina*, *T. tomentosus*, *T. welwitschii*, *T. fontanesii*).

Velenovsky (1906) focused on *Thymus* writing a monography on it. There he considered ten sections: *Coridothymus*; *Vulgares*; *Orientales*; *Anomali* (*T. antoninae*, *T. portae*); *Mastichina* (*T. fontanesii*); *Thymastra* (*T. algarbiensis*, *T. albicans*, *T. capitellatus*); *Pseudothymbra* with 2 groups (suffruticosi: *T. membranaceus*, *T. longiflorus*, *T. funkii*, and herbacei: *T. cephalotos*, *T. villosus*, and *T. granatensis*); *Piperella*; *Micantes*; and *Serpyllum* (includes *T. serpylloides*).

The most important Spanish author is Pau, whose interest in *Thymus* runs throughout his whole botanical work. In his important article published in 1929 entitled 'Introducción al estudio de los tomillos españoles', he analyzes the previous works of Linnaeus, Boissier, and Willkomm. In this article, many interesting details can be found. Further remarkable Spanish authors who worked in this genus were Huguet del Villar (1934), Vicioso (1974), and Elena-Rosselló (1976), and in recent years many contributions from Spanish authors are known.

Although Spain has always been a centre of the systematic, research on thyme, also outside the Iberian Peninsula several famous botanists were dedicated to *Thymus*. They are enumerated here in alphabetical order: Bonnet, Braun, Debray, Klokov, Lyka, Machule, Negre, Opiz, Podlech, Paulovsky, Ronniger, Roussine, Roux, Schmidt, Sennen. Two of them shall be emphasized: Ronniger, who left a very valuable herbarium (today in Vienna) and Jakko Jalas, a Finnish botanist, who edited the genus *Thymus* within the Flora Europaea (1972), Flora Iranica (1982), and the Flora of Turkey (1982).

Illustrations

A lot of old illustrations of *Thymus* are available, specially in the works of Hoffmannsegg and Link (1809), and Boissier (1838, 1839–1845, 1859). The early depictions were very primitive drawings like those of Laguna's translation (1555) of the Dioscorides or those of Barrellier (Figures 1.1 and 1.2). The herbariums from the seventeenth and the first half of the eighteenth century were bound like books and as we can see in the Linnean herbarium of the Hortus Cliffortianus, the plants were ornately arranged in vases (Figures 1.3 and 1.4). After Linnaeus, at the end of the 18th and in the 19th century, the drawings of plants spectacularly improved. Figure 1.7 shows beautifully coloured icons of *T. caespititius* with details of the calyx and the corolla. It is taken from the Portuguese Flora of Hoffmannsegg and Link. Another coloured icon showing *T. broussonetii* of North Africa is taken from Boissier's work (Figure 1.6). Although plant photography has reached a high standard, we must be aware that drawings can mediate more information on botanical details of plants than photographs. Therefore classification of plants can better be performed with drawings than with photographs.

Figure 1.7 T. *caespititius* (Hoffmannsegg and Link, 1809).

BOTANY – THE MORPHOLOGY AND BIOLOGY OF *THYMUS*

Thymus plants are morphologically characterised by their habit or life-forms. We can differentiate two groups, on the one hand little bushy plants, usually below 50 cm, only sporadically up to 1 m, e.g. *T. baeticus* and *T. hyemalis* in the south and southeast of Spain. On the other hand there are creeping life-forms sometimes with rooting twigs. The latter is very common among the species belonging to the section *Serpyllum* or

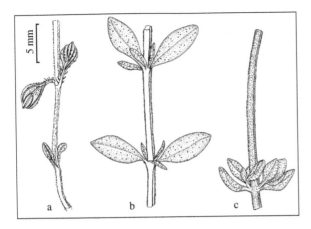

Figure 1.8 Stem morphology: (a) alelotrichous (*T. praecox*), (b) goniotrichous (*T. pulegioides*), (c) holotrichous (*T. piperella*).

Hyphodromi. T. caespititius is an exception with its caespitose habit which can have very long stems. Like most of the Lamiaceae, *Thymus* plants have quadrangular stems, the young being hirsute. The hairs can cover either all four faces of the stem (holotrichous) or only two faces alternating in each internode (alelotrichous). They also can be found only on the four ribs of the stems (goniotrichous). The function of the different types of hairs on the stems are not yet known. Figure 1.8 represents the different types of stems found within the genus, and Figures 1.17 to 1.23 show the plant morphology for different species of *Thymus*.

The leaves can be flat and more or less wide, or with revolute margins and almost acicular. All intermediates seem to be possible. The indumentum is very variable. Some species have leaves without hairs. The tector hairs in *Thymus* are always simple, but rarely single-celled. Leaves are very frequently ciliate at the margins, either at the whole margin or only at the base or on the petiole (Figure 1.9). The glandular trichomes are very important containing the essential oil. There exist two types of glandular trichomes: pedicellate glands with the upper cells full of essential oils, or big globose glands, typical of Lamiaceae, with some basal cells (Figure 1.10). Chapter 3 provides additional information on the anatomy and physiology of these glands.

The flowers grow more or less in clusters in the nodes. Few species have only two flowers per node (e.g. *T. antoninae*), but usually there are bigger clusters of flowers. Species with shorter internodes have globose and capituliform inflorescences. In these cases both leaves of the inflorescence node usually differentiate from the rest of the plant's leaves in form and size, and they are called bracts. This goes for *T. membranaceus*, *T. carnosus*, and other species belonging to the sections *Pseudothymbra* and *Thymus*. In some species the bracteoles can be extraordinarily long as in *T. satureioides*.

The calyx of thyme (2.5–8 mm) when dry plays an important role in the dispersion of the small fruits, or nuculas. Therefore its throat is closed by a hairy row and wind can take it over quite a big distance. The calyces of some species, like that of *T. mastichina*, have long ciliate teeth and seem to be a flying device like the pappus of the

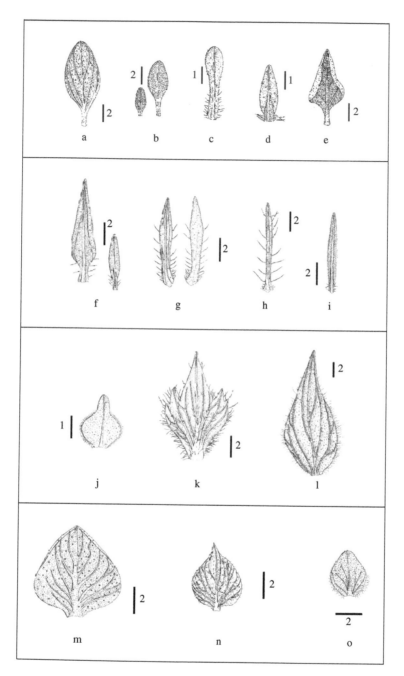

Figure 1.9 Leaf and bract morphology. **Leaves**: (a) *T. richardii*, (b) *T. albicans*, (c) *T. lacaitae*, (d) *T. hyemalis*, (e) *T. camphoratus*, (f) *T. longiflorus*, (g) *T. lotocephalus*, (h) *T. villosus*, (i) *T. zygis*. **Bracts**: (j) *T. lacaitae*, (k) *T. villosus*, (l) *T. lotocephalus*, (m) *T. camphoratus*, (n) *T. longiflorus*, (o) *T. albicans*.
Numbers beside the bars mean the length in mm.

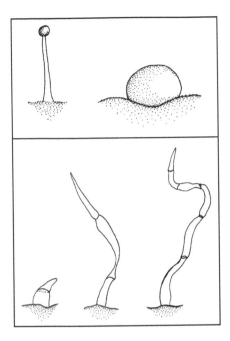

Figure 1.10   Morphology of essential oil glands (up) and hairs (down).

Asteraceae. Usually the calyx has five teeth; three upper and two lower, the latter always being longer and frequently curved upwards. They probably have to keep hold of the corolla's tube. The three upper teeth are shorter than the lower and sometimes reduced to one (*T. caespititius*). The corolla varies between 4 and 10 mm in length and finishes in one upper and three lower lobes, a typical structure to be pollinated by bees or similar insects. The production of pollen in the four stamens is low. Occasionally, the corolla can reach 2 cm like in *T. longiflorus*. Such long-tubed flowers are pollinated by insects with long trunks which can pollinate the flowers while they fly, like the flies of the *Bombilidae* family or crepuscular butterflies of the *Sphyngidae* and *Noctuidae* families do. Figure 1.11 presents examples of calyx and corolla morphology.

Thyme commonly presents gynodioecy, meaning that they produce two types of individuals, some with female flowers without stamens, and others with hermaphrodite flowers. It is proven that pollinators can pollinate female flowers faster than the hermaphrodites. The fruits are nutlets, up to four per flower, but usually some of them abort during early development. Seeds collected from wild populations germinate usually very easily and the seedlings grow relatively fast. Most of the species bloom in spring, others in summer like e.g. *T. serpyllum* or *T. praecox*. In the Mediterranean area, *T. vulgaris* subsp. *aestivus* and *T. piperella* flower in autumn, while *T. hyemalis* in winter. The latter inhabits the arid region of the southeast of the Iberian Peninsula.

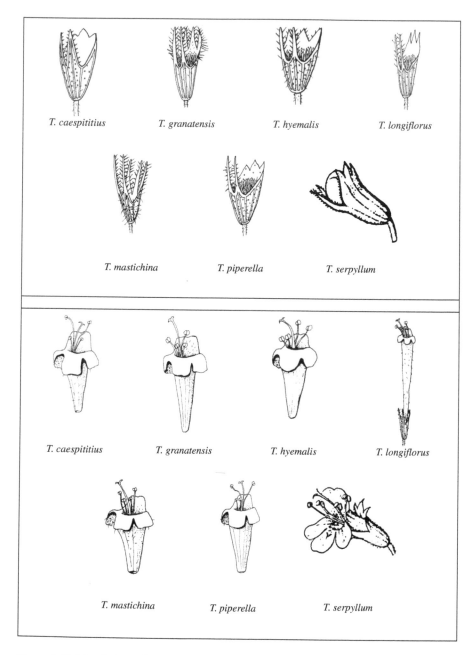

Figure 1.11 Morphology of calyx and corolla.

If we analyse some characteristics from the evolutionary point of view, flat leaves without hairs seem to be more primitive than leaves with revolute and hairy margins. The same occurs with spiciform inflorescences that present bracts similar to leaves. Globose inflorescences with special bracts seem to be more evolved. Woody species with erect life-forms may be phylogenetically older than herbaceous species with only woody parts at the base. An interpretation of the evolutionary relationships among the different sections within the genus is shown in Figure 1.12.

ECOLOGICAL ASPECTS

Thymes are heliophylous plants and like the sun, a fact which reflects the ecology of the genus. *Thymus* plants frequently live on rocks or stones and it is very important that the soils are well drained. But different *Thymus* species require very different substrata, e.g.

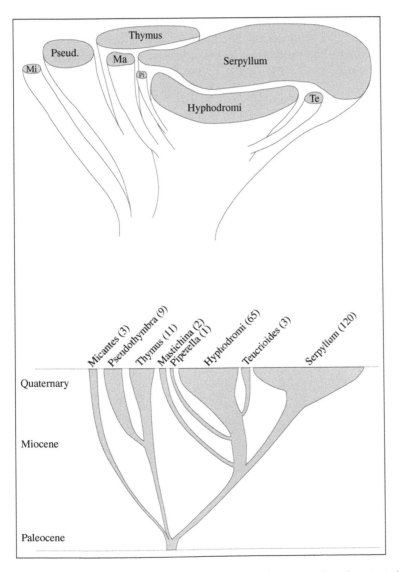

Figure 1.12 Evolutionary relationships in the genus *Thymus*. Number of species in brackets.

T. carnosus lives on sand dunes near the sea (Figure 1.13), *T. lacaitae* on gypsaceous soils, and *T. vulgaris* usually on calcareous soils.

Thymes are very resistant plants, which allows them to live under extreme climatic conditions concerning temperature and water supply. They do not avoid either cold or aridness. Dense and tomentose hairs as well as acicular leaves enable some species to support very dry conditions. The high production of essential oils can also be an adaptive characteristic for dry climate, because the volatile substances evaporate and produce a saturated atmosphere around the plant that makes the loss of water more difficult. Especially some species of the section *Serpyllum* can live in very cold climate, like *T. glacialis* in Siberia or *T. praecox* in Greenland. From an ecological point of view we can find the following correlation: bushy, woody, and erect plants are widely distributed in dry climates, whereas in more fresh and humid climates usually plants with flat leaves and woody only at the base are more common. The latter usually are herbaceous with creeping or lying stems. Such species mostly belong to the sections *Hyphodromi* and *Serpyllum*. The production of essential oils in this group is probably lower than in the first one.

SYSTEMATIC BOTANY

The genus *Thymus* is one of the most important genera of the Lamiaceae. It belongs to the tribe Mentheae within the subfamily Nepetoideae. The most related genera are *Origanum*, *Satureja*, *Micromeria* and *Thymbra*. *Thymus* is considered a well-defined genus, based on the morphological and chemical features of its species.

Figure 1.13 *T. carnosus* from Portugal.

General description

Perennial, subshrubs or shrubs, sometimes with herbaceous shape, but woody at the base, aromatic; stem erect to prostrate, sometimes caespititious and radicant, hairy in all the four sides, only in two alternating or only in the angles; leaves simple, entire or sometimes toothed, frequently revolute, glabrous or hairy, very variable in indumentum; inflorescence spiciform, interrupted in verticillasters or capituliform, bracts like the leaves or very different, lanceolate to broad ovate, usually coloured; flowers pedicellate or not, usually with little bracteoles (very small bracts at the pedicels' base); calyx two-lipped, sometimes nearly regular, more or less campanulate or cylindrical, ten-nerved; upper lip with three triangular teeth sometimes reduced to one, lower lip with two long triangular teeth curved upwards or widespread, throat barbate; corolla bilabiate, sometimes nearly regular, more or less tubular, sometimes with a very long tube, up to 20 mm, four-lobed, white, cream, pink or violet, frequently with clear spots in the throat or lower lobe; upper lobe more or less rounded, emarginate, straight, lower and lateral lobes rectangular to suborbicular, rounded, perpendicular to the tube; four stamens, sometimes reduced or not present (gynodioecy), inserted in the upper half of the tube, exserted or not; anthers with two parallel thecae; style apex branched; nutlets ovoid, smooth.

Biogeography

Thymus is widely distributed in the Old World (Figure 1.14). The Mediterranean region can be described as the centre of the genus – strictly speaking the West Mediterranean region. Only species of two sections occur outside the Mediterranean area. Seven sections are spread over the Iberian Peninsula and northwest Africa, five of them are endemic. In the Iberian Peninsula 35 species can be found, 24 of them endemic to the area. Two

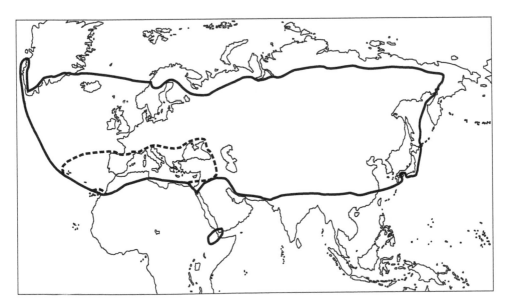

Figure 1.14 Distribution of the genus *Thymus* in the world. Dotted line represents all sections except sect. *Serpyllum* and sect. *Hyphodromi* subsect. *Serpyllastrum*.

species can be found in the Macaronesian region, one on the Canary Islands (*T. origanoides*) growing only at Riscos de Famara and surroundings, and the other one (*T. caespititius*) on Madeira and the Azores; the latter grows also in the western part of the Iberian Peninsula. Fifteen species (12 endemic) grow in northwest Africa, north of the Sahara desert (Morocco, Algeria, Tunis, and Libya), with only three of them also occurring in the Iberian Peninsula. Two species are common in the mountains of Ethiopia (*T. serrulatus*, *T. schimperi*) and one occurs in the southwest of the Arabian mountains (*T. laevigatus*). In Greece 18 species are recorded, 36 in Turkey and 17 in the Flora Iranica. Further eastwards *Thymus* can be found on the Sinai Peninsula (*T. bovei* and *T. decussatus*) and in the arid regions of West Asia up to the Himalayas reaching the limits of the tropical region up to East Asia and Japan. In China 11 species have been recorded. In the north it occurs in Siberia and northern Europe, the coasts of Greenland can be described as the most northern occurrence of *Thymus* (*T. praecox*). Introduced populations now growing wild are known to exist in regions as distant as Canada (*T. serpyllum* and *T. pulegioides*), Chile (*T. vulgaris*) or New Zealand (*T. pulegioides* and *T. vulgaris*).

We can suggest the origin of some taxa of the genus to be in the Mediterranean area, seeing that the sections *Serpyllum* and *Micantes* have been present there since the Paleocene. In the Miocene, some species of section *Thymus* and *Hyphodromi* developed. During the Quaternary the ancestors of the section *Serpyllum* and, to some extent, those within the section *Hyphodromi* have produced new speciation processes, colonizing all the ice-free land after the last Ice Age. These processes are not yet finished and may be the reason why all these species are difficult to be distinguished. We can assume that they are halfway in a process of speciation to produce clear species (Morales, 1989).

Pollen

The pollen grains of this genus have a very homogeneous morphology, both within the same species and among different taxa. According to Wunderlich (1967), it can be ascribed to the *Satureja* type. It has a radial isopolar symmetry and is usually hexacolpate (NPC 643) and three-celled. Octocolpate and tetracolpate grains are also known. The colpi are regularly disposed, and the mesocolpi usually are of the same width with one exception: the mesocolpi of *T. caespititius* are of varying wideness alternating a wider and a more narrow one. The pollen grains are more or less spheroidal and the index of Polar distance/Equatorial distance (P/E) varies between 0.9 and 1.3 (from prolate-spheroidal to oblate-spheroidal). The sizes of the pollen grains vary from 21 to 46 µm depending on the species and a correlation between ploidy level and size can be assumed. The ornamentation usually is suprareticulate, less frequently semitectate or reticulate. In the case of suprareticulate ornamentation, thick walls delimit in a lower level a net of narrower walls and pores. The wideness of walls and pores varies from one species to another, but it is homogeneous within each species. As an exception, pollen grains with cerebroid ornamentation can be found, which seems to be usual for tetracolpate pollen grains. Figure 1.15 illustrates the morphology of pollen grains from *T. hyemalis*.

Chromosomes

In the genus *Thymus* the chromosomes are very small. With 1–2 µm they appear like dots under the optical microscope. The following chromosome numbers are known: $2n = 24, 26, 28, 30, 32, 42, 48, 50, 52, 54, 56, 58, 60, 84$ and 90, corresponding to the

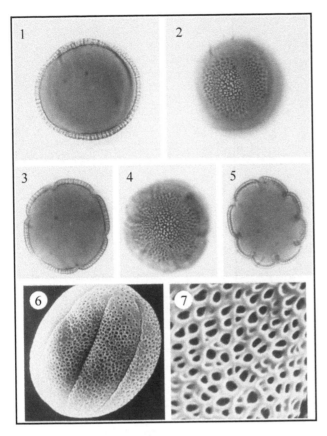

Figure 1.15 Pollen grains of *T. hyemalis* from Murcia (Spain). Images 1–5 are views from an optical microscope. Images 6 and 7 were obtained with a scanning electron microscope with a magnification of 1600x and 7000x respectively.

diploid, tetraploid and hexaploid levels. The secondary basic numbers x=14 and x=15 probably originate from a basic number x=7. The most frequent numbers are $2n=28$, 30, 56 and 60. Aneuploidy has been an important phenomenon during the evolution of this genus and is responsible for the other numbers. There are a lot of interesting cases of different levels within the same species. This is true for *T. mastichina* with $2n=30$, 60; *T. vulgaris* $2n=28$, 58; *T. zygis*, *T. leptophyllus*, *T. glabrescens*, *T. longicaulis*, *T. praecox* $2n=28$, 56; *T. algeriensis* $2n=30$, 56; *T. comptus* $2n=26$, 52; *T. zygioides* $2n=60$, 90; *T. longedentatus* $2n=30$, 90; *T. striatus* and *T. herba-barona* $2n=28$, 56, 84. The latter is most remarkable because the chromosome numbers studied in the West Mediterranean populations resulted to be $2n=28$ in Majorca, $2n=56$ in Corsica, and $2n=84$ in Sardinia. Chromosomes from different *Thymus* species are shown in Figure 1.16.

Other features

In *Thymus* hybridization is very common where two or more species live together. Up to date 60 hybrids have been detected among the 35 species living in the Iberian

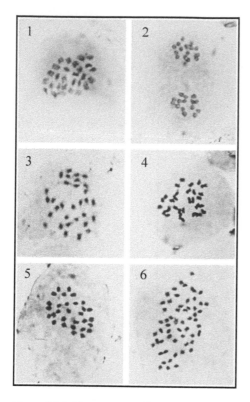

Figure 1.16 Chromosomes of some species of *Thymus*. (1): *T. mastigophorus*, 2n = 28 (Zaragoza, Spain). (2): *T. capitellatus*, n = 15 (Algarve, Portugal). (3): *T. camphoratus*, 2n = 30 (Alentejo, Portugal). (4): *T. camphoratus*, 2n = 30 (Algarve, Portugal). (5): *T. camphoratus*, 2n = 30 (Algarve, Portugal). (6): *T. carnosus*, 2n = 56 (Algarve, Portugal).

Peninsula, as we can see in the appendix (Morales, 1995). Some chemical studies show the genus to be homogeneous, in the comparison with others such as *Teucrium* or *Sideritis* both chemically heterogeneous (Morales, 1986). These two features are the evidence to consider *Thymus* to be a good taxonomical genus, probably monophyletic. Within the genus genetic incompatibility between species does not seem to exist, which makes taxonomic studies in this genus very difficult, especially in some taxonomical groups e.g. in the section *Hyphodromi* and particularly in the section *Serpyllum*, where the concept of species is more difficult to apply. If we impose synoptcal criteria, probably a lot of forms, sometimes ecological forms, would be included as simple populations into a given taxon. But when using analytical criteria we risk overlooking existing species considered as natural units. In case of doubt I recommend synoptical criteria. At the species level, there are a lot of names, more than 1 000, many of them of course are synonyms.

Popular names

In the whole area of distribution, *Thymus* is usually well known and used by the population as spice, medicinal plant or source of essential oils. Therefore a big variety of

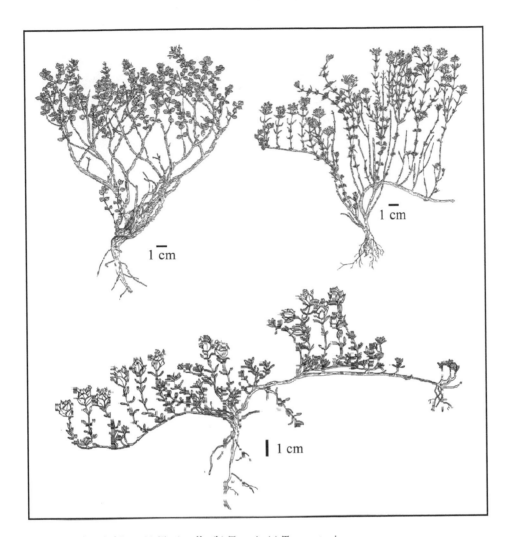

Figure 1.17 Plant habitus. (a) *T. piperella*, (b) *T. zygis*, (c) *T. granatensis*.

popular and vernacular names are known for the different species. If we begin in the west of its habitat *T. caespititius*, from the Azores, Madeira and the western part of the Iberian Peninsula has the portuguese names 'tomentelo' or 'tormentelo', and in Galicia 'tomelo do país', 'tomentelo do país' or 'tomillo'. The only species grown on the Canary Islands from this genus is *T. origanoides*, in the Riscos de Famara of Lanzarote island. An old name of this plant is 'tajosé'. 'Tomillo' is the popular name in other islands for *Thymus*-looking species of *Micromeria*. In continental Africa species of *Thymus* are found in Morocco, Algeria, Tunis and Libya. *T. algeriensis* is the most common in the four countries, and its popular names in arabic and berber languages are: 'azoukni', 'djertil', 'djoushshen', 'hamriya', 'hamzousha', 'khieta', 'mezoukesh', 'rebba', 'toushna'. *T. broussonetii* is named there 'zatar', 'za'atar el-hmir', 'za'ter el hmir', 'ze'itra', 'z'itra'. The Moroccan *T. maroccanus* has the name 'azukenni'. In the mountains of Ethiopia two *Thymus* species grow, *T. serrulatus* and *T. schimperi*, with the Abyssinian names for the first one: 'tausi',

Figure 1.18 T. mastichina (Spanish marjoram) very common in the Iberian Peninsula, from Central Spain.

'tazè', 'tenni', 'teschin', 'tessni', 'tesnè', 'thasne', and 'tessni', 'tosign', 'tosigne', 'tossign' or 'tossine' for the latter.

In Asia, in the Arabic Peninsula, the mountainous areas of Yemen are the southern-most localities in this continent, where *T. laevigatus* lives. It is named 'za'tar' or 'sa'tar'. In the far east in China the popular name of several species of thymes are 'bai li xiang' and the most used species, *T. quinquecostatus*, is called 'di jiao' or 'bian zhong'.

In the North of Europe *T. serpyllum* and other species of this group are widespread. The vernacular names in the nordic languages are 'timian' or 'timjan', 'stortimian' or 'backtimian'. And in Central Europe they are called 'Thymian', 'Feldthymian', 'Quendel', 'Kudelkraut', 'Kuttelkraut' in Germany; 'serpolet', 'piolet', 'piliolet', 'pignolet', 'pélevoué', 'pénévouet' in France; 'pepolino', 'sermollino selvatico' in Italy; 'erba pevarina', 'süsémbar' in Slavic (Puschlav); 'timian', 'masarón salvatg', 'pavradel', 'pavradella' in Rätorom. In English, the following names are known: 'thyme', 'wild thyme', 'penny mountain', 'hillwort', 'brotherwort', 'shepherds thyme', and in dutch 'tijm'.

In the different languages of the Iberian Peninsula, a lot of names are in use for the multitude of species of *Thymus* or 'tomillo' (Morales *et al.*, 1996):

T. baeticus: tomillo, tomillo basto, tomillo fino, tomillo gris, tomillo limonero.
T. granatensis: hierba luna, serpillo, serpol, tomillo, tomillo colorao, tomillo serpol.
T. hyemalis: tomillo, tomillo de invierno, tomillo fino, tomillo macho, tomillo morado, tomillo rojo.
T. lacaitae: tomillo lagartijero, tomillo de Aranjuez.
T. longiflorus: tomillo, tomillo real.

Figure 1.19 T. piperella from Valencia, Spain.

T. loscosii: ajedrea, tomillo sanjuanero.

T. lotocephalus: tomilho-cabeçudo.

T. mastichina: ajedrea de monte, almoradux, almoraú, amáraco, bela-luz, cantueso, escombrilla, marahú, marduix silvestre, mejorana, mejorana de monte, mejorana silvestre, mendaro, mendaroa, moraduix bord, salpurro, sarilla, tomilho-alvadio, tomillo, tomillo blanco, tomillo de las aceitunas, tomillo macho, tomillo salsero.

T. membranaceus: cantueso, escombrilla, mejorana, tomillo blanco, tomillo macho, tomillo terrero.

T. moroderi: cantahueso, cantueso, mejorana.

Figure 1.20 T. membranaceus from Murcia, Spain (Morales, 1986).

T. orospedanus: tomillo.

T. piperella L.: peberela, peberella, pebrella, pebrinella, piperesa, timó.

T. praecox: erva-ursa, farigola, farigoleta, folcó, herba de pastor, hierba luna, salia de pastor, samarilla, sarpoil, serpão, serpil, serpildo, serpilho, serpol, sérpol, serpolio, timó negre, tomillo de puerto, tomillo rastrero.

T. pulegioides: apiua, charpota, serpão glabro, serpol, te fino, te morado, té morau, tomelo, tomentelo, tomillo.

T. richardii: farigola de muntanya, farigola mascle, farigoleta, hierba luna, salsa de pastor, serfull, serpol, sèrpol, serpoll, timó negre, tomillo rojo.

T. serpylloides: samarilla, tomillo, tomillo de la sierra, tomillo de Sierra Nevada.

T. serpylloides subsp. *gadorensis*: samarilla, tomillo rojo, verbena.

T. villosus: azeitoneira, erva-azeitoneira, erva-das-azeitonas, tomilho-peludo, tomillo ansero.

T. vulgaris L.: arçã, arçanha, axedrea, boja, bojas, elar, elharr, ellbor, entremunsell, erle-bedarr, estremoncello, estremoncillo, estremonzillo, estremunsell, estremunzillo, ezkai, ezkaia, farigola, farigoleta, fariguala, forigola, frigola, frígola, friula, ispillu, lo timó, morquera, sajolida, senyorida, tem, timó, timó femella, timó mascle, timó normal, timón, timoncillo, timonet, timons, tomaní, tomelo, tomello, tomello vulgar, tomentelo, tomilho, tomilho-ordinário, tomilho-vulgar, tomilo, tomillo, tomillo ansero, tomillo blanco, tomillo borde, tomillo común, tomillo limonero, tomillo negrillo, tomillo royo, tomillo salsero, tomillo vulgar, tomillua, tomizo, tremoncillo, tremonsillo, tremontillo, tremonzillo, tumillo.

T. zygis: farigola salsera, morquera, paticas de mona, salsero salseta de pastó, serpão-do-monte, tomilhinha, tomillo, tomillo aceitunero, tomillo aceytunero, tomillo albar, tomillo ansero, tomillo blanco, tomillo del campo, tomillo español, tomillo fino, tomillo negrillo, tomillo risquero, tomillo rojo, tomillo salsero, tomillo sansero, tomillo sansero fino.

Figure 1.21 T. *vulgaris* (Hallier, 1884: 188, Tafel 1796).

T. *zygis* subsp. *gracilis*: tomillo, tomillo aceitunero, tomillo blanco, tomillo fino, tomillo rojo, tomillo salsero.

Sections of the genus *Thymus*

According to Jalas (1971), *Thymus* is divided into eight sections: *Micantes, Mastichina, Piperella, Teucrioides, Pseudothymbra, Thymus, Hyphodromi,* and *Serpyllum*. The sequence used here is that established by Jalas, but other ordinations would perhaps be more

Figure 1.22 T. lacaitae from Central Spain.

logical considering phylogenetic or evolutionary criteria. For distribution patterns of the sections and subsections see Morales (1997).

I.   Sect. *Micantes*
II.   Sect. *Mastichina*
III.   Sect. *Piperella*
IV.   Sect. *Teucrioides*
V.   Sect. *Pseudothymbra*
      1. Subsect. *Pseudothymbra*
      2. Subsect. *Anomalae*
VI.   Sect. *Thymus*
      1. Subsect. *Thymastra*
      2. Subsect. *Thymus*
VII.   Sect. *Hyphodromi*
      1. Subsect. *Subbracteati*
      2. Subsect. *Serpyllastrum*
      3. Subsect. *Thymbropsis*
VIII. Sect. *Serpyllum*
      1. Subsect. *Insulares*
      2. Subsect. *Kotschyani*
      3. Subsect. *Pseudopiperellae*
      4. Subsect. *Isolepides*
      5. Subsect. *Alternantes*

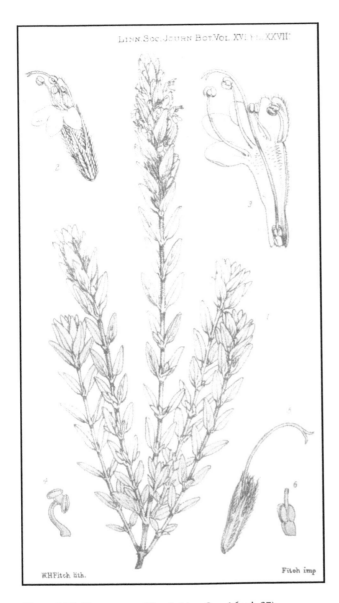

Figure 1.23 T. maroccanus (Bot. J. Linn. Soc., 16: pl. 27).

6. Subsect. *Pseudomarginati*
7. Subsect. *Serpyllum*

I. Sect. *Micantes* Velen., Bei. Bot. Centr. 19(B2): 278 (1906)

Typus: *T. caespititius* Brot.
Erect plants (North-African species) or caespitose; stems holotrichous; leaves flat, glabrous, long oblong-obovate; inflorescence spiciform, sometimes dense; lateral upper teeth of calyx very short or reduced.

It comprises three species, two of them are North African woody species, that occur in Morocco: *T. satureioides* and *T. riatarum*. The former inhabits the High Atlas region, while *T. riatarum* is a prostrate plant and lives in the Rif mountains. The Ibero-Macaronesian species *T. caespititius* occurs in the northwest of the Iberian Peninsula and also in Madeira and Azores. If we take into account their plesiomorphic features, like flat, non-revolute and glabrous leaves, and their geographical distribution, this section seems to be very old.

II. Sect. *Mastichina* (Miller) Bentham, Lab. Gen. Sp.: 340 (1834)

Mastichina Miller, Gard. Dict. ed. 4 (2) (1754).
Typus: *T. mastichina* (L.) L.
Erect plants with holotrichous stems, leaves flat, lanceolate to obovate; inflorescence capituliform; calyx very hairy, teeth similar and subulate, with long cilia.

This section is endemic to the Iberian Peninsula, and comprises *T. mastichina*, with two subspecies, and *T. albicans*. *T. mastichina* subsp. *mastichina* is a very common plant in Spain and Portugal. The subspecies *donyanae* occurs only in the southwest of the Iberian Peninsula around the 'Coto de Doñana' and in some locations in the 'Algarve'. The other species, *T. albicans*, is also living in the southwestern pinewoods of *Pinus pinea*. *T. mastichina* subsp. *donyanae* and *T. albicans*, with 2n=30 chromosomes, are probably the origin of the tetraploid apoendemic *T. mastichina* subsp. *mastichina*, a modern taxon that has spread throughout the entire Iberian Peninsula.

III. Sect. *Piperella* Willk., Prodr. Fl. Hisp. 2: 404 (1868).

Typus: *T. piperella* L.
Erect or decumbent plants, with holotrichous stems and leaves obovate, flat and glabrous; flowers growing in lax verticillasters.

T. piperella is found at Valencia province and surroundings, and it is the unique species of this monotypic section, endemic to this region.

IV. Sect. *Teucrioides* Jalas, Bot. J. Linn. Soc. 64(2): 201 (1971).

Typus: *T. teucrioides* Boiss. and Spruner.
Plants usually decumbent with leaves revolute, ovate or triangular-ovate; flowers in verticillasters. Endemic to the Balkan Peninsula. It inhabits the mountains of Greece and Albania. Three species can be recognised within this section: *T. teucrioides*, *T. hartvigi*, and *T. leucospermus*, that have been studied by Hartvig (1987). Chromosome numbers of these species are not yet known.

V. Sect. *Pseudothymbra* Bentham, Lab. Gen. Sp.: 341 (1834).

Typus: *T. lotocephalus* G. López and R. Morales (*T. cephalotos* auct. non L.)
Erect plants with holotrichous stems and linear revolute leaves, usually hairy and with cilia at the base; inflorescence capituliform with broad bracts; corolla very long.

In this section are included nine Iberian-North-African species, usually with long corollas, up to 2 cm, and bracts rather different from the leaves and subglobose inflorescence, except in subsection *Anomalae*. The North-African species are *T. munbyanus*, common and very variable, extending from the Middle Atlas and the Rif Mountains as far as the Algerian mountains. A difficult species with two subspecies and hybrids with *T. algeriensis* and *T. willdenowii*. *T. bleicherianus* is only known from three locations, one in Algeria and two more in the north of Morocco. The other species of this section are all Iberian.

We recognize two subsections:

V1. Subsect. *Pseudothymbra* (Bentham) R. Morales, *Ruizia* 3: 146 (1986).
Inflorescence capituliform and bracts are very different from the leaves.

V2. Subsect. *Anomalae* (Rouy) R. Morales, *Ruizia* 3: 146 (1986).
T. sect. *Anomalae* Rouy, Bull. Soc. Bot. France 37: 166 (1890).
Typus: *T. antoninae* Rouy and Coincy
Flowers in verticillasters, bracts with similar appearance as the leaves.

VI. Sect. *Thymus*

Erect or radicant plants with holotrichous stems, revolute leaves, usually hairy; flowers in spiciform or globose inflorescences.

Western Mediterranean section, with three most important species: *T. vulgaris, T. zygis* and *T. willdenowii*. The first usually occurs on basic soils and is distributed in northern Italy, south of France and east of Spain. *T. zygis* is a very common species in all the Iberian Peninsula and *T. willdenowii* is common in North Africa (Morocco and Algeria) and also grows only in Gibraltar area in the Iberian Peninsula.
We recognize two subsections:

VI1. Subsect. *Thymastra* (Nyman ex Velen.) R. Morales, *Ruizia* 3: 146 (1986).
T. sect. *Thymastra* Velen., Bei. Bot. Centr. 19(B2): 276 (1906).
Typus: *T. capitellatus* Hoffmanns. and Link.
Erect plants with triangular-ovate or lanceolate-ovate leaves, without cilia at the base; inflorescence more or less globose with bracts different from the leaves.

VI2. Subsect. *Thymus*
Erect or subtended plants with leaves usually hairy, and ciliate or not at the base, with revolute margins and more or less linear, bracts broader than the leaves, but not very different.

VII. Sect. *Hyphodromi* (A. Kerner) Halácsy, Denkschr. Akad. Wiss. Wien 61: 252 (1894).

Typus: *T. bracteosus* Vis. ex Bentham.
Plants usually subtended and rooting; stems holotrichous; leaves flat or revolute, usually not hairy; inflorescence frequently capituliform with bracts different from the leaves.

This section extends throughout the Mediterranean area and comprises around 60 species. From the three subsections, *Subbracteati* is characterized by more or less revolute or convolute leaves and seems to be Oriental. Only one species occurs in North Africa, from Morocco to Libya: *T. algeriensis*. Another species occurs in Central Spain: *T. mastigophorus*. *T. spinulosus* occurs in Sicily and Italy, and *T. striatus* in the Italian and Balkan Peninsulas. Both species are very variable. *T. argaeus, T. brachychilus, T. cappadocicus, T. cherlerioides, T. convolutus, T. pulvinatus,* and *T. revolutus* occur in Turkey; *T. boissieri, T. comptus, T. dolopicus,* and *T. plasonii* in the Balkan Peninsula; *T. atticus, T. parnassicus,* and *T. leucotrichus* inhabit Turkey and the Balkan Peninsula. The last species also grows in Syria and in the Lebanon. *T. integer* is only found on the island of Cyprus. This species is probably not different from *T. leucotrichus*. *T. samius* occurs in the Aegean islands. *T. borysthenicus* and *T. pallasianus* occur north of the Black Sea, *T. persicus* south of the Caucasus, but only one location for this species is known.

We recognize three subsections:

VII1. Subsect. *Subbracteati* (Klokov) Jalas, Bot. J. Linn. Soc. 64(2): 205 (1971), emend.
T. sect. *Subbracteati* Klokov, Not. Syst. (Leningrad) 16: 315 (1954) pro parte.
Typus: *T. pallasianus* H. Braun.

VII2. Subsect. *Serpyllastrum* Huguet del Villar, Cavanillesia 6: 124 (1934).
Lectotypus: *T. bracteosus* Vis. ex Bentham.

VII3. Subsect. *Thymbropsis* Jalas ex R. Morales, Anales Jard. Bot. Madrid 45(2): 562 (1989).
Typus: *T. maroccanus* Ball.

Subsection *Serpyllastrum* is a group of species characterized by the presence of prostrate stems and flat leaves more or less wide. Five species from this section are living in Spain: *T. bracteatus*, *T. leptophyllus*, *T. fontqueri*, *T. granatensis* and *T. lacaitae*. It is also well represented in the East, but no species occur in Italy and North Africa. *T. aznavourii* and *T. bracteosus* occur in the Balkan Peninsula; *T. canoviridis*, *T. haussknechtii*, *T. pectinatus* and *T. spathulifolius* are found in Turkey. *T. zygioides* extends from the Balkan Peninsula as far as the Crimean Peninsula and also in Turkey. This species and the Spanish endemic *T. lacaitae* are morphologically very similar. There is also a group of species that occur only in the Caucasus: *T. dagestanicus*, *T. hadzhievii*, *T. helendzhicus*, *T. karjagnii*, *T. ladjanuricus*, *T. lipskyi*, *T. majkopiensis*, and *T. sosnowskyi*. Seven more species from Central Asia are considered inside this subsection: *T. cuneatus*, *T. eremita*, *T. incertus*, *T. irtyschensis*, *T. kirgisorum*, *T. nerczensis*, *T. petraeus*.

Subsection *Thymbropsis* includes the North African *T. broussonetii*, *T. maroccanus*, *T. lanceolatus*, *T. numidicus*, *T. pallescens*, and the two endemic species from Greece *T. laconicus* and *T. holosericeus*. Five more species from this section are found in Turkey: *T. cariensis*, *T. cilicicus*, *T. eigii*, *T. leucostomus*, and *T. sipyleus*. *T. syriacus* occurs in Lebanon, Syria and a location in northern Irak; *T. bovei* lives in the Sinai Peninsula, Israel, Jordan, Irak and Saudi Arabia; and *T. decussatus* in Sinai and Saudi Arabia. This group has predominantly North-African and East-Asian species.

VIII. Sect. *Serpyllum* (Miller) Bentham, Lab. Gen. Sp.: 340 (1834).
Serpyllum Miller, Gard. Dict. ed. 4 (3) (1754).
Woody plants or only woody at the base, but with herbaceous appearance, usually subtended and rooting, with holotrichous stems or hairy only in two opposite sides or in the angles (goniotrichous or alelotrichous), leaves flat and usually ciliate at the base, with distinct lateral veins; inflorescence spiciform or more or less globose.

In this section there are around 120 species. They occur throughout the area of the genus, except in Madeira and the Azores. We find in the species of *Serpyllum* the widest chromosomal variation. There are also woody species that grow in the mountains in arid areas like *T. origanoides* on Lanzarote (Canary Islands); *T. serrulatus* and *T. schimperi* in Ethiopia, *T. laevigatus* in the southwest of the Arabian Peninsula. Another group of species are more or less herbaceous and occur in the Mediterranean mountains, and all of Eurasia and also along the coasts of Greenland. The species of the last group seem to be younger in evolutionary terms and have probably been actively evolving since the last glaciation when this group colonized the new lands free of ice. This group is also very difficult taxonomically and corresponds to the last three subsections. Few species of these subsections are present in the Mediterranean area. According to Jalas (1971), we divided this section into seven subsections.

Subsection *Insulares* comprises *T. willkommii*, an endemic species that occurs in the mountains of the provinces of Castellón and Tarragona (eastern Spain); *T. richardii*, with three subspecies: subsp. *richardii* from Majorca and Yugoslavia, subsp. *ebusitanus* from Ibiza and subsp. *nitidus* from Marettimo island near Sicily; the North-African *T. dreatensis* and *T. guyonii*, the Canary Island endemic *T. origanoides* and the endemic species to northwest Turkey *T. bornmuelleri*.

Subsection *Kotschyani* includes a lot of Asian species, but only *T. fallax* and *T. transcaucasicus* occur in Turkey. Other interesting species occurring outside the Mediterranean area are *T. laevigatus* from the mountains of Yemen or *T. schimperi* and *T. serrulatus* from the Ethiopian mountains.

Subsection *Pseudopiperellae* comprises *T. herba-barona* from Majorca, Corsica, and Sardinia (Mayol *et al.*, 1990) and *T. nitens* from the south of France.

Five species inhabiting the Balkan Peninsula belong to the subsection *Isolepides*: *T. bulgaricus*, *T. glabrescens*, *T. longedentatus*, *T. pannonicusm*, and *T. sibthorpii*.

Subsection *Alternantes* includes *T. linearis* from the Himalaya mountains; the European *T. pulegioides*, *T. froelichianus*, *T. alpestris*, *T. oehmianus*, *T. bihoriensis*, and *T. comosus*.

Subsection *Pseudomarginati* includes the species *T. longicaulis* and *T. praecox*, very common in Europe and also in Turkey; *T. nervosus*, an endemic of the Pyrenees and the French Massif Central; *T. ocheus*, *T. stojanovii*, and *T. thracicus* from the Balkan Peninsula and the East Mediterranean region.

Subsection *Serpyllum* includes *T. quinquecostatus* from Japan, the European *T. serpyllum* and *T. talijevii* and other Russian species.

LIST OF *THYMUS* SPECIES OF THE WORLD

I propose at the moment the following list of species. There are 214 species and 36 subspecies more: 250 taxa. When known, the chromosome numbers and the countries are given (Ag=Algeria, Al=Albania, An=Asian Turkey, Az=Azores, Bl=Balearic Islands, Bu=Bulgaria, Co=Corsica, Cy=Cyprus, E=East Aegean Islands, Ga=France, Gr=Greece, Hs=Spain, It=Italy, Ju=former Jugoslavia, Li=Libya, LS=Lebanon and Syria, Lu=Portugal, Ma=Morocco, Ru=Romania, Sa=Sardinia, Si=Sicily, Tn=Tunis, Tu=European Turkey, URSS=former Soviet Union).

I. Sect. *Micantes* Velen.
T. caespititius Brot. 2n=30 Hs Lu Az Madeira
T. satureioides Cosson subsp. *satureioides* Ma
   subsp. *commutatus* Batt. 2n=30 Ma
T. riatarum Humbert and Maire Ma

II. Sect. *Mastichina* (Miller) Bentham
T. mastichina (L.) L. subsp. *mastichina* 2n=56, 58, 60 Hs Lu

subsp. *donyanae* R. Morales 2n=30 Hs Lu
T. albicans Hoffmanns. and Link 2n=30 Hs Lu

III. Sect. *Piperella* Willk.
T. piperella L. 2n=28 Hs

IV. Sect. *Teucrioides* Jalas
T. teucrioides Boiss. and Spruner subsp. *teucrioides* Gr Al
  subsp. *alpinus* Hartvig Gr
  subsp. *candilicus* (Beauverd) Hartvig Gr
T. hartvigi R. Morales subsp. *hartvigi* Gr
  subsp. *macrocalyx* (Hartvig) R. Morales Gr
T. leucospermus Hartvig Gr

V. Sect. *Pseudothymbra* Bentham
V1. Subsect. *Pseudothymbra* (Bentham) R. Morales
T. lotocephalus G. López and R. Morales 2n=30 Lu
T. villosus L. subsp. *villosus* Lu
  subsp. *lusitanicus* (Boiss.) Coutinho 2n=54 Lu Hs
  subsp. *oretanicus* Hs
T. longiflorus Boiss. 2n=28 Hs
T. membranaceus Boiss. 2n=28 Hs
T. moroderi Pau ex Martínez 2n=28 32 Hs
T. munbyanus Boiss. and Reuter subsp. *munbyanus* Ma Ag
  subsp. *coloratus* (Boiss. and Reuter) Greuter and Burdet Ma Ag
T. bleicherianus Pomel Ma Ag
T. funkii Cosson 2n=28 Hs

V2. Subsect. *Anomalae* (Rouy) R. Morales
T. antoninae Rouy and Coincy 2n=56 Hs

VI. Sect. *Thymus*
VI1. Subsect. *Thymastra* (Nyman ex Velen.) R. Morales
T. capitellatus Hoffmanns. and Link 2n=30 Lu
T. camphoratus Hoffmanns. and Link 2n=30 Lu

VI2. Subsect. *Thymus*
T. carnosus Boiss. 2n=56 Lu Hs
T. vulgaris L. subsp. *vulgaris* 2n=28, 30 Hs Ga It
  subsp. *aestivus* (Willk.) O. Bolós and A. Bolós 2n=58, 60 Hs Bl
T. orospedanus Huguet del Villar 2n=28 Hs
T. hyemalis Lange subsp. *hyemalis* 2n=58 Hs
  subsp. *millefloris* (Rivera and al.) R. Morales 2n=58 Hs
  subsp. *fumanifolius* (Pau) R. Morales Ma Ag
T. zygis Loefl. ex L. subsp. *zygis* 2n=28 Hs Lu
  subsp. *gracilis* (Boiss.) R. Morales 2n=28 Hs Ma
  subsp. *sylvestris* (Hoffmanns. and Link) Coutinho 2n=56, 58 Hs Lu
T. baeticus Boiss. ex Lacaita 2n=58 Hs
T. willdenowii Boiss. 2n=30 Hs Ma Ag

T. loscosii Willk. 2n=54 Hs
T. serpylloides Bory subsp. *serpylloides* 2n=58 Hs
   subsp. *gadorensis* (Pau) Jalas 2n=56, 58 Hs

VII. Sect. *Hyphodromi* (A. Kerner) Halácsy

VII1. Subsect. *Subbracteati* (Klokov) Jalas

T. algeriensis Boiss. and Reuter 2n=30, 56 Ma Ag Tn Li
T. argaeus Boiss. and Bal. An
T. atticus Celak. An Bu Gr Tu
T. boissieri Hal. Al Gr Ju
T. borysthenicus Klokov and Shost.
T. brachychilus Jalas An
T. cappadocicus Boiss. An
T. cherlerioides Vis. 2n=28 An
T. comptus Friv. 2n=26, 28, 52 Gr Tu
T. convolutus Klokov An
T. dolopicus Form. Gr
T. integer Griseb. Cy
T. leucotrichus Halácsy An Gr Ju LS
T. mastigophorus Lacaita 2n=28 Hs
T. pallasianus H. Braun subsp. *pallasianus* north to Black Sea
   subsp. *brachyodon* (Borbás) Jalas
T. parnassicus Halácsy An Gr Ju
T. persicus (Ronniger ex Rech. fil.) Jalas
T. plasonii Adamovic Gr
T. pulvinatus Celak. An
T. revolutus Celak. An
T. spinulosus Ten. 2n=56 It Si
T. striatus Vahl 2n=26, 28, 42, 54, 56, 84 Al Bu It Gr Ju Tu

VII2. Subsect. *Serpyllastrum* Huguet del Villar

T. aznavourii Velen. Tu
T. bracteatus Lange ex Cutanda 2n=56, 58 Hs
T. bracteosus Vis. ex Bentham Ju
T. canoviridis Jalas An
T. cuneatus Klokov Central Asia
T. dagestanicus Klokov and Shost. 2n=28 Caucasus
T. eremita Klokov Central Asia
T. fontqueri (Jalas) Molero and Rovira 2n=56 Hs
T. granatensis Boiss. subsp. *granatensis* 2n=28 Hs
   subsp. *micranthus* (Willk.) O. Bolós and Vigo Hs
T. hadzhievii Grossh. 2n=28 Caucasus
T. haussknechtii Velen. An
T. helendzhicus Klokov and Shost. Caucasus
T. incertus Klokov Central Asia
T. irtyschensis Klokov Altai
T. karjaginii Grossh. Caucasus
T. kirgisorum Dubjanski 2n=26 South of Russia and wide area until Siberia

T. lacaitae Pau 2n=28 Hs

T. landjanuricus Kem. Caucasus

T. leptophyllus Lange subsp. *leptophyllus* 2n=28 Hs
  subsp. *paui* R. Morales 2n=56 Hs
  subsp. *izcoi* (Rivas Martínez and al.) R. Morales Hs

T. lipskyi Klokov and Shost. Caucasus

T. majkopiensis Klokov and Shost. 2n=28 Caucasus

T. nerczensis Klokov N Mongolia

T. pectinatus Fisch. and Meyer An

T. petraeus Serg. Central Asia

T. samius Ronniger and Rech. fil. AE

T. sosnowskyi Grossh. 2n=60 Caucasus

T. sphatulifolius Hausskn. and Velen. An

T. zygioides Griseb. 2n=56, 60, 62, 90 An Bu Gr Ru Tu Crimea

VII3. Subsect. *Thymbropsis* Jalas ex R. Morales

T. bovei Bentham Sinai IJ Irak Saudi Arabia

T. broussonetii Boiss. subsp. *broussonetii* Ag Ma Tn
  subsp. *hannonis* (Maire) R. Morales Ma

T. cariensis Hub.-Mor. and Jalas An

T. cilicicus Boiss. and Bal. An AE

T. decussatus Bentham Sinai Saudi Arabia

T. eigii (Zohary and Davis) Jalas An

T. holosericeus Celak. 2n=28 Gr

T. laconicus Jalas Gr

T. lanceolatus Desf. Ag

T. leucostomus Hausskn. and Velen. An

T. maroccanus Ball. subsp. *maroccanus* Ma
  subsp. *rhombicus* Huguet del Villar Ma

T. numidicus Poiret Ag

T. pallescens de Noé (*T. fontanesii*) Ag

T. sipyleus Boiss. subsp. *sipyleus* An AE
  subsp. *rosulans* (Borbás) Jalas

T. syriacus Boiss. An LS Iraq

VIII. Sect. *Serpyllum* (Miller) Bentham

VIII1. Subsect. *Insulares* Jalas

T. bornmuelleri Velen. An

T. dreatensis Batt. Ma Ag

T. guyonii De Noe Ag

T. origanoides Webb and Berthelot 2n=28 Canary Islands

T. richardii Pers. subsp. *richardii* 2n=28, 30 Bl Ju
  subsp. *ebusitanus* (Font Quer) Jalas 2n=30 Bl
  subsp. *nitidus* (Guss.) Jalas 2n=28 Si

T. willkommii Ronniger 2n=56 Hs

VIII2. Subsect. *Kotschyani* (Klokov) Jalas

T. ararati-minoris Klokov and Shost.

T. armeniacus Klokov

T. binervulatus Klokov and Shost.
T. carmanicus Jalas
T. collinus M. Bieb.
T. daenensis Celak. subsp. *daenensis*
  subsp. *lancifolius* (Celak.) Jalas
T. desjatovae Ronniger
T. eriocalyx (Ronniger) Jalas
T. eriophorus Ronniger
T. fallax Fisch. and Meyer An
T. fedtschenkoi Ronniger
T. fominii Klokov and Shost.
T. guberlinensis Iljin
T. intercedens (H. Braun) Rech. fil.
T. kjapazi Grossh.
T. koeieanus Ronniger
T. kotschyanus Boiss. and Hohen.
T. laevigatus Vahl Yemen
T. migricus Klokov and Shost.
T. pubescens Boiss. and Kotschy ex Celak.
T. punctatus Vis.
T. roseus Schipcz.
T. schimperi Ronniger subsp. *schimperi* 2n=30 Ethiopia
  subsp. *hedbergianus* Sebsebe Ethiopia
T. serrulatus Hochst. ex Bentham Ethiopia
T. squarrosus Fisch. and Meyer
T. transcaspicus Klokov
T. transcaucasicus Ronniger An
T. trautvetteri Klokov and Shost.
T. ziaratinus Klokov and Shost.

VIII3. Subsect. *Pseudopiperellae* Jalas
T. herba-barona Loisel. subsp. *herba-barona* 2n=56, 84 Co Sa
  subsp. *bivalens* 2n=28 Bl
T. nitens Lamotte 2n=28 Ga

VIII4. Subsect. *Isolepides* (Borbás) Halácsy
T. bulgaricus (Dom and Podp.) Ronniger
T. coriifolius Ronniger
T. czernjaevii (*tschernjajevii*) Klokov and Shost.
T. dimorphus Klokov and Shost.
T. elisabethae Klokov and Shost.
T. glabrescens Willd. subsp. *glabrescens* 2n=28, 32, 56, 58
  subsp. *decipiens* (H. Braun) Domin 2n=52
  subsp. *urumovii* (Velen.) Jalas 2n=28, 56
T. karamarianicus Klokov and Shost.
T. klokovii (Ronniger) Shost.
T. latifolius (Bess.) Andr.
T. lavrenkoanus Klokov
T. longedentatus (Degen and Urum.) Ronniger 2n=30, 90 Gr

T. markhotensis Malejev
T. pannonicus All. 2n=28, 35 URSS, China
T. podolicus Klokov and Shost.
T. przewalskii Kom.
T. sibthorpii Bentham 2n=28
T. tiflisiensis Klokov and Shost.
T. turczaninovii Serg.

VIII5. Subsect. *Alternantes* Klokov
T. alpestris Tausch ex A. Kerner 2n=28
T. alternans Klokov
T. bihoriensis Jalas
T. buschianus Klokov and Shost.
T. caucasicus Willd. ex Ronniger
T. comosus Heuffel ex Griseb. and Schenk 2n=28, 58
T. disjunctus Klokov URSS, China
T. froelichianus Opiz 2n=56
T. komarovii Serg. 2n=24, 26
T. nummularius M. Bieb.
T. oehmianus Ronniger and Soska
T. pseudonummularius Klokov and Shost.
T. pseudopulegioides Klokov and Shost.
T. pulchellus C. A. Meyer
T. pulegioides L. 2n=28, 30
T. semiglaber Klokov

VIII6. Subsect. *Pseudomarginati* (Braun ex Borbás) Jalas
T. linearis Bentham subsp. *linearis*
   subsp. *hedgei* Jalas
T. longicaulis C. Presl. subsp. *longicaulis* 2n=26, 28, 30, 50, 56, 58
   subsp. *chaubardii* (Boiss. and Heldr. ex Reichenb. fil.) Jalas
T. nervosus Gay ex Willk. 2n=28 Ga Hs
T. ocheus Heldr. and Sart. ex Boiss. An Bu Gr Ju
T. praecox Opiz subsp. *praecox* 2n=24, 50, 54, 56, 58
   subsp. *skorpilii* (Velen.) Jalas 2n=28, 56
   subsp. *polytrichus* (A. Kerner ex Borbas) Jalas 2n=28, 50, 54, 55, 56
   subsp. *britannicus* (Ronniger) Holub 2n=28, 50, 54, 56
   subsp. *zygiformis* (H. Braun) Jalas
   subsp. *grossheimii* (Ronniger) Jalas
T. pulcherrimus Schur subsp. *pulcherrimus* 2n=28, 56
   subsp. *carpathicus* (Celak.) Mártonfi
T. stojanovii Degen. Bu Gr Ju
T. thracicus Velen. 2n=28, 56, 58 Al An Bu Gr Ju Tu

VIII7. Subsect. *Serpyllum*
T. alatauensis (Klokov and Shost.) Klokov
T. altaicus Klokov and Shost. URSS, China
T. amurensis Klokov URSS, China
T. arsenijevii Klokov
T. aschurbajevii Klokov

T. asiaticus Serg. 2n=26
T. bituminosus Klokov
T. bucharicus Klokov
T. cerebrifolius Klokov
T. chancoanus Klokov
T. crenulatus Klokov
T. curtus Klokov URSS, China
T. diminutus Klokov
T. diversifolius Klokov
T. eravinensis Serg.
T. eubajcalensis Klokov
T. extremus Klokov
T. flexilis Klokov
T. glacialis Klokov
T. iljinii Klokov and Shost.
T. inaequalis Klokov URSS, China
T. jenisseensis Iljin
T. mandschuricus Ronniger 2n=24 China
T. minussinensis Serg.
T. mongolicus Klokov URSS, China
T. narymensis Serg.
T. nervulosus Klokov URSS, China
T. ochotensis Klokov
T. oxyodontus Klokov 2n=24, 28
T. phyllopodus Klokov 2n=24
T. proximus Serg. URSS, China
T. quinquecostatus Celak. 2n=24, 26 China, Japan
T. reverdattoanus Serg.
T. schischkinii Serg.
T. seravshanicus Klokov
T. serpyllum L. subsp. *serpyllum* 2n=24, 26
   subsp. *tanaensis* (Hyl.) Jalas 2n=24
T. sibiricus (Serg.) Klokov and Shost.
T. sokolovii Klokov
T. talijevii Klokov and Shost.
T. tonsilis Klokov
T. ussuriensis Klokov

Appendix

List of hybrids in the Iberian Peninsula. Hybridization occurs frequently in the Iberian Peninsula, where 60 hybrids have been detected and some of them described.

T. baeticus Boiss. ex Lacaita x *T. hyemalis* Lange
T. x indalicus Blanca, Cueto, Gutiérrez and Martínez, Folia Geobot. Phytotax 28: 138 fig. 1 (VIII-1993)
T. x garcia-martinoi Sánchez Gómez and Sáez in Saéz, Sánchez Gómez and Morales, Anales Jard. Bot. Madrid 51(1): 158 (XII-1993)

T. baeticus Boiss. ex Lacaita x *T. mastichina* (L.) L. subsp. *mastichina*
T. x arundanus Willk., Oesterr. Bot. Z. 41: 52 (1891), pro sp.
T. x fontquerianus Pau, Mem. Mus. Ci. Nat. Barcelona, Ser. Bot. 1(1): 61 (1922)

T. baeticus Boiss. ex Lacaita x *T. zygis* subsp. *gracilis* (Boiss.) R. Morales
T. x arcanus G. López and R. Morales, Anales Jard. Bot. Madrid 41(1): 94 (1984)

T. bracteatus Lange ex Cutanda x *T. mastichina* (L.) L. subsp. *mastichina*
T. x bractichina R. Morales, Anales Jard. Bot. Madrid 43(1): 39 (1986)
T. x pectinatus R. Morales, Anales Jard. Bot. Madrid 41(1): 94 (1984) non Fischer and Meyer, nom. illeg.
T. x rivas-molinae Mateo and M. B. Crespo, Rivasgodaya 7: 130 (1993)
T. x sennenii Pau var. *leucodonthus* Pau, Bol. Soc. Aragonesa Ci. Nat. 15: 160 (1916), nom. inval.
T. x sennenii auct. non Pau

T. bracteatus Lange ex Cutanda x *T. pulegioides* L.

T. bracteatus Lange ex Cutanda x *T. zygis* Loefl. ex L. subsp. *zygis*
T. x borzygis Mateo and M. B. Crespo, Thaiszia, Kosice 3(1): 7 fig. 2 (1993)

T. caespititius Brot. x *T. mastichina* (L.) L. subsp. *mastichina*
T. x henriquesii Pau, Brotéria, Sér. Bot. 22: 121 (1926)

T. camphoratus Hoffmanns. and Link x *T. mastichina* (L.) L. subsp. *mastichina*
T. x ramonianus Paiva and Salgueiro, Anales Jard. Bot. Madrid 52(1): 114 fig. 2 (1994)

T. carnosus Boiss. x *T. mastichina* (L.) L. subsp. *mastichina*
T. x welwitschii Boiss., Diagn. Pl. Orient. 3(4): 9 (1859), pro sp.
T. noeanus Rouy, Bull. Soc. Bot. France 52: 507 (1905)

T. funkii Cosson x *T. vulgaris* L. subsp. *vulgaris*
T. x lainzii Sánchez Gómez, Fernández Jiménez and Sáez in Sánchez Gómez and Fernández Jiménez, Anales Jard. Bot. Madrid 54

T. funkii Cosson x *T. zygis* subsp. *gracilis* (Boiss.) R. Morales
T. x paradoxus Rouy, Bull. Soc. Bot. France 20: 78 (1883)

T. granatensis Boiss. subsp. *granatensis* x *T. longiflorus* Boiss.
T. x almijarensis Ruiz de la Torre and Ruiz del Castillo, Ecología 6: 103 fig. 2(1992), pro sp.

T. granatensis Boiss. subsp. *granatensis* x *T. serpylloides* subsp. *gadorensis* (Pau) Jalas

T. granatensis Boiss. subsp. *granatensis* x *T. orospedanus* Huguet del Villar
T. x mariae Socorro, Arrébola and Espinar, Lagascalia 16(1): 121 (1991)

T. hyemalis Lange x *T. mastichina* (L.) L. subsp. *mastichina*
T. x mastichinalis Sánchez Gómez and Sáez in Sáez, Sánchez Gómez and Morales, Anales Jard. Bot. Madrid 51(1): 158 (1993)

T. hyemalis Lange x *T. moroderi* Pau ex Martínez
T. x diazii Alcaraz, Rivas Martínez and Sánchez Gómez, Itinera Geobot. 2: 118 (1989)

T. hyemalis Lange x *T. vulgaris* subsp. *aestivus* (Reuter ex Willk.) O. Bolós and A. Bolós

T. hyemalis Lange x *T. zygis* subsp. *gracilis* (Boiss.) R. Morales
T. x enicensis Blanca, Cueto, Gutiérrez and Martínez, Folia Geobot. Phytotax. **28**(2): 138 fig. 2 (VIII-1993)

T. x sorianoi Sáez and Sánchez Gómez in Sáez, Sánchez Gómez and Morales, Anales Jard. Bot. Madrid 51(1): 158 (XII-1993)

T. lacaitae Pau x *T. vulgaris* L. subsp. *vulgaris*
T. x armuniae R. Morales, Anales Jard. Bot. Madrid 41(1): 94 (1984)

T. lacaitae Pau x *T. zygis* subsp. *sylvestris* (Hoffmanns. and Link) Coutinho
T. x arcuatus R. Morales, Anales Jard. Bot. Madrid 41(1): 93 (1984)

T. leptophyllus subsp. *izcoi* (Rivas Martínez, Molina and Navarro) R. Morales x *T. mastichina* (L.) L. subsp. *mastichina*
T. x celtibericus Pau, Mem. Real. Soc. Esp. Hist Nat. 15: 71 (1929)

T. leptophyllus subsp. *izcoi* (Rivas Martínez, Molina and Navarro) R. Morales x *T. vulgaris* L. subsp. *vulgaris*
T. x moralesii nothosubsp. *navarroi* (Mateo and M. B. Crespo) R. Morales, Anales Jard. Bot. Madrid 53(2): 208 (1995)
T. x navarroi Mateo and M. B. Crespo, Rivasgodaya 7: 132 (1993)

T. leptophyllus Lange subsp. *leptophyllus* x *T. mastichina* (L.) L. subsp. *mastichina*
T. x celtibericus nothosubsp. *bonichensis* (Mateo and M. B. Crespo) R. Morales, Anales Jard. Bot. Madrid 53(2): 202 (1995)
T. x bonichensis Mateo and M. B. Crespo, Thaiszia, Kosice 3(1): 5 fig. 1 (1993)

T. leptophyllus Lange subsp. *leptophyllus* x *T. vulgaris* L. subsp. *vulgaris*
T. x moralesii nothosubsp. *cistetorum* Mateo and M. B. Crespo, Anales Jard. Bot. Madrid 49(2): 288 (1992)

T. leptophyllus Lange subsp. *leptophyllus* x *T. zygis* Loefl. ex L. subsp. *zygis*
T. x xilocae Mateo and M. B. Crespo, Anales Jard. Bot. Madrid 49(2): 289 (1992)

T. leptophyllus subsp. *paui* R. Morales x *T. pulegioides* L.
T. x benitoi Mateo, Mercadal and Pisco, Bot. Complutensis 20: 70 fig. 1 (1996)

T. leptophyllus subsp. *paui* R. Morales x *T. vulgaris* L. subsp. *vulgaris*
T. x moralesii Mateo and M. B. Crespo in Mateo, Cat. Fl. Teruel: 234 (1990)

T. longiflorus Boiss. x *T. zygis* subsp. *gracilis* (Boiss.) R. Morales
T. x ruiz-latorrei C. Vicioso in Ruiz del Castillo, Anales Inst. Nac. Invest. Agrar., Ser. Rec. Nat. 1: 31 lam. 16 (1974), pro sp.

T. loscosii Willk. x *T. mastichina* (L.) L. subsp. *mastichina*
T. x riojanus Uribe-Echebarría, Est. Mus. Ci. Nat. Alava 5: 67 fig. 1 (1990)

T. loscosii Willk. x *T. vulgaris* L. subsp. *vulgaris*
T. x rubioi Font Quer, Treb. Mus. Ci. Nat. Barcelona, Ser. Bot. 3: 215 (1920)

T. lotocephalus G. López and R. Morales x *T. mastichina* subsp. *donyanae* R. Morales
T. x mourae Paiva and Salgueiro, Anales Jard. Bot. Madrid 52(1): 114 fig. 1 (1994)

T. mastichina (L.) L. subsp. *mastichina* x *T. mastigophorus* Lacaita
T. x ibericus Sennen and Pau in Sennen, Bull. Acad. Int. Géogr. Bot. 18 (229): 461 (1908)

T. mastichina (L.) L. subsp. *mastichina* x T. *orospedanus* Huguet del Villar
T. x mixtus Pau, Carta Bot. 3: 7 (1906)

T. mastichina (L.) L. subsp. *mastichina* x *T. praecox* subsp. *britannicus* (Ronniger) Holub
T. x genesianus Galán Cela, Anales Jard. Bot. Madrid 45(2): 562 fig. 1 (1989)

T. mastichina (L.) L. subsp. *mastichina* x *T. pulegioides* L.
T. x sennenii Pau, Bol. Soc. Aragonesa Ci. Nat. 6: 29 (1907)
T. jovinieni Sennen and Pau in Pau, op. cit.

T. mastichina (L.) L. subsp. *mastichina* x *T. serpylloides* subsp. *gadorensis* (Pau) Jalas
T. x hieronymi Sennen, Diagn. Nouv. Pl. Espagne Maroc: 92 (1936)

T. mastichina (L.) L. subsp. *mastichina* x *T. serpylloides* Bory subsp. *serpylloides*
T. x hieronymi nothosubsp. *hurtadoi* (Socorro, Molero Mesa, Casares and Pérez Raya)
    R. Morales, Anales Jard. Bot. Madrid 43(1): 39 (1986)
T. x hurtadoi Socorro, Molero Mesa, Casares and Pérez Raya, Trab. Dep. Bot. Univ.
    Granada 6: 109 (1981)

T. mastichina (L.) L. subsp. *mastichina* x *T. villosus* subsp. *lusitanicus* (Boiss.) Coutinho
T. x toletanus Ladero, Anales Inst. Bot. Cavanilles 27: 97 fig. 6 (1970)

T. mastichina (L.) L. subsp. *mastichina* x *T. vulgaris* L. subsp. *vulgaris*
T. x eliasii Sennen and Pau in Sennen, Bol. Soc. Ibérica Ci. Nat. 32: 79 (1933); in Pau,
    Cavanillesia 4: 55 (1931), nom. inval.

T. mastichina (L.) L. subsp. *mastichina* x *T. zygis* subsp. *sylvestris* (Hoffmanns. and Link)
    Coutinho
T. x brachychaetus (Willk.) Coutinho, Bol. Soc. Brot. 23: 79 (1907), pro var.

T. mastichina var. *brachychaetus* Willk. in Willk. and Lange, Prodr. Fl. Hispan. 2: 400 (1968)
T. x mixtus var. *toletanus* Pau, Bol. Soc. Aragonesa Ci. Nat. 15: 160 (1916)

T. mastichina (L.) L. subsp. *mastichina* x *T. zygis* Loefl. ex L. subsp. *zygis*

T. mastigophorus Lacaita x *T. vulgaris* L. subsp. *vulgaris*
T. x severianoi Uribe-Echebarría, Est. Mus. Ci. Nat. Alava 5: 69 figs. 3a y 4b (1990)

T. mastigophorus Lacaita x *T. zygis* Loefl. ex L. subsp. *zygis*
T. x zygophorus R. Morales, Anales Jard. Bot. Madrid 41(1): 93 (1984)

T. membranaceus Boiss. x *T. moroderi* Pau ex Martínez

T. membranaceus Boiss. x *T. orospedanus* Huguet del Villar
T. x beltranii Socorro, Espinar and Arrébola, Lagascalia 17(1): 186 (1993)

T. membranaceus Boiss. x *T. vulgaris* L. subsp. *vulgaris*
T. x guerrae Sáez and Sánchez Gómez in Sáez, Sánchez Gómez and Morales, Anales Jard.
    Bot. Madrid 51(1): 157 (1993)

T. membranaceus Boiss. x *T. zygis* subsp. *gracilis* (Boiss.) R. Morales
T. x almeriensis G. López and R. Morales, Anales Jard. Bot. Madrid 41(1): 94 (1984)

T. moroderi Pau ex Martínez x *T. vulgaris* L. subp. *vulgaris*
T. x carrionii Sáez and Sánchez Gómez in Sáez, Sánchez Gómez and Morales, Anales
    Jard. Bot. Madrid 51(1): 157 (1993)

T. moroderi Pau ex Martínez x *T. zygis* subsp. *gracilis* (Boiss.) R. Morales
T. x martinezii Pau ex Martínez, Mem. Real Soc. Esp. Hist. Nat. 14: 467 fig. 7 (1934), pro sp.

T. funkii var. *martinezii* (Pau ex Martínez) C. Vicioso, Anales Inst. Nac. Invest. Agrar., Ser. Rec. Nat. 1: 19 (1974)
T. capitatus Lag., Elench. Pl.: 18 (1816), non (L.) Hoffmanns. and Link (typus: MA 106457)
T. villosus sensu Willk., Suppl. Prodr. Fl. Hispan.: 146 (1893)

T. orospedanus Huguet del Villar x *T. zygis* subsp. *gracilis* (Boiss.) R. Morales
T. x jimenezii Socorro, Arrébola and Espinar, Lagascalia 16(1): 122 (1991)

T. piperella L. x *T. vulgaris* subsp. *aestivus* (Reuter ex Willk.) O. Bolós and A. Bolós
T. x josephi-angeli Mansanet and Aguilella, Mediterránea, Ser. Biol. 8: 84 (1985)

T. piperella L. x *T. vulgaris* L. subsp. *vulgaris*
T. x josephi-angeli nothosubsp. *edetanus* Mateo, M. B. Crespo and Laguna, Anales Jard. Bot. Madrid 49(1): 140 fig. 1 (1991)

T. pulegioides L. x *T. vulgaris* L. subsp. *vulgaris*
T. x carolipaui Mateo and M. B. Crespo in Mateo, Cat. Fl. Teruel: 232 (1990)

T. pulegioides L. x *T. zygis* subsp. *gracilis* (Boiss.) R. Morales

T. pulegioides L. x *T. zygis* Loefl. ex L. subsp. *zygis*
T. x viciosoi Pau ex R. Morales, Anales Jard. Bot. Madrid 53(2): 210 (1995)
T. x viciosoi (Pau) R. Morales, Anales Jard. Bot. Madrid 43(1): 41 (1986), comb. inval.
T. bracteatus f. *viciosoi* Pau, Bol. Soc. Aragonesa Ci. Nat. 15: 159 (1916), nom. inval.

T. serpylloides subsp. *gadorensis* x *T. vulgaris* subsp. *aestivus*
T. x aitanae nothosubsp. *dominguezii* (Socorro and Arrébola) R. Morales, Anales Jard. Bot. Madrid 53(2): 200 (1995)
T. x dominguezii Socorro and Arrébola, Lagascalia 17(2): 355 (1995)

T. serpylloides subsp. *gadorensis* x *T. vulgaris* subsp. *vulgaris*
T. x aitanae Mateo, M. B. Crespo and Laguna, Anales Jard. Bot. Madrid 49(1): 142 fig. 3 (1991)

T. serpylloides subsp. *gadorensis* (Pau) Jalas x *T. zygis* subsp. *gracilis* (Boiss.) R. Morales
T. x pastoris Socorro and Arrebola, Lagascalia 17(2): 353 (1995)

T. vulgaris L. subsp. *vulgaris* x *T. zygis* Loefl. ex L. subsp. *zygis*
T. x monrealensis Pau ex R. Morales, Anales Jard. Bot. Madrid 41(1): 93 (1984)
T. x monrealensis Pau, Mem. Real Soc. Esp. Hist. Nat. 15: 71 (1929), nom. inval. sine descr.

T. vulgaris L. subsp. *vulgaris* x *T. zygis* subsp. *gracilis* (Boiss.) R. Morales
T. x monrealensis nothosubsp. *garcia-vallejoi* Sánchez Gómez, Alcaraz and Sáez, Anales Jard. Bot. Madrid 49(2): 289 (1992)

ACKNOWLEDGEMENTS

Thanks are given to Juan Castillo and Leopoldo Medina for their drawings. This work was made in part under the financial support of the Project Flora Iberica V PB96-0849 of the DGICyT, Spain, that has transferred some unpublished drawings. Part of this work has been possible thanks to the 'Acciones integradas hispano-austríacas HU96-13 and HU1997-34' from Subdirección General de Cooperación Internacional, Spain.

REFERENCES

Barrelier, J. (1714) *Plantae per Galliam, Hispaniam et Italiam observatae iconibus aeneis exhibitae*, Paris.

Bauhin, C. (1623) *Pinax theatri botanici*, Basel.

Bentham, G. (1834) *Lab. Gen. Sp.: Thymus*, London.

Boissier, E. (1838) *Elenchus Plantarum Novarum minusque cognitorum quas in itinere hispanico*, Geneve, pp. 73–76.

Boissier, E. (1839–1845) *Voyage botanique dans le Midi de l'Espagne*, Paris.

Boissier, E. (1859) *Diagn. Pl. Orient. Novarum*, Ser. 2, Vol. 3, Lipsiae and Parisiis, p. 9.

Boissier, E. (1867–1884) *Flora orientalis*, Basel, Geneva, Lyons.

Briquet, J. (1897) Labiatae. In A. Engler and K. Prantl (eds), *Die Natürlichen Pflanzenfamilien* IV 38(a), Leipzig.

Brotero, F. A. (1804) *Flora Lusitanica*, Lisbon.

Clusius, C. (1576) *Rariorum aliquot stirpium per Hispanias observatarum historia*, Antwerp.

Dodonaeus, R. (1616) *Stirpium Historiae Pemptadis secundae liber tertius*, Antwerp.

Elena-Rosselló, J.A. (1976) *Projet d'une etude de taxonomie expérimentale du genre Thymus*. Doctoral thesis, Montpellier.

Hallier, E. (1884) *Flora von Deutschland*, p. 188.

Hartvig, P. (1987) A taxonomical revision of *Thymus* sect. *Teucrioides (Lamiaceae). Pl. Syst. Evol.*, 155, 197–213.

Hoffmannsegg, J.C. and Link, H.F. (1809) *Flore portugaise*, Vol. 1, Berlin, 123–138.

Huerta, G. (1629) *Historia Natural de Cayo Plinio Segundo* II, Madrid.

Huguet del Villar, E. (1934) Quelques *Thymus* du Sud-est Ibérique. *Cavanillesia*, 6, 104–125.

Jalas, J. (1971) Notes on *Thymus* L. *(Labiatae)* in Europe. I. Supraspecific classification and nomenclature. *Bot. J. Linn. Soc.*, 64, 199–215.

Jalas, J. (1972) Thymus L. In T. Tutin, V.H. Heywood, N.A. Burges, D.M. Moore and D.H. Valentine (eds), *Flora Europaea*, Vol. 3, University Press, Cambridge, pp. 172–182.

Jalas, J. (1982) Thymus. In K.H. Rechinger (ed.), *Flora Iranica*, Graz, pp. 532–551.

Jalas, J. (1982) Thymus. In P.H. Davis (ed.), *Flora of Turkey and the East Aegean Islands*, Vol. 7, Edinburgh, pp. 349–382.

Laguna, A. (1555) *Pedacio Dioscórides Anazarbeo. Acerca de la materia medicinal y de los venenos mortíferos*, Salamanca.

Linné, C. (1738) *Hortus Cliffortianus*, Amstelaedami.

Linné, C. (1748) *Hortus Upsaliensis* I, Stockholm.

Linné, C. (1753) *Species Plantarum*, (ed. 1), Holmiae.

Linné, C. (1754) *Genera Plantarum*, Holmiae.

Linné, C. (1762–1763) *Species Plantarum*, (ed. 2), Stockholm.

Linné, C. (1767) *Systema Naturae*, (ed. 12), reformata, Holmiae.

Lobelius, M. (1581) *Elenchus Plantarum fere congenerum*, Antwerp.

Martínez, M. (1936) Sobre algunas plantas valencianas citadas en los "icones" de Barrelier. *Bol. Soc. Esp. Hist. Nat.*, 36, 199–204.

Mayol, M., Rosselló, J. A., Mus, M. and Morales, R. (1990) *Thymus herba-barona* Loisel., novedad para España, en Mallorca. *Anales Jard. Bot. Madrid*, 47, 516.

Morales, R. (1986) Taxonomía de los géneros *Thymus* (excluida la sección *Serpyllum*) y *Thymbra* en la Península Ibérica. *Ruizia*, 3, 1–324.

Morales, R. (1989) El género *Thymus* L. en la región mediterránea occidental (Lamiaceae). *Biocosme Mésogéen*, 6, 205–211.

Morales, R. (1995) Híbridos de *Thymus* L. (Labiatae) en la Península Ibérica. *Anales Jard. Bot. Madrid*, 53, 199–211.

Morales, R., Maciá, M.J., Dorda, E., and García-Villaraco, A. (1996) *Nombres vulgares* II. Archivos de Flora Iberica 7. Real Jardín Botánico. Madrid.

Morales, R. (1997) Synopsis of the genus *Thymus* L. in the Mediterranean area. *Lagascalia*, 19, 249–262.

Pau, L. (1929) Introducción al estudio de los tomillos españoles. *Mem. Real. Soc. Esp. Hist. Nat.*, 15, 65–71.

Tournefort, J.P. (1719) *Institutiones Rei Herbariae*, (ed. 3), Paris.

Velenovsky, J. (1906) Vorstudien zu einer Monographie der Gattung Thymus L. *Bot. Zentralbl. Beih.*, 19 B2, 271–287.

Vicioso, C. (1974) Contribución al conocimiento de los tomillos españoles. In J. Ruiz del Castillo (ed.), *Anales Inst. Nac. Invest. Agrar. ser. Recursos Nat.*, 1, 11–63.

Willkomm, M. (1868) Labiatae. In M. Willkomm and J. Lange (eds), *Prodr. Fl. Hisp.*, 96. vol. 2, Stuttgart, pp. 389–480.

Wunderlich, R. (1967) Ein Vorschlag zu einer natürlichen Gliederung der Labiaten auf Grund der Pollenkörner, der Samenentwicklung und des reifen Samens. *Österr. Bot. Z.*, 114, 383–483.

2 Population structure and the spatial dynamics of genetic polymorphism in thyme

John D. Thompson

THE SPATIAL STRUCTURE OF GENETIC DIVERSITY IN THYME

Ecological and evolutionary significance of the population structure

Many plant species occur as a mosaic of local populations in discrete patches dispersed across the landscape. In many populations, pollen and seed dispersal are highly localised, increasing the tendency for reproduction to occur within spatially localised groups. Spatial structure is thus a characteristic of plant populations. Where natural selection favours particular genotypes in particular sites or spatial isolation permits random genetic drift, genetic differentiation can be marked and highly localised.

Evolution at the level of the local population can also be influenced by regional processes associated with the arrangement of populations in the landscape and the rates of colonisation and extinction of such populations. In some cases, the colonisation of new populations may involve a small number of genetically related individuals that are only a subset of the genetic variation contained by source populations. Such founder events can have a marked effect on the genetic variation within populations and also the spatial organisation of the genetic variation in the landscape. A critical point here is that such founder events may produce spatial variation in phenotypic traits which cause their evolution to follow new directions in colonist populations.

Polymorphic variation in thyme

The genus *Thymus* provides a particularly interesting situation to study the ecological and evolutionary significance of the spatial population structure. Since the early 1960s, one species, *Thymus vulgaris* has been at the heart of ecological and genetic research on the evolutionary dynamics of not just one but two genetic polymorphisms (see review by Thompson *et al.*, 1998). First, like most labiates, thyme is an aromatic plant: glandular trichomes on the leaves and floral parts contain monoterpenoid essential oils. Thyme plants vary in the monoterpene composition of their essential oils, one monoterpene being present in a high percentage for a particular plant. In the south of France, one of six different monoterpenes may dominate the essential oil produced by a plant species, and thus six different chemical forms can be detected. This secondary compound variation has a genetic basis and the presence of six distinct genetically-based forms has thus provided an attractive system to explore the ecological genetics of secondary compound variation, in particular the role of this variation in mediating inter-specific interactions and the determinants of variation in relative abundance across the landscape.

The second of the two polymorphisms is gynodioecy, a sexual polymorphism in which natural populations contain two types of plants: females and hermaphrodites. Hermaphrodites bear only perfect flowers whilst females bear smaller flowers that lack anthers or have only rudimentary anthers that do not bear pollen. All flowers on a given plant are either female or hermaphrodite. Darwin (1877) first remarked on the occurrence of gynodioecy in the genus *Thymus*, and since 1963, when Professor Valdeyron first started counting and observing females in the garrigues landscape around Montpellier in southern France, the functional significance and evolutionary dynamics of gynodioecy in wild thyme has greatly intrigued researchers. The purpose of this research has been to elucidate why females are so abundant in many populations and what causes variation in their abundance among populations.

The purpose of this chapter is to review work on the ecology and evolution of the chemical and sexual polymorphisms, with focus on the importance of the spatial population structure in *T. vulgaris*, the species which has received by far the most attention.

THE ESSENTIAL OIL OF THYME: THE ECOLOGICAL GENETICS OF A CHEMICAL POLYMORPHISM

The Mediterranean is full of aromatic plants. Data on the ecological role of the monoterpenes which characterise the essential oils of Mediterranean aromatic plants remains however rather thin on the ground. *Thymus vulgaris* shows genetic variation in the production of monoterpenes, providing a fascinating opportunity to study the ecological role and evolutionary significance of monoterpene production.

The genetic basis of thyme monoterpenes

In southern France, natural populations of *T. vulgaris* contain one or several of six genetically distinct chemical forms (hereafter chemotypes) that can be distinguished on the basis of the dominant monoterpene produced in glandular trichomes on the surface of the leaves and calyces (Passet, 1971; Vernet *et al.*, 1986). Each of the six chemotypes, geraniol (G), α-terpineol (A), *tr*-sabinene hydrate or thuyanol-4 (U), linalool (L), carvacrol (C) and thymol (T), is named after the dominant monoterpene in the essential oil of a plant. Each monoterpene is at the end of a branch in a common reaction chain that has as precursor, geranyl pyrophosphate (Figure 2.1). The six monoterpenes have different molecular structures (Figure 2.1), with an important difference being the phenolic nature of carvacrol and thymol and the non-phenolic nature of the other four monoterpenes. In Spanish populations the geraniol chemotype has not been observed, whereas a seventh chemotype, 1,8-cineole is present (Adzet *et al.*, 1977). Some of the chemotypes are discernible to the human nose in the field. This is particularly so for geraniol which often has a lemon smell and the two phenolics which have the characteristic thyme odour which makes them readily distinguishable from the four non-phenolic types.

The first description of the essential oil variation in *T. vulgaris* occurred in the early 1960s (Granger *et al.*, 1963) with a more comprehensive study arriving a few years later in the form what was to be the first of a long line of PhD theses on this species in Montpellier (Passet, 1971). The construction of a chromatograph capable of determining the chemical phenotype of a plant from a small sample (3–4 leaves), permitted the

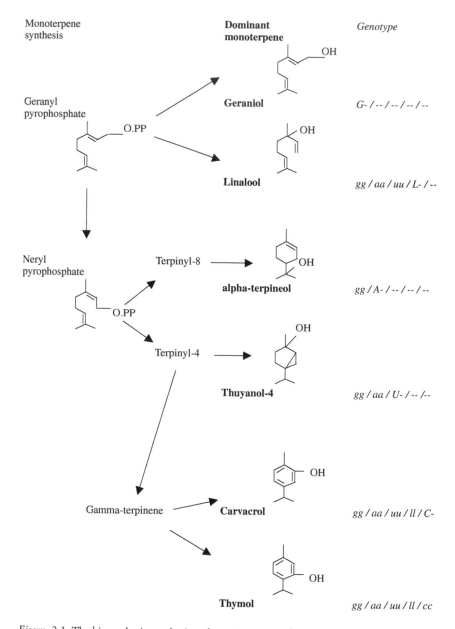

Figure 2.1 The biosynthetic synthesis and genetic control of the dominant monoterpenes in *T. vulgaris* in southern France (based on Passet, 1971; Vernet *et al.* 1986).

rapid determination of the chemotype for a large number of individuals (Passet, 1971; Gouyon *et al.*, 1981). This method of analysis made it possible to study the genetic control and spatial distribution of the chemical forms, work which required an extensive program of controlled crosses and natural population sampling, i.e. thousands of plants (Gouyon *et al.*, 1986; Vernet *et al.*, 1986).

The presence of the dominant monoterpene in *T. vulgaris* is controlled by an epistatic series of five biosynthetic loci that has the following sequence: $G \rightarrow A \rightarrow U \rightarrow L \rightarrow C \rightarrow T$ (Vernet *et al.*, 1986). As can be seen in Figure 2.1, a plant with a dominant G allele will have the G phenotype, regardless of whether it has dominant or recessive alleles at the other loci (Vernet *et al.*, 1986). Two loci probably code for the G phenotype, otherwise a single pair of alleles at each locus codes for the remaining chemotypes. If a plant is homozygous recessive at the G loci (i.e. gg) and has a dominant A allele then it will have the A phenotype, again regardless of whether it has dominant or recessive alleles at the other loci. If the plant is homozygous recessive at the G and A loci (i.e. gg, aa), but has a dominant U allele then it will have the U phenotype, and so on down the chain (Figure 2.1). A plant homozygous for recessive alleles at all five loci has the T phenotype. Figure 2.1 also illustrates (see Passet, 1971 for details) that the metabolic pathway leading to the production of the two phenolic chemotypes is much longer than that of the non-phenolic and that there is an almost perfect fit between the genetic chain and the metabolic chain, only linalool is "out of place". The basis of this non-concordance remains unknown.

The most plausible explanation for the relation between genetic constitution and dominant monoterpene (i.e. genotype and phenotype) is that there is a series of regulatory proteins coded by alleles at the G, A, U, L and C loci, each of which can interrupt, at different stages, the sequence of reactions that would normally lead to the synthesis of the T phenotype (i.e. homozygous recessive for all loci). An alternative possibility is that enzyme G may consume all the substrate for a particular reaction causing chemotype G to be produced; likewise for the other chemotypes down the chain.

Spatial structure and the adaptive value of thyme monoterpenes

The ecological role of secondary compounds in thyme can be addressed under three headings: adaptation to the abiotic environment; competitive interactions with other plants; and chemical defence against herbivores and pathogens. Clearly, these factors are unlikely to act in isolation from one another, indeed the dynamics of the secondary compound polymorphism are most likely influenced by the combined and interactive effects of the different features of the abiotic and biotic environment.

Spatial distribution of chemotypes: an adaptation to the abiotic environment?

There are several pieces of evidence which indicate that the monoterpene variation may represent an adaptive strategy in relation to environmental variation. The first piece of evidence for adaptive variation concerns the geographic and localised distribution of chemotypes in *T. vulgaris* in southern France. Based on bulk samples of plants, it is clear that phenolic chemotypes (C and T) dominate thyme populations in hot dry sites close to the Mediterranean sea, whereas the non-phenolic (G, A, U and L) chemotypes dominate sites further inland, particularly above 400 m elevation, i.e. in wetter, cooler climates (Passet, 1971; Granger and Passet, 1973). This trend has been confirmed by a recent study which compared the actual frequency of plants of each chemotype in approximately 12 sites at low altitude (<300 m) close to the Mediterranean with a similar number of sites above 400 m in the Drome valley (J. Thompson and J.-C. Chalchat, unpublished data). Above 400 m no phenolic plants were found.

Spatial differentiation in the distribution of the different chemotypes in *T. vulgaris* also occurs on a very localised scale of 8×10 km in and around the St Martin-de-Londres valley roughly 25 km north of Montpellier in southern France (see Vernet *et al.*, 1977a, b; Gouyon *et al.*, 1986). This valley is reputed for the temperature inversion that can occur in winter due to the accumulation of cold moist air in the valley. In winter, temperatures are often several degrees colder than in the hills that surround the valley. In the valley, soils are less stony, deeper and moister than in the surrounding hills. Again based on bulk samples of plants, the climatic and soil gradient has been found to be correlated with differences in the distribution of different chemotypes (Figure 2.2a and b). Phenolic chemotypes predominate over large areas of hot, dry limestone plateau areas on shallow, stony iron-rich soils around the valley above 250 m elevation (Figure 2.2b). Below 200 m elevation inside the valley, where populations are often fragmented by agricultural land-use, there is a mosaic of smaller thyme populations where one or two non-phenolic chemotypes are most abundant (Figure 2.2). This differentiation occurs in about 1 km where altitude drops from 250 to 200 m.

In the 336 populations analysed by Gouyon *et al.* (1986) roughly 20 per cent (72 populations) only contained a single chemotype and 50 per cent (165) contained a mixture of two chemotypes (Figure 2.3). For the 165 populations with two chemotypes, 72 had the two non-phenolic chemotypes, 69 had two phenolic chemotypes and only 24 had one non-phenolic and one phenolic chemotype. So, although populations with two chemotypes are the most common (Figure 2.3), those with mixtures of phenolic and non-phenolic chemotypes are relatively rare. In fact, populations containing both phenolic and non-phenolic chemotypes (i.e. a combination of those with 2, 3, 4, or 5 chemotypes) are rare (Figure 2.2b). All populations with appreciable percentages of both phenolic and non-phenolic chemotypes occur either inside the elevation transition between 200 m and 250 m (Figure 2.2b) or adjacent to it. Most mixed populations out-side this transition zone tend to be dominated by non-phenolic chemotypes only and are inside the bassin (Figure 2.2a). Selection or drift must be driving the evolution of chemotype frequencies such that populations tend to have either phenolic or non-phenolic forms, but not both. This differentiation between populations has also been reported over a very small spatial scale of several metres (Mazzoni and Gouyon, 1985).

What is the cause of this sharp pattern of differentiation? Two main patterns of correlation were commented on by Gouyon *et al.* (1986). First, marked variation in soil type across the study region is correlated with the shift from phenolic to non-phenolic types. Phenolic populations occur on drier, more stony fersiallitic soils than non-phenolic chemotypes which predominate on deeper, moist soils. Within the non-phenolic types, it is the α-terpineol chemotype which is most abundant on the wettest soils. For the two phenolic types, carvacrol is limited to the driest conditions, whereas thymol is less specialised and can occur on moist soils.

A second clear-cut correlation concerns the absence of phenolic types, particularly carvacrol, from sites which experience sub-freezing temperatures, i.e. inside the valley or as mentioned above, at elevations above 400 m further inland. As one drops into the valley the shift is towards non-phenolic types, as one moves northwards around the valley, carvacrol is replaced by thymol, with the shift occurring as one passes the Pic St Loup where freezing temperatures increase in frequency and intensity in winter. Although Varinard (1983) found no effect of freezing on the survival of seedlings of the C, T and U chemotypes in controlled conditions, it is still possible that the non-phenolics suffer less effects due to temperatures several degrees below zero than do phenolic types.

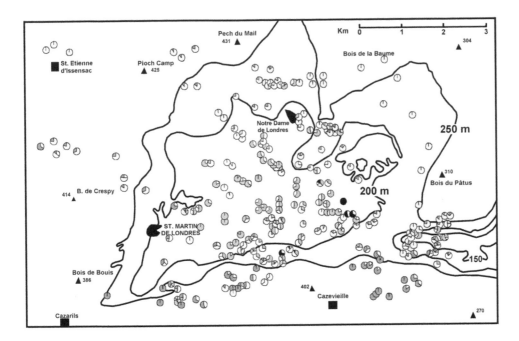

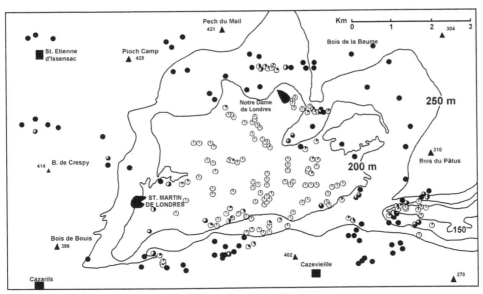

Figure 2.2 The spatial structure of *T. vulgaris* chemotypes in and around the St Martin-de-Londres valley in southern France. Data are from the original records compiled by P.-H. Gouyon and used in Vernet *et al.* (1977a,b) and Gouyon *et al.* (1986). In (a) all six chemotypes are shown, black – geraniol, blue – α-terpineol, green – thuyanol-4, white – linalool, red – carvacrol and yellow – thymol. In (b) the data are simplified to contrast the distribution of the two phenolic chemotypes (black circles) to that of the four non-phenolics (yellow circles). Each circle represents a population in which a bulk sample of ca. 30 plants was obtained (see the above cited authors for details). Three contour lines, 150 m, 200 m and 250 m are shown to illustrate the altitudinal segregation of the phenolic and non-phenolic types. (*See Color Plate 1*)

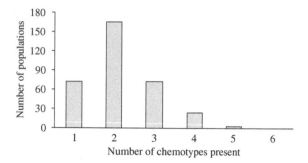

Figure 2.3 The frequency of populations with one or more chemotypes in and around the St Martin-de-Londres valley in the south of France. Data are from Gouyon *et al.* (1986) and are the same as those in Figure 2.2.

In *Trifolium repens* the genes responsible for the cyanogenesis polymorphism show a latitudinal and altitudinal cline and one hypothesis proposed to explain this cline is that freezing temperatures may, via the rupture of cell membranes, allow the release of hydrogen cyanide (HCN) that is toxic to the plant (reviewed by Briggs and Walters, 1997). The possibility that phenolic chemotypes of thyme may be excluded from areas with severe cold temperatures in winter due to the greater toxicity of the phenolic molecules that may, following freezing and the rupture of cell membranes, cause mortality during harsh winters is currently being investigated. Why plants with a carvacrol phenotype suffer such effects more than those with a thymol chemotype is not known.

A comparative study of plant species in Mediterranean-type ecosystems has shown that roughly 49 per cent of the species produce aromatic volatile oils and that these are predominantly the evergreen, xeromorphic woody shrubs (like thyme) and not the drought-avoiding annuals and deciduous species (Ross and Sombrero, 1991). Hence the presence of essential oils is correlated with the persistence during the Mediterranean summer "drought". In *T. vulgaris*, data obtained in controlled conditions by Couvet (1982) suggest that the non-phenolic A chemotype is significantly less resistant to drought and hot temperature stress than the C, T and L chemotypes. Phenolic types may thus be better adapted to dry (and hot) conditions. Non-phenolic essential oils are vapourised at lower temperatures than the phenolic oils (Couvet, 1982), hence it is possible that they are less able to withstand high temperatures due to a toxic effect of vapourised oil in the summer. However, it is difficult to conceive of an ecological role of the oils in relation to drought stress. There is no evidence that the vapours of such essential oils permit the regulation of leaf temperature or transpiration rate (Audus and Cheetham, 1940). What is more, the actual compounds that are vapourised may not be in the same proportions as those detected in the plant (Seufert *et al.*, 1995).

The spatial pattern of chemotype distribution in southern France primarily concerns the identity of the two most abundant chemotypes at a given site and whether the site is dominated by either phenolic or non-phenolic chemotypes (Vernet *et al.*, 1977a,b; Gouyon *et al.*, 1986; Figure 2.2). The presence of one of the six chemotypes has only a weak contribution to the spatial structure. This is not surprising, in a population dominated by non-phenolic chemotypes, the genes which cause the carvacrol phenotype (C-) and the thymol phenotype (cc) can be present even though their phenotype is not detected since their expression is masked by dominant alleles at the non-phenolic loci. Crosses among plants heterozygous at the non-phenolic loci will produce offspring with a phenolic chemotype. This may be why one may sometimes observe phenolic plants at very low frequency (<5 per cent) in populations dominated by non-phenolic plants, but not vice versa (J.D. Thompson and J.-C. Chalchat, unpublished data).

Geraniol is the rarest of the six chemotypes in southern France (Granger and Passet, 1973; Gouyon *et al.*, 1986), perhaps in part because it is the only gene that cannot "hide" behind other phenotypes hence may be more frequently "lost" during episodes of colonisation and extinction. In the study by Gouyon *et al.* (1986), none of the populations contained all six chemotypes, three populations contained five chemotypes and 24 populations were represented by four chemotypes (Figure 2.3). All three populations with five chemotypes lacked the geraniol chemotype. Of the 24 populations with four chemotypes, 22 also lacked the geraniol chemotype. In fact, when one distinguishes phenolic chemotypes from non-phenolic chemotypes the frequency of absence of particular chemotypes depends on their position in the epistatic chain: those chemotypes whose dominant genes prevent the expression of genes at subsequent loci in the chain are the chemotypes most frequently absent. For example, when one examines which non-phenolic chemotypes are absent from populations with four chemotypes, geraniol is absent from 22 of the 24 populations, α-terpineol from six and thuyanol from five of the 24 populations.

In sharp contrast, linalool, the last of the non-phenolic chemotypes in the genetic chain, is present in all 24 of these populations. A similar pattern is observed for the two phenolic chemotypes, carvacrol is absent from 13 of these 24 populations and thymol is absent from only 2 of the 24 populations. The combination of two chemotypes which are the most often jointly absent from populations with four chemotypes is geraniol and carvacrol (i.e. the most dominant gene for non-phenolic and phenolic chemotypes), which are concomitantly absent from 11 of the 24 populations. Finally, for the 72 monomorphic populations, 16 were linalool, 16 carvacrol and 28 thymol, the three chemotypes which are the most abundant in southern France (Gouyon *et al.*, 1986; Granger and Passet, 1973; J. Thompson and J.-C. Chalchat, unpublished data).

The data suggest that a combination of natural selection, which also acts at the level of phenolic–non-phenolic distinction, and chance effects within the two groups of chemotypes linked to the epistatic mechanism of genetic determination jointly impinge on the abundance of the six chemotypes. The challenge will be to demonstrate where and how natural selection acts on chemotype frequency by carefully replicating transplantation in the field (see also Boursot and Gouyon, 1983) and controlled experimentation of particular factors. It is also possible, as will be discussed in the rest of this section, that interactions with the biotic environment also contribute to the spatial and evolutionary dynamics of the chemical polymorphism.

Interactions on a single trophic level

In the context of potential interactions with other species, volatile oils may have a negative "allelopathic" effect on the germination and growth of associated plant species and in this way reduce competition with other species. The potential allelopathic effects of monoterpenes and their role in structuring plant communities have been the centre of much interest and critical discussion (Muller, 1969; Harper, 1977; Rice, 1979; Williamson, 1990; Fisher, 1991). Bare zones under and around aromatic shrubs have been remarked in different aromatic plant species (Muller *et al.*, 1964; Katz *et al.*,

1987) and much work on the effects of monoterpenes on the germination and growth of associated species has involved labiates in Mediterranean communities, e.g. *Salvia* (Muller *et al.*, 1964), *Coridothymus* (Vokou and Margaris, 1982) and *Calamintha* (Tanrisever *et al.*, 1988).

In the genus *Thymus*, there have been several investigations of the potential effects of monoterpene exudates on seed germination and plant growth. In *T. serpyllum*, Paul (1970) found that an aqueous extract of leaves and litter differentially inhibits the germination and growth of different species. For four studied species, the extract from *T. serpyllum* leaves had the strongest effect on *Plantago ramosa* which had significantly reduced germination in the presence of the extract and which also was least abundant where *T. serpyllum* occurred. Tarayre *et al.* (1995) tested the hypothesis that the different essential oils produced by *T. vulgaris*, have different effects on the germination of the grass *Bromus phoenicoides*, a common grass species in southern France. In a series of controlled germination trials in petri dishes, these authors found that the two phenolic chemotypes (C and T), when present in the form of leaves or as pure essence, have a significantly greater inhibitory effect on the germination of the grass than do the non-phenolic chemotypes. In the presence of phenolic leaves, the percentage germination of the grass was around 75 per cent, and in the presence of non-phenolic leaves from 85–90 per cent. Germination in the absence of thyme leaves was 90–95 per cent. Phenolic chemotypes may thus be able to better resist competition from associated grasses than non-phenolic chemotypes.

Using soil collected under thyme plants and away from thyme plants in phenolic and non-phenolic populations Y. Linhart, P. Gauthier and J. Thompson (unpublished data) studied the germination and growth of several different species that also occur with thyme with and without a cover of thyme leaves from the same sites. In general, germination and growth tended to be lower on soil collected from under phenolic thyme plants or in the presence of phenolic leaf litter. However, the effects of different chemotypes on germination and growth varied across the range of associated species (one *Nigella*, one *Medicago*, two *Bromus*, one *Crepis* and one *Daucus* species) used in the study. So there clearly exists a potential for inhibitory effects on associated species germination and growth and for variation in such effects depending on the identity of the associated species. In this context, it is interesting to note that in the study by Tarayre *et al.* (1995) the percentage germination of the grass in the presence of thyme leaves always exceeded 75 per cent, i.e. even in the presence of phenolic leaves this grass germinates well. It would thus be most interesting to compare populations or species that occur in association with thyme with others that do not occur in association with thyme, to examine whether the latter shows greater inhibition in the presence of thyme leaf litter.

The different monoterpenes may also affect the germination of thyme seeds and subsequent plant growth. Although the term "auto-allelopathy" has been used to describe the effect of thyme monoterpenes on seed germination one should refrain from using this term since there is some evidence that a temporal inhibition of germination may in fact be an adaptive response to irregular germination cues in the form of a short-term dormancy mechanism and not a toxicity phenomenon. It has been reported that aqueous extracts from *T. vulgaris* (Tarayre *et al.*, 1995) and *Thymbra capitata* (*Coridothymus capitatus*) (once *Thymus*) (Vokou and Margaris, 1986; Thanos *et al.*, 1995) can significantly delay their own seed germination.

In *Coridothymus*, germination was significantly slower in the presence of calyces (which are the unit of dispersal and which contain the essential oil) than when seeds were germinated alone, an effect relieved by leaching of the essential oil over time (Thanos *et al.*, 1995). In *T. vulgaris* all six of the oils cause an inhibition of seed germination to roughly 50 per cent that of seeds germinated in the absence of thyme monoterpenes (Tarayre *et al.*, 1995). At the end of the experiment (which was stopped due to fungal growth in the petri dishes) germination in controls had finished whereas germination in the presence of the different chemotypes continued (Tarayre *et al.*, 1995). Hence the inhibition effects may gradually wear off. This delayed rather than completely inhibited germination has been suggested (see Tarayre *et al.*, 1995; Thanos *et al.*, 1995) to represent an adaptive response to the irregular germination cues experienced by mature seeds of such Mediterranean shrubs in late summer when rainfall may be particularly erratic and interspersed by extreme drought stress for small plants and seedlings. Such inhibition could represent an evolved response to variation in cues for seed germination (see Angevine and Chabot, 1979; Fenner, 1985).

Interactions across trophic levels: escape in space and thyme?

As a defence against the strong pressure imposed on them by herbivores, parasites and pathogens, plants have evolved an immense diversity of chemical defences (Jones, 1962; Ehrlich and Raven, 1964; Bryant *et al.*, 1991). The more diverse, the partners at different trophic levels, the more important it may be to have a diverse, defence system. Herein lies a clue to the reason why there may be so many chemotypes in *T. vulgaris*: spatio–temporal variation in the abundance of different potential herbivores, parasites, etc. may lead to disruptive selection on chemical phenotype and thus contribute to the maintenance of several forms.

The first piece of work which investigated the possible role of thyme monoterpenes as a chemical defence against herbivores was that of Gouyon *et al.* (1983). These authors found marked variation between chemotypes in their palatability to slugs: U was the least palatable and A and C were the most palatable. Experiments with all six chemotypes by Linhart and Thompson (1995) showed that snails (*Helix aspersa*) have a preference for non-phenolics, particularly the L chemotype, and a marked distaste for the two phenolics C and T (Table 2.1). What is more, the most deterrent monoterpenes to snails, the phenolic C chemotype, caused snails fed on a diet of exclusively thyme plants of this chemotype to lose weight. Interestingly, when L genotypes are at the seedling stage (1–3 months old) their leaves do not have an L phenotype, they have a phenolic (C or T) phenotype, and they only develop their "true" phenotype after this very young seedling stage (Vernet *et al.*, 1986). As suggested by Linhart and Thompson (1995), the chemotype most preferred by snails (L) may thus "hide" behind a less palatable phenotype (C or T) during early seedling development – a stage in the life cycle that is likely to be critical for survival in the face of snail herbivory.

Investigation of feeding preferences in a range of herbivores have shown marked differences in the rank order of feeding preferences in *T. vulgaris*, i.e. different chemotypes vary in their ability to deter herbivore feeding, and different herbivores respond differently to the different chemotypes (Table 2.1). If one compares the palatability of the chemotypes to molluscs and grasshoppers (experiments done almost simultaneously in similar conditions) the deterrence ranks are completely reversed. What is tasty for a snail is unpalatable for a grasshopper and vice versa. For micro-organisms, Simeon de Bouchberg (1976) observed a similar reversal of deterrence; whereas the T chemotype had the most severe effects on bacterial population growth, it was the G chemotype

Table 2.1 The rank order of deterrent effects of the six chemotypes of *Thymus vulgaris* on different potential herbivores and inhibitory effects on microbial population growth and seed germination of an associated grass species. Chemotypes are ranked from (1) least to (6) most deterrent; for chemotype codes see text

Herbivore	Rank order of deterrence					
	1	2	3	4	5	6
Helix (snail)[1]	L	A	U	G	T	C
Deroceras (slug)[2]	G	A	L	U=C=T		
Leptophyes (grasshopper)[2]	T	C	A	G=U		L
Arima (chrysomelid beetle)[2]	L	A=U		G	C	T
Capra (goat)[2]	A	C	L	T	G	U
Ovis (sheep)[2]	U	T	A	G	C	L
Agriolimax (slug)[3]	A=C		T	U		
Fungi[4]	U	L	C	A	T	G
Bacteria[4]	U	A	G	L	T	C
Brachypodium (grass)[5]	L	U	G	A	T	C

Notes
1 Linhart and Thompson (1995);
2 Linhart and Thompson (1999);
3 Gouyon *et al*. (1983) for four chemotypes;
4 Simeon de Bouchberg (1976);
5 Tarayre *et al*. (1995).

that had the greatest impact on fungal growth. Elsewhere, the closely related *Thymbra capitata*, which has similar monoterpene oils, has been reported to significantly influence soil microorganism activity in the soil (Vokou and Margaris, 1984). It would thus be most worthwhile to examine how different chemotypes interact with soil organisms and the potential feedback effects on plant growth.

A glance down Table (2.1) indicates that every chemotype can be the preferred chemotype depending on the component of the environment studied and that all but the α-terpineol chemotype can be the most deterrent. A key point is that no one chemotype provides the best defence across the spectrum of potential herbivores, pathogens, etc. although in general the phenolic chemotypes do tend to be more deterrent than non-phenolic chemotypes. There are differences in the abundance of different snail species in phenolic and non-phenolic populations (Linhart and Thompson, 1995). Hence variation in the deterrence effects of the different chemotypes combined with variation in the spatio–temporal abundance of different herbivores, parasites and pathogens could influence the maintenance of the polymorphism in secondary compounds and contribute to the spatial variation in their relative abundance (Linhart and Thompson, 1999), as has been illustrated for secondary compounds in other species (Linhart, 1989).

In fact, the different facets of the biotic environment may act in concert with spatial variation in the abiotic environment (see above) to influence the maintenance of the spatial structure in the chemotype polymorphism. In one of the 12 doctoral theses on thyme done in Montpellier, Pomente (1987) reported that phenolic (T and C) plants had a better tolerance of drought stress, that the U chemotype was particularly favoured in humid conditions, and that the presence of an associated grass was correlated with a decrease in the survival of thyme plants in conditions of drought stress. This effect of grass presence was not an effect of competition, but rather because the grass maintained a more humid environment in which slugs sheltered and subsequently caused a greater mortality of thyme plants. The U chemotype grew best in humid conditions (Pomente, 1987) and is also the chemotype most deterrent to slugs (Gouyon *et al.*, 1983).

Gene Flow Versus Selection

If one wishes to study the spatial structure of plant populations it is necessary to investigate, and distinguish between, potential gene flow, i.e. the dispersal of pollen and seeds, and effective gene flow, which depends on the fertilisation of seeds and seed establishment (Levin and Kerster, 1974). When effective gene flow is significant, the development of the spatial structure of the genetic variation is limited, populations will show more genetic homogeneity. This is unless natural selection is strong enough to overcome the effects of such gene flow.

What do we know about gene flow among thyme populations dominated by different monoterpenes? To examine this question Tarayre and Thompson (1997) conducted a study of the spatial structure of several polymorphic isozyme loci by protein electrophoresis of 25–50 individuals in each of 23 populations of *T. vulgaris* representing a range of the phenolic and non-phenolic populations in Figure 2.2. Despite the significant differentiation of chemotype abundance across populations, isozymes showed relatively low levels of population differentiation. A mean F_{st} value of 0.038 indicated that less than 4 per cent of the genetic variation for the loci studied was due to population differentiation. Even along three transects which ran from 100 per cent non-phenolic to 100 per cent phenolic populations, F_{st} values did not exceed 0.06.

Pollen transfer distances are likely to be very small in the study region, most insect visitation is by honey bees which fly almost exclusively between adjacent plants (Brabant *et al.*, 1980; Rolland, 1999). However, many butterfly species (Lepidoptera) visit thyme in the study region and since they travel larger distances between plants than do bees, the potential for pollen flow among populations exists. Indeed, Tarayre *et al.* (1997) show that gene flow in pollen is much greater than that via seed, attesting to at least occasionally important pollinator movements.

The data of Tarayre and Thompson (1997) provide strong support for previous work (Gouyon *et al.*, 1987) which illustrated that, although limited migration can occur between phenolic and non-phenolic populations, there is little "effective" gene flow for the genes coding for the monoterpenes. When gene flow occurs, the chemical genes will be transported in either the pollen and seed that moves among populations. If genetic drift were responsible for the patterns of chemotype distribution, then we would have expected isozymes to show higher levels of genetic differentiation among populations, since the random nature of genetic drift is likely to act equally on all genes. Selection, in contrast, only acts on those genes of adaptive significance (although it may secondarily cause variation in the frequency of other genes linked to those under selection). The data, at present, strongly suggest that natural selection on the chemical phenotypes maintains the spatial pattern observed in the study region. What is more, selection appears to be particularly strong.

Future directions

There are clearly several avenues of work that are urgently needed for a clear understanding of the ecological significance of the chemical polymorphism in *T. vulgaris*.

First, the different chemotypes show marked variation in how they interact with the complexity of factors that determine the biotic and abiotic environment in which plants grow. What may be crucial to the dynamics of the chemical polymorphism is that the interaction of thyme monoterpenes with the biotic environment (herbivores, parasites, competitors) varies depending on the feature of the environment studied (Linhart and Thompson, 1999). Thus there is the potential for spatial (and temporal) variation in such biotic interactions to contribute to the maintenance of this genetic polymorphism. Documenting the extent to which such interactions occur and vary in the field represents a major challenge for future work. Not only could such interactions play a key role in the dynamics of the chemical polymorphism, they could also translate into effects on the structure and diversity of garrigues plant communities. The study of the chemical polymorphism in the field is thus a model system to do both population biology and community ecology, illustrating the often under-appreciated link between these two fields.

Second, several potentially important factors remain completely unstudied. One that comes immediately to mind is that the different monoterpenes may not have the same energy requirements for their production. The evolution of many polymorphisms that involve resistance to a particular environmental feature can be greatly influenced by what is known as the "cost of resistance". In the absence of the selective feature of the environment, a resistant genotype incurs a fitness cost associated with the presence of the resistance gene it carries. In the absence of the particular feature of the environment that favours resistance genes, non-resistant types will be favoured. In thyme, although phenolic chemotypes may be favoured because of a more generalised toxicity to herbivores, the phenolic molecules may be more costly to produce since they are further down the biosynthetic chain of production and thus requires more enzyme and precursor synthesis.

In fact, plants with a phenolic phenotype have only 50–70 per cent of their oil dominated by thymol or carvacrol, whereas the non-phenolic chemotypes G, L and A regularly have >80 per cent of their oil composed of their dominant monoterpene (Passet, 1971; J. Thompson and J.-C. Chalchat, unpublished data). The cost of production could thus be greater for the two phenolic molecules, causing a fitness cost to phenolic-based chemotypes where their selective agents are absent. Alternatively, the resistance cost incurred by phenolic plants may as mentioned above involve a lack of freezing tolerance imposed by the toxicity of the phenolic molecules. Future research should thus consider the possible importance of a cost to monoterpene production and the variation in this cost among chemotypes.

Third, since aromatic plants are such an essential component of the current day Mediterranean flora, a feature which would well merit attention in future research is the possible role of the essential oil variation in speciation and adaptation of new species to different environments. The dominant monoterpenes in *Thymus* vary (a) among the populations of individual species in different environments and (b) among different species across the range of the genus. In *T. vulgaris* it is even possible to observe variation on this theme of six chemotypes, a seventh form, based on 1,8-cineole occurs in Spain, where the geraniol chemotype has not been observed (Passet, 1971). Whether this pattern is due to hybridisation with other species containing this molecule in Spain or to a selective elimination of the geraniol chemotype in Spain and the 1,8-cineole chemotype in France merits close attention, as does the position of 1,8-cineole in the genetic and metabolic pathways.

Finally, in addition to the dominant monoterpene which characterises their essential oil, thyme plants may also contain a second monoterpene in a non-negligeable proportion. There are several examples of this phenomenon. The oil of carvacrol plants may contain up to 15 per cent thymol where plants are heterozygous at the C locus (Vernet *et al.*, 1986). In a recent survey, of approximately 100 plants having a linalool or geraniol phenotype, four were found to have equal amounts of the two compounds, while three out of twenty α-terpineol plants also contained 15–30 per cent thymol (J. Thompson and J.-C. Chalchat, unpublished data). The oil of both carvacrol and thymol chemotypes frequently contains up to 30 per cent of their two precursors, α-terpinene and *p*-cymene (Passet, 1971; J. Thompson and J.-C. Chalchat, unpublished data). Finally, the thuyanol-4 chemotype is actually a mixture of several compounds (Passet, 1971).

Such variation is not just background noise, it no doubt reflects developmental variation and adjustments, the study of which could provide a more concrete basis for understanding the precise link between genes and physiology. What triggers the co-occurrence of different molecules in some plants compared to a more pure oil in others? What is the relative resource cost of producing different molecules? Where and how do genes switch on and off the different elements of the monoterpene chain of synthesis? The answer to such questions may not only provide more precise information on the genetic control of chemotype in relation to the metabolic pathway of monoterpene synthesis, but could, as more species are studied, provide interesting insights into the genetics of adaptation and diversification in aromatic plants.

GYNODIOECY IN THYME: THE POPULATION GENETICS OF A SEXUAL POLYMORPHISM

Introduction: high and variable sex-ratios

In his book entitled "The different forms of flowers on plants of the same species" Charles Darwin (1877) remarked on the occurrence of two types of plants in a species of the genus *Thymus* in southern England. One type of plant bore only perfect flowers, with both male and female functions. The second type had smaller flowers that had no or completely reduced anthers and were thus completely male-sterile. This co-existence of hermaphrodite and female plants in a single population is called gynodioecy.

The presence of gynodioecy in the genus *Thymus* is associated with two intriguing observations: first, there is enormous variation in female frequency; and second, mean female frequency is very high, considerably higher than in populations of most other gynodioecious species. In *T. vulgaris*, populations contain 5–95 per cent females (Assouad *et al.*, 1978; Dommée *et al.*, 1978; Manicacci *et al.*, 1998) with a mean around 63 per cent. The frequency of females is extremely variable and on an average very high in *T. zygis* (mean: 51 per cent, range 17–87 per cent) and *T. mastichina* (mean 72 per cent, range 34–88 per cent) in central Spain (Manicacci *et al.*, 1998). In the clonal *T. serpyllum* a mean of 67 per cent females and a range of 30–95 per cent has been observed in six populations in south-central France (J.D. Thompson, unpublished data). Female frequencies above 50 per cent have also been observed in *T. sibthorpii* in Greece (K. Kateradi pers. comm.), *T. albicans* in South-West Spain (Valdés *et al.*, 1998) and *T. x arundanus* in South-East Spain (J. Arroyo, pers. comm.). So in several different species, which occur in different ecological conditions, in different parts of the geographic range of the genus and in different taxonomic sections of the genus (Morales, 1986), mean female frequency is very high (above 50 per cent) and can be extremely variable... why?

Why so many females: the importance of the spatial population structure

In a gynodioecious population, females do not produce pollen and therefore do not transmit genes to the next generation via male function. In thyme, as in many gynodioecious species, sexual phenotype is governed by a complex interaction between nuclear and cytoplasmic (mitochondrial) genes. The maternally inherited cytoplasmic genes inhibit male function, causing the sexual phenotype to be female. The effect of these cytoplasmic genes can be repressed by nuclear alleles at particular loci which restore male function, causing the individual to be hermaphrodite. The sexual phenotype of an individual will thus depend on the combination of its nuclear and cytoplasmic genome. In *T. vulgaris*, the genetic determination of sexual phenotype is extremely complex and involves several cytoplasmic types restored by a range of dominant and recessive restorer alleles which may show epistatic interactions (Belhassen *et al.*, 1991; Charlesworth and Laporte, 1998).

Theoretical models predict the maintenance of a nucleo-cytoplasmic polymorphism at loci involved in sex determination as a result of either a cost of restoration caused by negative pleiotropic effects of restorer genes in cytoplasmic backgrounds they do not restore (Gouyon *et al.*, 1991) or, in the absence of such a cost, due to the spatial structure of cytoplasmic and restorer genes (Couvet *et al.*, 1998). In the extreme case of spatial structure, a cytoplasmic male-sterility type may occur in a population where the nuclear alleles that restore its male-fertility are absent. This situation causes sex determination to be cytoplasmic and all the offspring of the female will be female. As the classic models of Lewis (1941) predict, this cytoplasmic sex determination will allow female frequency to rise as soon as females produce slightly more offspring than hermaphrodites. Couvet *et al.* (1986, 1998) propose (see scenario in Figure 2.4) that nucleo-cytoplasmic sex determination may vary in space and time. Sex determination may be locally cytoplasmic due to founder events during colonisation of new sites causing the absence of restorer alleles for the cytoplasmic types that are present. Subsequent arrival of nuclear restorer genes, via pollen or seeds, will later permit a decline in female frequency (Figure 2.4).

Strong evidence supporting this hypothesis came with the observation that female frequencies often attain very high values in young populations that are actively colonising new sites after either disturbance (e.g. forest fire) or the abandon of agricultural practice (Dommée *et al.*, 1983; Belhassen *et al.*, 1989). Seeds can resist up to 75 °C, hence populations can rapidly establish after a fire (Belhassen *et al.*, 1987). After this period of population establishment (usually around 10–15 years), the frequency of females appears to decline (Belhassen *et al.*, 1989). In young populations actively colonising new sites, the presence of dense patches (3–4 m in diameter) each composed of exclusively female plants are often observed (Manicacci *et al.*, 1996). This development of female patches causes female frequency to be locally very high, attaining 95 per cent in extreme cases.

The individuals within a patch tend to have a single cytoplasmic type, both for mitochondrial (Manicacci *et al.*, 1996) and chloroplast (Tarayre *et al.*, 1997) DNA, but patches less than 10 m apart can have different cytoplasmic DNA profiles. It is thus probable that each patch originates from a single female or sibling females from a

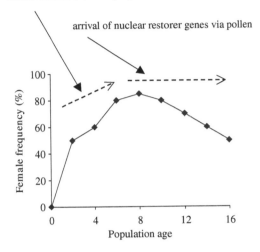

"mismatch" of cytoplasmic sterility genes and
nuclear restorer aleles = cytoplasmic sex determination

arrival of nuclear restorer genes via pollen

Figure 2.4 A schematic illustration of how a shift in the genetic control of sex may be related to female frequency variation in thyme populations.

common mother plant. The hermaphrodite that pollinated these mother plants must have lacked the restorer alleles for the maternal cytoplasm, allowing female patches to develop around the original mother plant (note that there is no mechanism for long-distance seed dispersal in thyme). Females produce 2–3(5) times more seed than herm-aphrodites in natural populations (Assouad *et al.*, 1978; Couvet *et al.*, 1986) allowing the rapid development of such patches in young populations. So there is evidence that founder events may occur and cause reduced cytoplasmic diversity, which in turn permits high female frequency in colonist populations.

A comparison of population differentiation for cytoplasmic and nuclear genes has shown that pollen migration is many times more frequent than seed migration among thyme populations (Tarayre *et al.*, 1997), hence the possibility that nuclear restorer genes may arrive via pollen and thus cause female frequency to decline as populations become older. The number of migrants per generation (Nm) among populations (estimated from F_{St} values) was 1.6 and 11.65 for the cytoplasmic DNA (cpDNA) and allozyme markers respectively, indicating that gene flow through pollen is roughly 14 times that for genes dispersed in seeds among the studied *T. vulgaris* populations. Within a single population where female patches are monomorphic for their cpDNA haplotype, Nm values for gene flow among patches and the surrounding (roughly 10 m away) more continuous cover of thyme are 0.42 and 12.91 for cpDNA and allozymes respectively. Hence, even at the scale of several metres, pollen migration greatly exceeds seed migration. It is therefore reasonable to assume that restorer genes will arrive via migration in pollen whilst the founder effect on cytoplasmic genes may persist longer during the life of a population.

There are several lines of evidence for spatial variation in the frequency of nuclear restorer genes. First, experimental pollination of plants in an insect-free glasshouse by

60   *John D. Thompson*

Couvet *et al.* (1985a) and Belhassen *et al.* (1991) have shown that male fertility restoration is greatest when females are pollinated with pollen from a hermaphrodite present in their original population compared to when the pollen source is a hermaphrodite from a different population. This result suggests that the restorer gene frequency is variable among populations and that hermaphrodites carry restorer genes adapted to local cytoplasmic male-sterility types.

Second, Manicacci *et al.* (1997) cloned five different females that had previously been found to show different rates of male fertility restoration when pollinated with a range of hermaphrodites, i.e. they have functionally distinct cytoplasmic male-sterilities. These authors placed the clonal replicates back into the five original populations from which the cytoplasmic types had originally been sampled. The females were allowed to flower and were then returned to the experimental garden where the seeds produced by pollination in each population were collected and sown. The sex ratio of the offspring of each female in each population (25 sex ratios in total) was determined the following flowering season. Marked variation in percentage restoration between populations was observed for three of the females (Figure 2.5), suggestive of spatial variation in the abundance of different restorer genes. Two of these three females had a percentage restoration that was greatest when transplanted into their original population, in agreement with the results of Couvet *et al.* (1985a) and Belhassen *et al.* (1991) who suggest that restorer genes are selected in populations that contain the associated male-sterility.

Interestingly, female *ph* was relatively well restored in her home population PH and, to a lesser extent, in the PB population, the closest other population (ca. 1 km away). In the three other populations distant by more than 10 km the restorer genes for this cytoplasm were virtually absent, and thus appear to have a very localised spatial occurrence

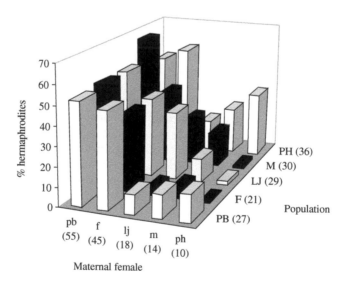

Figure 2.5 The percentage of male fertility restoration in the offspring of five females reciprocally transplanted among five original populations in southern France (redrawn with permission from Manicacci *et al.* 1997). Values in parentheses represent the mean percentage of hermaphrodites in the offspring of either each maternal female (pooled across the five populations) or for each population (pooled across the five females).

in and around the PH population. A different pattern was however observed in two of the females (*pb* and *f*) which showed a high percentage restoration in all five populations (Figure 2.5). There are several potential causes for the lack of variation in the restoration rates for these two females. First, the cytoplasmic male-sterility carried by these females may be restored by different restorer alleles which are present in different populations. Second, some restorer alleles may be very common across populations in the absence of their associated male-sterility cytoplasm.

Third, the cytoplasmic male-sterility represented by these two females may be present in the other populations (note that only one cytoplasm was investigated per population) and as a result their nuclear restorer alleles are also present. Fourth, some cytoplasmic male-sterilities may be restored by generalist restorer alleles. A fifth interpretation of this result is particularly appealing. Sex expression is generally observed as a qualitative phenomenon in thyme, there being females and hermaphrodites. However, the restoration of male fertility may also have a quantitative component since several restorer genes are probably involved in the determination of male function. If this is the case, plants of each sex may be more or less close to a threshold which determines the functional sexual phenotype. The two females in the study by Manicacci *et al*. (1997) that show consistently high restoration rate across populations may be closer to this threshold than the three other females.

It has long been known in thyme that some female flowers bear reduced male structures and that it is possible to classify females according to the phenotypic expression of stamen reduction. Some females have no visible stamens (type D); some have a small swelling on the corolla with a very short filament (type C). A third group has a more or less well-developed filament and anther (type B) (Assouad, 1972, Thompson *et al*., 2002). All flowers on a given plant are the same, none of the females produce pollen, and female offspring are dominated by females with the same type of male fertility reduction as the mother plant after controlled crosses and in a field population (Dommée, 1973).

What is more, there is a gradient in flower size and rate of restoration in the offspring of the different types of females from those with the most reduced male structures which have the smallest flowers and produce a more female-biased offspring than those with the most well-developed stamens that have flowers similar in size to those of hermaphrodites and which produce much less female-biased progenies in the field and in controlled pollination (J. Thompson and B. Dommée, unpublished data, Thompson *et al*., 2002). The two females in the study by Manicacci *et al*. (1997) with a high and stable rate of restoration could thus be females of type B while those with variable rates of restoration in their offspring may be females of types C or D.

The idea that female frequency is high in young populations due to founder events causing a mis-match of the cytoplasmic sterility genes and nuclear restorer genes plus high levels of seed set on females, but declines as populations age following the immigration of nuclear restorer alleles that restore the male fertility of the cytoplasmic types present in the founding population thus has some support. Other factors may however also contribute to the pattern of sex ratio variation in natural populations of *T. vulgaris*. First, at high female frequency females may suffer pollen limitation and reduced seed set due to the low numbers of hermaphrodites present, this causing their frequency to subsequently decline. Second, if colonist females form patches of females via the establishment of their exclusively female offspring, then their offspring, are likely to be pollinated by the same hermaphrodite which pollinated the colonist female, i.e. their father. Such mating among related individuals could cause female offspring to suffer inbreeding depression.

A study of rates of biparental inbreeding in females that occur in four populations with very different sex ratios has not however found any evidence that biparental inbreeding actually occurs on females (M. Tarayre and J. Thompson, unpublished manuscript). However, Thompson and Tarayre (2000) found a negative effect of biparental inbreeding on female seed set, hence this hypothesis cannot be completely dismissed.

Third, since females produce many more seeds than hermaphrodites (see below), their abundance may decline if they suffer a reproductive cost that causes them to have shorter life span than the hermaphrodites that are also present during colonisation. We are currently quantifying the survival of related females and hermaphrodites to test this possibility. The issue of what causes variation in female frequency is thus complex, with local variation in sex determination being of paramount importance. The observed sex ratio variation may nevertheless be greatly enhanced by differences in the production of viable offspring by females and hermaphrodites. This so-called female fertility advantage is the subject of the next section.

The advantages of being female

The nature of female fitness advantage

In thyme, as in most gynodioecious species, females produce more viable seeds than hermaphrodites, 2–3(5) times than that of hermaphrodites, although this female advantage may vary among populations and years (Assouad *et al.*, 1978; Couvet *et al.*, 1986). Female offspring can also be more vigorous than that of hermaphrodites (Assouad *et al.*, 1978). Survival of the two forms appears to be equal (Assouad, 1972), although this needs verification over the full life cycle of thyme plants in natural populations. Differences in the production of viable offspring by females and hermaphrodites may result from an "outcrossing advantage" due to the fact that females cannot self and/or a "resource compensation" advantage due to the fact that females may be able to re-allocate resources (not spent on producing pollen) to seed production.

The outcrossing advantage

Females cannot self, hence their offspring will not suffer inbreeding depression due to selfing. Hermaphrodites are self-compatible, hence, although they bear protandrous flowers in which the anthers dehisce before the stigmas are receptive (Assouad, 1972), may self-pollinate due to pollen movement between flowers on a given plant. An individual thyme plant can bear many hundreds of open flowers at a given moment and since the most important pollinator of *T. vulgaris*, the honey bee *Apis mellifera*, tends to visit many flowers on a plant during each visit (Brabant *et al.*, 1980; Rolland, 1999), high levels of selfing may occur. In fact, selfing rates vary among populations (Valdeyron *et al.*, 1977; M. Tarayre and J. Thompson, unpublished manuscript) with the highest rates of selfing in populations with the highest female frequencies (Table 2.2).

Such a positive correlation between female frequency and hermaphrodite selfing rate could, as suggested by Sun and Ganders (1986) be interpreted as evidence that gynodioecy may be maintained due to its positive effect on outcrossing. However, as we

Table 2.2 Variation in hermaphrodite selfing rate in relation to female frequency observed in two studies of natural populations of *T. vulgaris* in southern France

Population	Sex ratio (% females)	Selfing rate
Valdeyron *et al.* (1977)		
Le Vigan	46	0.10
Les Chênes	66	0.36
Pic St Loup	–	0.35
Viols le Fort	76	0.49
Thompson *et al.* (2002)		
Pompignan-haut	12	0.09
Courège	62	0.12
Tourrière	80	0.33

point out elsewhere (M. Tarayre and J. Thompson, unpublished manuscript), it is also possible that hermaphrodites show higher rates of selfing when females are abundant simply because of problems of pollen transfer resulting from the reduced abundance and spatial isolation of hermaphrodite plants in such populations. An important point here is that the selfing rate on hermaphrodites is very low (<10 per cent) in populations with female frequencies of up to 60 per cent. Only above 60 per cent females does the selfing rate on hermaphrodites become significant (Table 2.2). Female frequencies thus reach their mean value in the study region (ca. 60 per cent) in the absence of significant rates of selfing on hermaphrodites. Problems associated with selfing would thus not appear to be driving the evolution of female frequencies in thyme populations, in sharp contrast to what many authors have argued for other gynodioecious species. Nevertheless, once female frequencies do attain high levels, selfing on hermaphrodites may contribute to the maintenance of high female frequencies, if selfing is followed by inbreeding depression.

Where selfing does occur on hermaphrodites in thyme, females may have a fitness advantage due to inbreeding depression (the reduced fitness of selfed progeny relative to outcrossed progeny). Assouad *et al.* (1978), Bonnemaison *et al.* (1979), Perrot *et al.* (1982) and Thompson and Tarayre (2000) have all found that inbreeding depression negatively affects the performance of selfed offspring in thyme. A re-analysis of previous data (Thompson *et al.*, 1998) showed that, when quantified over several stages of the life-cycle, i.e. seed production, seed germination and seedling size, inbreeding depression can be extremely high (0.818). This would suggest that selfed progeny contributes little to the next adult generation. In the Tourrière population where we have recently detected significant selfing on hermaphrodites (Table 2.2), the inbreeding coefficient (F_{is} = 0. 05) is not significantly different from zero (Tarayre and Thompson, 1997), indicating that selfing contributes few offspring to the mature adult generation.

This result has important bearings on the heterozygosity levels and on the spatial structure of natural populations. The theoretical models of Gouyon and Couvet (1987) predict that, for a constant hermaphrodite selfing rate and (at least some degree of) cytoplasmic inheritance of sex, as female frequency increases, the heterozygosity of local populations should increase. A study of isozyme variation for one enzyme system along

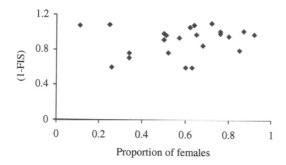

Figure 2.6 Heterozygosity values ($1 - F_{IS}$) for 23 populations of *T. vulgaris* in and around the St Martin-de-Londres valley in southern France. The populations occur across the area shown in Figure 2.2. Redrawn with permission from Tarayre and Thompson (1997).

a series of adjacent successional populations of thyme has illustrated that heterozygosity levels do decline as female frequency declines along the succession (Dommée *et al*. 1983). Using the chemotype genes as markers, Gouyon and Vernet (1982) found that females are more frequently heterozygous than hermaphrodites at the C locus in a single population.

However, Bonnemaison (1980) found that this result was not common to three studied populations and that females and hermaphrodites had similar heterozygosity values for the L locus in all three populations. In a more recent study by Tarayre and Thompson (1997) of several isozyme loci in 23 populations, no evidence for a correlation between heterozygosity and sex ratio was found (Figure 2.6), and in general adult females and hermaphrodites had similar levels of heterozygosity. The lack of a positive correlation between female frequency and population heterozygosity in this study was primarily due to high levels of heterozygosity in two populations with very low (<20 per cent) female frequencies (Figure 2.6), probably due to low levels of selfing on hermaphrodites (Table 2.2).

In populations with many hermaphrodites, females produce high percentages of hermaphrodites (Manicacci *et al*., 1997), hence most hermaphrodites will be produced by females and thus may have heterozygosity values similar to those of females. So, as a result of variation among populations in both the rate of selfing and male fertility restoration, heterozygosity values do not decline at low female frequencies. Furthermore, any selfing that does occur can be followed by inbreeding depression, allowing heterozygosity levels of adult hermaphrodites to remain high and similar to those for females.

It must also not be forgotten that although females cannot self, they may incur biparental inbreeding, at levels similar to hermaphrodites, due to mating with related individuals. Since pollination is primarily by honey bees which fly among adjacent plants and since seed dispersal is low (Belhassen *et al*., 1987; Tarayre *et al*., 1997), pollination in natural populations is often likely to be among related plants. In a recent study, females in three populations out of four showed biparental inbreeding depression on viable seed production when crossed with hermaphrodites of the same family (Thompson and Tarayre, 2000). This only occurred for hermaphrodites from one of the populations, hence female seed production may be more negatively influenced by biparental inbreeding than that of hermaphrodites. In other words, the effects of inbreeding may not simply be to produce a female outcrossing advantage, but in fact may reduce the female seed fertility advantage in population contexts where biparental inbreeding occurs.

So although outcrossing is likely to be important in the ecological adaptation of thyme to different environments around the Mediterranean basin (Bonnemaison, 1980; Gouyon and Vernet, 1982) and may contribute to the female fitness advantage in some populations, it is clearly not the principal cause of female fertility advantage. What is more, the evolution of high female frequencies in thyme populations is not likely to be driven by an outcrossing advantage of females, since hermaphrodites only self at significant rates when female frequencies are already at very high levels. This provides further support for the hypothesis that stochastic effects on sex determination are behind the pattern of sex ratio variation observed in thyme populations.

Resource compensation and sexual specialisation

A second possible cause of female seed fertility advantage is that females may re-allocate resources (otherwise used to produce pollen) to seed production and provisioning. In *T. vulgaris*, Atlan *et al*. (1992) reported negative correlations between the male (full pollen grains per flower) and female (germinating seeds per fruit) fertility of hermaphrodites grown in uniform garden conditions. In the plants sampled from one population, hermaphrodites that were the progeny of females produced small amounts of pollen but relatively high numbers of seed, whilst hermaphrodites that were the progeny of hermaphrodites produced more pollen but fewer seeds. Pomente (1987) has documented genetic variation in pollen production by hermaphrodites, hence the potential for the evolution of male fertility exists.

Based on a comparison of seed number per fruit in open-pollinated plants ad controlled crosses, Couvet *et al*. (1985b) illustrate that hermaphrodites abort 45 per cent more seeds than females, and that less than 20 per cent of seed abortion can be attributed to selective embryo maturation. Low fruit set in hermaphrodites would appear to be primarily determined by sexual selection and subsequent specialisation in male function. What is more there is a genetically-based variation in the degree of sexual specialisation and relative female fertility for a range of families from four populations (Thompson and Tarayre, 2000).

Selection on the functional gender of hermaphrodites may also be imposed via the female frequency in a population. If female frequency in a population remains stable long enough to act as a selective force on hermaphrodite resource investment; resource allocation theory (Lloyd, 1976; Charnov, 1982) makes two predictions. First, for hermaphrodites, relative allocation to male function (e.g. number of viable pollen grains as a function of seed-set) would be positively correlated with female frequency in a population. As there are more females, hermaphrodites with greater male function would be favoured. Second – and following the previous prediction – viable seed production of females relative to hermaphrodites should also increase with female frequency.

In a study of three *Thymus* species, *T. vulgaris*, *T. mastichina* and *T. zygis*, Manicacci *et al*. (1998) found evidence for a correlation between female frequency and either hermaphrodite male function or the relative fecundity of females across six populations of each species. However, on an average hermaphrodites were better males, and females better females, in the species with the highest female frequency. In other words, among the three species, an increase in sex ratio was correlated with both an increase in seed-set on females and an increase in the relative allocation of resources to pollen (compared to seeds) in hermaphrodites.

The lack of any correlation across populations may have several explanations. First, relative seed set on females may not increase with their frequency because of frequency-dependent effects on seed-set, notably pollen limitation. In a recent study of female fertility advantage of females and hermaphrodites from four populations of *T. vulgaris* grown and pollinated under controlled conditions, no evidence for the predicted evolution of gender at high female frequencies was observed (Thompson and Tarayre, 2000), confirming the data for natural populations in Manicacci *et al.* (1998). The lack of greater female fertility advantage at high female frequency is thus not likely due to a problem associated with pollen limitation in the wild. Second, Manicacci *et al.* (1998), propose that thyme populations are often subject to disturbances which cause extinction and re-colonisation to be particularly frequent. Selection may thus not have time to precisely adjust the functional gender of hermaphrodites at the population level.

Third, Manicacci *et al.* (1998) also suggest that where sex ratios are female-biased this may in fact cause hermaphrodites with a female-biased gender to be maintained. The argument for this is that since hermaphrodites with a male-biased gender will pollinate more females they will not contribute to the maintenance of hermaphrodites in the population since they are unlikely to restore male function on these females (otherwise female frequency would already be low!). Hence the only hermaphrodites contributing to the next generation of hermaphrodites will be those with a female-biased gender that set seed and in doing so produce the next generation of hermaphrodites.

Finally, the results of Gigord *et al.* (1999) which suggest a positive correlation between the frequency of hermaphrodites in a given progeny and the male (and female) function of hermaphrodites of that progeny may contribute to the lack of increased male function for hermaphrodites in populations with a female-biased sex ratio. These authors report that hermaphrodites in families with many hermaphrodites may produce more pollen per flower than hermaphrodites in families with a high female frequency. Selection due to a high female frequency would thus be constrained by lower male performance of hermaphrodites in progeny with a high female frequency. Important to note here is that progeny sex ratios are positively correlated with population sex ratios (Manicacci *et al.*, 1997). If this genetic constraint on gender variation is real, it could also contribute to prevent the evolution of dioecy from gynodioecy in *Thymus*, a genus where there are many gynodioecious but no dioecious species.

Inherent differences in seed fertility exist between females and hermaphrodites. The differences do not concern inflorescence and flower production which is similar for the two sexes (Assouad, 1972; Bonnemaison, 1980). The differences do not vary in relation to sex ratio and are not just due to inbreeding depression on the seed fertility of hermaphrodites. Assouad (1972) first observed that female seed fertility exceeded that of hermaphrodites even on outcrossing. He also observed that pollen adherence on female stigmas was greater than that on hermaphrodite stigmas, suggesting that the difference is not just a resource compensation effect but more the result of specialisation of sex functions in the two morphs.

In a recent study of females and hermaphrodites from four populations of *T. vulgaris* that vary in sex ratio from 11–80 per cent females, it has also been found that even when plants are outcrossed with pollen from hermaphrodites of a different population, female seed fertilty is at least twice that of hermaphrodites (Thompson and Tarayre, 2000). Given that selfing rates on hermaphrodites in three of these populations are close to zero (Table 2.2). It would appear that the seed fertility advantage of females has little to do with inbreeding depression and is more likely the result of sexual specialisation and/or resource compensation. An interesting feature of the results of this study is that the seed fertility advantage of females relative to hermaphrodites showed variation across families. In some families this seed fertility advantage was fairly high, while in others much lower. The precise combination of cytoplasmic and nuclear genes responsible for sex determination may thus also impinge on quantitative variation in sexual function within each of the two sexual phenotypes.

Flower size dimorphism: from pollination biology to the genetics of sex determination

As in most gynodioecious species (see review by Delph, 1996), female flowers of *T. vulgaris* are smaller than hermaphrodite flowers (Assouad, 1972). This flower size dimorphism shows a variation which has implications for both the pollination ecology of thyme and our understanding of the genetics of sex determination in this species.

In natural populations, flower size varies significantly between the two sexual phenotypes, among populations and also as a result of an interaction between population and sexual phenotype (Thompson *et al.* 2002). In other words, the difference in flower size between the sexual morphs varies in extent across populations, primarily because female flower size varies to a greater extent across populations than does hermaphrodite flower size. For the same populations, the relative proportion of bees and butterflies visiting thyme also varies markedly suggesting that pollinator-mediated selection may contribute to differences among populations (Rolland, 1999). A study of the F_2 progeny of three populations by Thompson *et al.* (2002) confirmed that there is a population–sex interaction in homogeneous conditions, due to greater genetic variation among females in different populations. Within populations, these authors also detected a family–maternal sex interaction, due to greater genetic variation of females among families than among hermaphrodites of different families.

Although Rolland (1999) found that hermaphrodites are more attractive to bee pollinators than females, she did not detect any effect of flower size within females on bee attraction. Although a range of populations and different pollinators should be studied, it would appear reasonable to conclude that the functional significance of variation in female flower size is not related to pollinator-mediated selection.

An alternative explanation for the variation in female flower size among families and populations is that this variation is related to the genetic determination of sex, in particular to differences in the degree of restoration of females in different families and populations. As mentioned above, several types of female phenotypes can be observed depending on the degree of stamen reduction. Thompson *et al.* (2002) found that females with larger flowers have a more or less well-developed filament and anther (type B females), whilst females with the smallest flowers have no visible trace of stamens (type D females). Females which have a small swelling on the corolla where the filament would normally be fixed (type C females) have intermediate flower size. These are the different female phenotypes previously described above which give different sex ratios in their offspring. It is thus possible that populations or families with large-flowered females may contain predominantly females of type B whilst populations and families with small-flowered females are dominated by females of type D.

Perspectives: the importance of variation within sexes

Gynodioecy in thyme is controlled by a complex interaction of nuclear and cytoplasmic genes. The difference in inheritance of cytoplasmic and nuclear genes causes the selective pressures acting on the genes they contain to fundamentally differ: selection for feminising genes in the maternally-inherited cytoplasm is in conflict with selection for restorer genes in the nucleus (Couvet *et al.*, 1990). A key point here is that the precise determination of sex in a local population may vary in space and time causing the sex ratio to show high levels of variation and allowing for unusually high female frequencies. In thyme populations, the random impact of founder events on the relative distribution of cytoplasmic and nuclear genes and differences in gene flow via pollen and seeds among established populations may greatly contribute to this pattern of sex ratio variation.

The seed fertility advantage of females is primarily due to sexual specialisation and has very little to do with any outcrossing advantage. This female seed fertility advantage may contribute to the local population structure by allowing the rapid development of high female frequencies in colonist situations. Gynodioecy may thus be a key parameter in the colonising ability of *T. vulgaris*, a species which rapidly establishes in early successional habitats.

We do not know how many functionally different male-sterilities exist in natural populations nor whether nuclear restorer genes are highly specific to particular cytoplasms or whether they can restore different cytoplasmic male-sterility types. Particularly an interesting point concerns whether or not male fertility restoration involves a quantitative effect of restorer genes on male function. The presence of a range of different female phenotypes that vary in the degree of stamen reduction, flower size and rate of offspring restoration represents particularly an intriguing aspect of variation in sex expression in this species. The study of these different female types will provide a useful tool for advancing our knowledge of the genetics of sex expression and the spatial dynamics of gynodioecy in thyme.

Unfortunately, we only have a faint inkling of the relationship between spatial variation in restorer gene frequency and how selection may act on such genes. The variation in the frequency of some restorer genes among thyme populations (Figure 2.5) is higher than what one would predict given that population differentiation for nuclear isozyme variation among populations in the same zone ($F_{st}=0.038$) is relatively low (Tarayre and Thompson, 1997). The greater population differentiation for some restorer alleles may be due to selection on particular restorer alleles in certain environments. The marked population structure for cytoplamsic genes (Tarayre *et al.*, 1997) could provide the selective context for particular restorer genes once they arrive in a population (Couvet *et al.*, 1998).

Why then do some restorer genes (for example those that restore the *f* and *pb* females in Figure 2.5) have a more widespread distribution than others (i.e. those that restore the *pb*, *m* and *lj* females in Figure 2.5)? Is it because that the latter three cytoplasmic types are in fact present in the different populations or because these cytoplasmic types are close to the threshold necessary for the restoration of male fertility? This is a distinct possibility: females that have well-developed anthers do tend to produce more hermaphrodites than females that have no visual male structures. The former may thus incur a fitness cost since the latter are producing (female) offspring that will set at least twice as many seeds as the (hermaphrodite) offspring of the former. In fact females with no trace of any anther development are the most common in natural populations (Assouad, 1972; J.D. Thompson, unpublished data).

One would also predict that females that bear reduced anthers would set less viable seed than females with no trace of male organs. Furthermore, females with no trace of male organs are the most abundant female type, such populations should have the highest female frequency and a smaller mean flower size than populations with a low female frequency (and a higher frequency of type B females). Whether or not the different female types are caused by the expression of different cytoplasmic male-sterility types, the expression of a gradient of male fertility restoration, or the result of an interaction between the cytoplasmic type and the nuclear gene is the subject of ongoing experimental work.

Hermaphrodites also vary in their sex expression, both among each other (Thompson and Tarayre, 2000) and over time, plants having a male-biased pattern of resource allocation early in flowering and a more female-biased pattern of resource allocation later in flowering (Manicacci, 1993). The functional significance of such genetic and environmental (developmental) variation will be interesting to study. So a shift in emphasis, from studying differences between females and hermaphrodites, towards an appreciation of the relevance of sex variation in sexual phenotype and functional gender will be a key component of the future work necessary for us to make further advances in our understanding of the population structure and genetics of gynodioecy in thyme.

CONCLUDING REMARKS: SEX, MONOTERPENES AND THYME

In the early 1960s, chemical variation in *T. vulgaris* began to attract the attention of ecologists and geneticists with the report that this species had a variety of different chemical forms (Granger *et al.*, 1963). At roughly the same time, a discussion between L. Emberger and G. Valdeyron led the latter to start counting the frequency of female plants in natural populations around Montpellier. Since then, as this chapter attests, the two kinds of polymorphism have been the focus of continued research on the spatial dynamics of polymorphic variation in thyme.

A question that is often asked when one talks about the two kinds of polymorphism in thyme concerns whether or not there is a link between sexual and chemical phenotypes or more subtly between the patterns of variation observed for each polymorphism. There is in fact evidence for a subtle relationship. Gouyon *et al.* (1986) found that although 61 per cent of populations with more than 50 per cent hermaphrodites are predominantly phenolic populations, only 34 per cent of populations with less than 50 per cent hermaphrodites are populations dominated by phenolics.

What may cause populations with many hermaphrodites to tend to be dominated by phenolic chemotypes and populations with high female frequencies to be predominantly of the non-phenolic type? Gouyon and Vernet (1982) provide data which suggests that hermaphrodites are more homozygous than females and that the correlated high frequencies of hermaphrodites and phenolic chemotypes may be due to inbreeding. Recent work however shows no evidence for greater homozygosity of hermaphrodites (Tarayre and Thompson, 1997). Another possibility, currently under study, is that females with a phenolic phenotype may have lower fertility and/or lower survival than both hermaphrodites with a phenolic chemotype and females with a non-phenolic chemotype due to the greater resource cost of phenolic molecules.

Finally, the study of the spatial population structure has necessitated the development of two different approaches. To understand the spatial dynamics of gynodioecy has required a population genetics approach in which stochastic effects on gene frequency play a key role. In contrast, the chemical polymorphism continues to provide a classic example of how an ecological genetic approach (the study of selection versus drift and variation across populations) can be used to elucidate the multiple selective factors that may influence the dynamics of a genetic polymorphism. But in both cases, understanding the dynamics of polymorphic variation has required recognition that plants occur in patches, which form mosaics of local populations.

Each local population experiences the selective forces of the environment and the regional processes of gene flow dynamics associated with the colonisation and extinction of individual populations and occasional migration between established populations. The dynamics of the two thyme polymorphisms is the result of an intricate balance between the processes acting at these different levels of the spatial population structure. Other *Thymus* species probably show similar patterns of variation as those discussed here. Further study of these species will provide interesting comparative examples with which to examine the general significance of the spatial population structure for the evolution of genetic polymorphism in thyme.

ACKNOWLEDGEMENTS

I am particularly grateful to Isabelle Litrico who compiled Figure 2.2, and to Domenica Manicacci, Yan Linhart, Anne-Gaëlle Rolland, and Perrine Gauthier for their helpful discussion of the manuscript.

REFERENCES

Adzet, T., Granger, R., Passet, J. and San Martin, R. (1977) Le polymorphisme chimique dans le genre *Thymus*: sa signification taxonomique. *Biochem. Syst. Ecol.*, 5, 269–272.

Angevine, M.W. and Chabot, B.F. (1979) Seed germination syndromes in higher plants. In O. T. Solbrig, S. Jain, G.B. Johnson and P.H. Raven (eds), *Topics in Plant Population Biology*, Columbia University Press, New York, pp. 188–206.

Assouad, M.W. (1972) *Recherches sur la génétique écologique de Thymus vulgaris L. Etude expérimentale du polymorphisme sexuel.* Ph.D. thesis, Université des Sciences et Techniques du Languedoc, Montpellier, France.

Assouad, M.W., Dommée, B., Lumaret, R. and Valdeyron, G. (1978) Reproductive capacities in the sexual forms of the gynodioecious species *Thymus vulgaris* L. *Biol. J. Linn. Soc.*, 77, 29–39.

Atlan, A., Gouyon, P.H., Fournial, T., Pomente, D. and Couvet, D. (1992) Sex allocation in an hermaphroditic plant: the case of gynodioecy in *Thymus vulgaris* L. *J. Evol. Biol.*, 5, 189–203.

Audus, L.J. and Cheetham, A.H. (1940) Investigations on the significance of ethereal oils in regulating leaf temperatures and transpiration rates. *Ann. Bot.*, 4, 465–483.

Belhassen, E., Pomente, D., Trabaud, L. and Gouyon, P.H. (1987) Recolonisation après incendie chez *Thymus vulgaris* L.: résistance des graines aux températures élevées. *Oecol. Plant.*, 8, 135–141.

Belhassen, E., Trabaud, L., Couvet, D. and Gouyon, P.H. (1989) An example of nonequilibrium processes: gynodioecy of *Thymus vulgaris* L. in burned habitats. *Evolution*, 43, 662–667.

Belhassen, E., Dommee, B., Atlan, A., Gouyon, P.H., Pomente, D., Assouad, M.W. and Couvet, D. (1991) Complex determination of male sterility in *Thymus vulgaris* L.: genetic and molecular analysis. *Theor. Appl. Genet.*, 82, 137–143.

Bonnemaison, F. (1980) *Etude stationnelle de la dynamique du maintien d'un polymorphisme génétique: cas de quatre populations naturelles de Thymus vulgaris L.* Ph.D. thesis, Université des Sciences et Techniques du Languedoc, Montpellier, France. pp. 527–536

Bonnemaison, F., Dommée, B. and Jacquard, P. (1979) Etude expérimentale de la concurrence entre formes sexuelles chez le thym, *Thymus vulgaris* L. *Oecol. Plant.*, 14, 85–101.

Boursot, P. and Gouyon, P.H. (1983) Mortalité et sélection chez les plantules de *Thymus vulgaris* L. *Oecol. Plant.*, 4, 53–60.

Brabant, P., Gouyon, P.H., Lefort, G., Valdeyron, G. and Vernet, P. (1980) Pollination studies in *Thymus vulgaris* L. (Labiatae). *Oecol. Plant.*, 15, 37–44.

Briggs, D. and Walters, S.M. (1997) *Plant Variation and Evolution.* Cambridge University Press, Cambridge.

Bryant, J.P., Provenza, F.D., Pastor, J., Reichardt, P.B., Clausen, T.P. and Toit (du), J.T. (1991) Interactions between woody plants and browsing mammals mediated by secondary metabolites. *Annu. Rev. Ecol. Syst.*, 22, 431–446.

Charlesworth, D. and Laporte, V. (1998) The male-sterility polymorphism of *Silene vulgaris*: analysis of genetic data and comparison with *Thymus vulgaris*. *Genetics*, 150, 1267–1282.

Charnov, E.L. (1982) *The Theory of Sex Allocation*, Princeton University Press. Princeton, New Jersey.

Couvet, D. (1982) Contribution à l'étude des polymorphismes chémotypique et sexuels, D.E.A., U.S.T.L., Montpellier.

Couvet, D., Gouyon, P.-H., Kjellberg, F. and Valdeyron, G. (1985a) La différenciation nucleo-cytoplasmique entre populations: une cause de l'existence de males-steriles dans les populations naturelles de thym. *C. R. Acad. Sci. Paris*, 300, 665–668.

Couvet, D., Henry, J.P. and Gouyon, P.H. (1985b) Sexual selection in hermaphroditic plants: the case of gynodioecy. *Am. Nat.*, 126, 294–299.

Couvet, D., Bonnemaison, F. and Gouyon, P.H. (1986) The maintenance of females among hermaphrodites: the importance of nuclear-cytoplasmic interactions. *Heredity*, 57, 325–330.

Couvet, D., Ronce, O. and Gliddon, C. (1998) The maintenance of nucleocytoplasmic polymorphism in a metapopulation: the case of gynodioecy. *Am. Nat.*, 152, 59–70.

Couvet, D., Atlan, A., Belhassen, E., Gliddon, C., Gouyon, P.H. and Kjellberg, F. (1990) Co-evolution between two symbionts: the case of cytoplasmic male-sterility in higher plants. In D. Futuyma and J. Antonovics (eds), *Oxford Surveys Evolutionary Biol.*, Oxford University Press, pp. 225–248.

Darwin, C.R. (1877) *The Different Forms of Flowers on Plants of the Same Species*, John Murray, London.

Delph, L.F. (1996) Flower size dimorphism in plants with unisexual flowers. In D.G.B. Lloyd (ed.), *Floral Biology: Studies on Floral Evolution in Animal-Pollinated Plants*, Chapman & Hall, New York, pp. 217–237.

Dommée, B. (1973) *Recherches sur la génétique écologique de Thymus vulgaris L. Déterminisme génétique et répartition écologique des formes sexuelles.* Ph.D. thesis, Université des Sciences et Techniques du Languedoc, Montpellier, France.

Dommée, B., Assouad, M.W. and Valdeyron, G. (1978) Natural selection and gynodioecy in *Thymus vulgaris* L. *Bot. J. Linn. Soc.*, 77, 17–28.

Dommée, B., Guillerm, J.L. and Valdeyron, G. (1983) Régime de reproduction et hétérozygotie des populations de *Thymus vulgaris* L., dans une succession postculturale. *C. R. Acad. Sci. Paris*, 296, 111–114.

Ehrlich, P.R. and Raven, P.H. (1964) Butterflies and plants: a study in coevolution. *Evolution*, 18, 586–608.

Fenner, M. (1985) *Seed Ecology*, Chapman & Hall, London.

Fisher, W.H. (1991) Plant terpenoids as allelopathic agents. In J.B. Harborne and F.A.T. Barberán (eds), *Ecological Chemistry and Biochemistry*, Clarendon Press, Oxford, pp. 377–397.

Gigord, L., Lavigne, C., Shykoff, J.A. and Atlan, A. (1999) Evidence for effects of restorer genes on male and female reproducive functions of hermaphrodites in the gynodioecious species *Thymus vulgaris* L. *J. Evol. Biol.*, 12, 596–604.

Gouyon, P.H. and Vernet, P. (1982) The consequences of gynodioecy in natural populations of *Thymus vulgaris* L. *Theor. Appl. Genet.*, 61, 315–320.

Gouyon, P.H. and Couvet, D. (1987) A conflict between two sexes, females and hermaphrodites. In S.C. Stearns (ed.), *The Evolution of Sex and its Consequences*, Birkhäuser Verlag, Basel, pp. 245–261.

Gouyon, P.H., Jaoul, R., Maladiere, H., Milhomme, M. and Vernet, P. (1981) Introduction automatique d'échantillons solids dans une chromatographe. *Analysis*, 9, 305–310.

Gouyon, P.H., Fort, P. and Caraux, G. (1983) Selection of seedlings of *Thymus vulgaris* by grazing slugs. *J. Ecol.*, 71, 299–306.

Gouyon, P.H., Vernet, P., Guillerm, J.L. and Valdeyron, G. (1986) Polymorphisms and environment: the adaptive value of the oil polymorphism in *Thymus vulgaris* L. *Heredity*, 57, 59–66.

Gouyon, P.H., King, E.B., Bonnet, J.M., Valdeyron, G. and Vernet, P. (1987) Seed migration and the structure of plant populations: an experimental study on *Thymus vulgaris* L. *Oecologia*, 72, 92–94.

Gouyon, P.H., Vichot, F. and Van Damme, J.M.M. (1991) Nuclear-cytoplasmic male sterility: single-point equilibria versus limit cycles. *Am. Nat.*, 137, 498–514.

Granger, R. and Passet, J. (1973) *Thymus vulgaris* L. spontané de France: races chimiques et chemotaxonomie. *Phytochemistry*, 12, 1683–1691.

Granger, R., Passet, J. and Teulade-Arbousset, G. (1963) Diversité des essences de *Thymus vulgaris* L. *La France et ses Parfums*, 6, 225–230.

Harper, J.L. (1977) *The Population Biology of Plants*. Academic Press, London.

Jones, D.A. (1962) Selective eating of the acyanogenic form of the plant *Lotus corniculatus* L. by various animals. *Nature*, 193, 1109–1110.

Katz, D. A., Sneh, B. and Friedman, J. (1987) The allelopathic potential of *Coridothymus capitatus* L. (Labiatae). Preliminary studies on the roles of the shrub in the inhibition of the annuals germination and/or to promote allelopathically active Actinomycetes. *Plant and Soil*, 98, 53–66.

Levin, D.A. and Kerster, H.W. (1974) Gene flow in seed plants. *Evol. Biol.*, 7, 139–220.

Lewis, D. (1941) Male sterility in natural populations of hermaphrodite plants. *New Phytol.*, 40, 56–63.

Linhart, Y.B. (1989) Interactions between genetic and ecological patchiness in forest and their dependent species. In J.H. Bock and Y.B. Linhart (eds), *Evolutionary Ecology of Plants*, Westview Press, Boulder, Colorado, pp. 393–430.

Linhart, Y.B. and Thompson, J.D. (1995) Terpene-based selective herbivory by *Helix aspersa* (Mollusca) on *Thymus vulgaris* (Labiatae). *Oecologia*, 102, 126–132.

Linhart, Y.B. and Thompson, J.D. (1999) Thyme is of the essence: biochemical variability and multi-species deterence. *Evol. Ecol. Res.*, 1, 151–171.

Lloyd, D.G. (1976) The transmission of genes via pollen and ovules in gynodioecious angiosperms. *Theor. Popul. Biol.*, 9, 299–316.

Manicacci, D. (1993) *Evolution et maintien de la gynodioecie: allocation sexuelle et structuration spatiale du polymorphisme nucléo-cytoplasmique*. Etude théorique et approches expérimentales dans le genre *Thymus*. Ph.D. thesis, Université des Sciences et Techniques du Languedoc, Montpellier, France.

Manicacci, D., Couvet, D., Belhassen, E., Gouyon, P.H. and Atlan, A. (1996) Founder effects and sex ratio in the gynodioecious *Thymus vulgaris* L. *Mol. Ecol.*, 5, 63–72.

Manicacci, D., Atlan, A. and Couvet, D. (1997) Spatial structure of nuclear factors involved in sex determination in the gynodioecious *Thymus vulgaris* L. *J. Evol. Biol.*, 10, 889–907.

Manicacci, D., Atlan, A., Elena-Rosselló, J.A. and Couvet, D. (1998) Gynodioecy and reproductive trait variation in three *Thymus* species (Lamiaceae). *Int. J. of Plant Sci.*, 159, 948–957.

Mazzoni, C. and Gouyon, P.H. (1985) Horizontal structure of populations: migration, adaptation and chance. An experimental study on *Thymus vulgaris* L. In P. Jacquard, G. Heim, and J. Antonovics (eds), *Genetic Differentiation and Dispersal in Plants*, NATO ASI Series, Berlin Heidelberg, pp. 395–412.

Morales R. (1986) Taxonomía de los géneros *Thymus* (excluida la sección *Serpyllum*) y *Thymbra* en la península Ibérica. *Ruizia*, 3, 1–324.

Muller, C.H. (1969) Allelopathy as a factor in ecological processes. *Vegetatio*, 106, 348–357.

Muller, C.H., Muller, W.H. and Haines, B.L. (1964) Volatile growth inhibitors produced by aromatic shrubs. *Science*, 143, 471–473.

Passet, J. (1971) *Thymus vulgaris* L.: *Chémotaxonomie et biogénèse monoterpénique*. Ph.D. thesis, Faculté de Pharmacie, Montpellier, France.

Paul, P. (1970) Mise en évidence de l'action inhibitrice sur la germination *par Thymus serpyllum* ssp. *serpyllum* (L.) Briq. *Bull. Soc. Bot. Fr.*, 117, 325–334.

Perrot, V., Dommée, D. and Jacquard, P. (1982) Etude expérimentale de la concurrence entre individus issus d'autofécondation et d'allofécondation chez le thym (*Thymus vulgaris* L.). *Oecol. Plant.*, 3, 171–184.

Pomente, D. (1987) *Etude expérimentale génétique, écologique et écophysiologique du polymorphisme végétal: chémotypes et formes sexuelles du thym*. Ph.D. thesis, Université des Sciences et Techniques du Languedoc, Montpellier, France.

Rice, E.L. (1979) Allelopathy – an update. *Bot. Rev.*, 45, 15–109.

Rolland, A.-G. (1999). Maintien de la variation de la taille des fleurs chez une espèce gynodioïque (*Thymus vulgaris* L.): approche théorique et empirique. D.E.A. Université de Montpellier 2, Montpellier.

Rolland, A.-G. and Thompson, J.D. Genetic variation in the expression of flower size dimorphism within and among populations of gynodioecious *Thymus vulgaris*: why do females vary more than hermaphrodites? Unpublished manuscript.

Ross, J.D. and Sombrero, C. (1991) Environmental control of essential oil production in Mediterranean plants. In J.B. Harborne and F.A.T. Barberán (eds), *Ecological Chemistry and Biochemistry of Plant Terpenoids*, Clarendon Press, Oxford, pp. 83–94.

Seufert, G., Kotzias, D., Spartà, C. and Versino, B. (1995) Volatile organics in Mediterranean shrubs and their potential role in a changing environment. In J. M. Moreno and W.C. Oechel (eds), *Global Change and Mediterranean Type Ecosytems*, Springer, New York, pp. 343–370.

Simeon de Bouchberg, M., Allegrini, J., Bessiere, C., Attisso, M., Passet, J. and Granger, R. (1976) Proprietès microbiologiques des huiles essentielles de chimiotypes de *Thymus vulgaris* L. *Riv. Ital.*, 58, 527–536.

Sun, M. and Ganders, F.R. (1986) Female frequencies in gynodioecious populations correlated with selfing rates in hermaphrodites. *Amer. J. Bot.*, 73, 1645–1648.

Tanrisever, N., Fischer, N.H. and Williamson, G.B. (1988) Calaminthone and other menthofurans from *Calamintha aschei*: their germination and growth regulatory effects on *Schizachyrium scoparium* and *Lactuca sativa*. *Phytochemistry*, 27, 2523–2526.

Tarayre, M., Thompson, J.D., Escarré, J. and Linhart, Y.B. (1995) Intra-specific variation in the inhibitory effects of *Thymus vulgaris* (Labiatae) monoterpenes on seed germination. *Oecologia*, 101, 110–118.

Tarayre, M., Saumitou-Laprade, P., Cuguen, J., Couvet, D. and Thompson, J.D. (1997) The spatial genetic structure of cytoplasmic (cpDNA) and nuclear (allozyme) markers within and among populations of the gynodioecious *Thymus vulgaris* (Labiatae) in southern France. *Amer. J. Bot.*, 84, 1675–1684.

Tarayre, M. and Thompson, J.D. (1997) The population genetic structure of the gynodioecious *Thymus vulgaris* (Labiateae) in southern France. *J. Evol. Biol.*, 10, 157–174.

Tarayre, M. and Thompson, J.D. Variation in selfing rates in populations of the gynodioecious *Thymus vulgaris* (Lamiaceae). Unpublished manuscript.

Thanos, C.A., Kadis, C.C. and Skarou, F. (1995) Ecophysiology of germination in the aromatic plants thyme, savory and oregano (Labiatae). *Seed Sci. Res.*, 5, 161–170.

Thompson, J.D., Manicacci, D. and Tarayre, M. (1998) Thirty five years of thyme: a tale of two polymorphisms. *BioScience*, 48, 805–815.

Thompson, J.D. and Tarayre, M. (2000) Exploring the genetic basis and causes of variation in female fertility advantage in gynodioecious *Thymus vulgaris*. *Evolution*, 54, 1510–1520.

Thompson, J.D., Rolland, A.-G and Pugnolle, F. (2002) genetic variation for sexual dimorphism in flower size within and between populations of gynodioecious *Thymus vulgaris*. *J. Evol. Biol.*, in press.

Valdés, B., Diaz Lifante, Z. and Parra, R. (1998) Nutlet production and germination in female and hermaphrodite plants of *Thymus albicans* Hoffmanns & Link (Labiatae). Poster presented at OPTIMA IX meeting, May 1998. Paris.

Valdeyron, G., Dommée, B. and Vernet, P. (1977) Self-fertilisation in male-fertile plants of a gynodioecious species: *Thymus vulgaris* L. *Heredity*, 39, 243–249.

Varinard, O. (1983) Contribution à l'étude du polymorphisme chimique et sexuel chez les espèces végétales, D.E.A., Université de Montpellier 2, Montpellier.

Vernet, P., Guillerm, J.L. and Gouyon, P.H. (1977a) Le polymorphisme chimique de *Thymus vulgaris* L. (Labiée) I. Repartition des formes chimiques en relation avec certains facteurs écologiques. *Oecol. Plant.*, 12, 159–179.

Vernet, P., Guillerm, J.L. and Gouyon, P.H. (1977b) Le polymorphisme chimique de *Thymus vulgaris* L. (Labiée) II. Carte à l'echelle 1/25000 des formes chimiques dans la région de Saint-Martin-de-Londres (Herault-France). *Oecol. Plant.*, 12, 181–194.

Vernet, P., Gouyon, P.H. and Valdeyron, G. (1986) Genetic control of the oil content in *Thymus vulgaris* L.: a case of polymorphism in a biosynthetic chain. *Genetica*, 69, 227–231.

Vokou, D. and Margaris, N.S. (1982) Volatile oils as allelopathic agents. In N. Margaris, A. Koedam, and D. Vokou (eds), *Aromatic Plants : Basic and Applied Aspects*. Martinus Nijhoff publishers, The Hague, Boston, London, pp. 59–72.

Vokou, D. and Margaris, N.S. (1984) Effects of volatile oils from aromatic shrubs on soil micro-organisms. *Soil Biol. Biochem.*, 16, 509–513.

Vokou, D. and Margaris, N.S. (1986) Autoallelopathy of *Thymus capitatus*. *Oecol. Plant.*, 7, 157–163.

Williamson, G.B. (1990) Allelopathy, Koch's Postulates, and the Neck Riddle. In J.B. Grace and T. Tilman (eds), *Perspectives on Plant Competition*. Academic press, Harcourt Brace Jovanovich, Publishers, pp. 143–162.

3 Essential oil chemistry of the genus *Thymus* – a global view

Elisabeth Stahl-Biskup

INTRODUCTION

The subject of plant chemistry has developed enormously in the last four decades and this has been due to the increasingly successful identification of organic molecules in minor quantities by means of sophisticated chemical techniques. It has also been due to the awareness that secondary metabolites have a significant role in the complex interaction occurring between plants and animals or plants and plants in their exposition to the environment. Economic and medicinal interests as well as taxonomical studies, all three in quest of new natural products, have always been the strongest stimulants for research in plant chemistry.

Concerning the genus *Thymus*, we can state that its chemistry is fairly well known at least concerning the two main classes of secondary products, the volatile essential oils on the one hand and the polyphenols, especially the flavonoids, on the other hand. Both, essential oils and flavonoids, are mainly responsible for the pharmacological activities of *Thymus* plants (Simeon de Bouchberg *et al.*, 1976, Van den Broucke, 1983).

Traditionally essential oils have been regarded as the relatively toxic waste products of plant metabolic processes with no practical value to the plant. Nowadays it is thought that they possess properties that assist the plant in repelling leaf-eating insects and in preventing microbial attack. There is also evidence that terpenes leached from the leaves contribute to the allelopathic effects on the ground inhibiting the germination and growth of competitors. It has been suggested although not proven, that oil vapours near the leaf surface may reduce water loss, and the oils in the flowers might release odours attractive to pollinating agents.

In Lamiaceae, essential oils are widespread (Hegnauer, 1966; Richardson, 1992) and many species are used as aromatic herbs for flavouring foods. The essential oils themselves are products of great demand in the manufacture of perfumes and cosmetics, and they are also used for medicinal purposes. This fact also holds good for the genus *Thymus*. Indeed all the *Thymus* species produce essential oils, and several representatives are important herbs and spices used in all parts of the world. As will be shown the oils of *Thymus* species have been studied extensively.

In Lamiaceae, essential oils are stored in glandular peltate trichomes. They are situated on the epidermal surface on both sides of the leaves and show a very typical anatomy (Figure 3.1). Bruni and Modenesi (1983) intensively studied the trichomes and their development in *Thymus vulgaris* by conventional, fluorescent and electron scanning microscopy. The glandular peltate trichomes are composed of one basal stalk cell, an

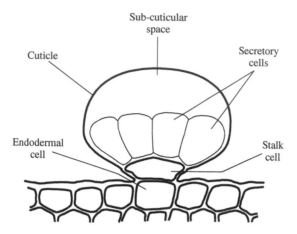

Figure 3.1 Anatomy of the glandular peltate trichome of *Thymus vulgaris* L.

"endodermal" cell that prevents backflow of secreted substances through the apoplast, and a gland head formed by 10–14 secretory cells whose prominent cytological characteristics are a relatively large nucleus and a great number of small osmiophilic vacuoles. The essential oils are produced in the secretory cells and are secreted into the sub-cuticular space, where they are stored.

When the essential oil begins to penetrate into the sub-cuticular space, a separation between the outermost and innermost layers of the secretory walls occurs. The outer-most layer of the cell wall raises together with the cuticle and forms a framework which the cuticle lies on. Furthermore, it was found that the mature peltate trichomes possess a dehiscence mechanism whereby stored essential oils are released. It ends in forming a crescent-shaped pore from which the sub-cuticular secreted material is released, demonstrated in living field-collected leaves after a sunny day (Bruni and Modenesi, 1983).

To date, the essential oils of 162 taxa of the genus *Thymus* have been chemically investigated revealing about 360 different volatile components in total. Among these the terpenes lead by almost 75 per cent, the monoterpenes being the most prominent group (43 per cent). Sesquiterpenes cover 32 per cent of the volatiles, although there are some ubiquitous sesquiterpenes present in most of the oils. Besides the terpenes a small group of non-terpenoid aliphates (17 per cent) occur in many oils but in very low concentrations. Simple benzene derivatives (6 per cent) and phenylpropanoids (2 per cent) have been found only very sporadically.

After revision of about 270 papers dealing with the essential oil composition of *Thymus* species it became obvious that they are not all of the same calibre. One has to take into account that within the last three decades the analytical methods have developed enormously. In the field of essential oils, Gas Chromatography (GC) was most concerned being the most frequently applied analytical method. Especially the on-line coupling with mass spectometry (GC/MS) nowadays allows effective oil analyses. Today identifi-cation of 90–95 per cent of the essential oil constituents is the standard, while prior to the 1960s only the main compounds could be identified.

CONSTITUENTS OF THE ESSENTIAL OILS

Monoterpenes and sesquiterpenes

Most of the terpenoid volatiles detected in *Thymus* oils belong to the monoterpene group. In the oils the monoterpenes usually make up more than 90 per cent. Sesquiterpenes are always present, but with only few exceptions in minor percentages. About 270 terpenes occur in *Thymus* oils, but their single presence is not significative when characterizing the genus, thus, quantitative aspects also have to be considered. Only constituents with concentrations above 10 per cent in at least one *Thymus* taxon will be mentioned here in order to enhance clarity and manageability. Fifty-two terpenes are concerned, 34 of them were selected as the most important volatiles within the genus *Thymus*. Their skeletons are presented in Figure 3.2.

Further classification of the 52 individual terpenes can be made by evaluating the number of *Thymus* taxa in whose oils they occur in concentrations beyond 10 per cent. As a result, in Figure 3.3 the terpenes are arranged in order of their importance within the genus *Thymus* showing the 34 most significant constituents (y-axis). How many *Thymus* taxa present the compound going beyond 10 per cent can be seen from the x-axis. Thus, the diagram reflects with clarity the chemical character of the genus *Thymus*. It once more shows the prominence of monoterpenes also among the most important volatile compounds in *Thymus*.

The phenolic terpenes, thymol and carvacrol, rank highest in importance. They occur in the oils of 77 (thymol) resp. 73 (carvacrol) different *Thymus* taxa in percentages beyond 10 per cent. Analysing Table 3.2 (see below), which has served as a basis for this diagram, one can gather that both phenols often amount to between 20 and 50 per cent of the oils. Their characteristic strong smell has always been closely associated with the genus *Thymus*, being *T. vulgaris* the most famous representative. Indeed, the 162 *Thymus* taxa investigated can be classified into phenolic and non-phenolic taxa. More than a half of the *Thymus* taxa (89 taxa=55 per cent) belong to the phenolic group, while 73 (45 per cent) to the non-phenolic group. Among the phenolic taxa both can be found: taxa with thymol plus carvacrol (46 taxa) and taxa with either thymol (27 taxa) or carvacrol (16 taxa).

In the plant kingdom, outside the genus *Thymus*, thymol and carvacrol occur quite restrictedly. Among the Lamiaceae some species of *Coridothymus*, *Origanum*, *Satureja*, and *Monarda* are known to contain thymol and carvacrol as main components of their essential oils. Aside from the Lamiaceae family it was only found in *Trachyspermum copticum* (Apiaceae). Therefore the genus *Thymus* is the most common source for the monoterpenoid phenols as is the genus *Mentha* for menthol. The limited occurrence of the phenols is one of the reasons why *Thymus* oils containing thymol or carvacrol have always been of great interest. The search for phenol-containing species has always been a great impetus for the chemical examination of the volatiles within the genus *Thymus*.

The high rank of the monoterpene hydrocarbons *p*-cymene (56 taxa) and γ-terpinene (38 taxa) can not be considered independently of the presence of thymol and carvacrol. All four terpenes are closely connected by biogenetical processes. As will be shown later *p*-cymene and γ-terpinene are the precursors in the biochemical pathway of the phenols. As a result they always occur simultaneously in the essential oils. Usually the hydroxylated phenols are more abundant in the oils than the hydrocarbons. But this is not obligatory as in few oils also the opposite is realised, with one of the hydrocarbons,

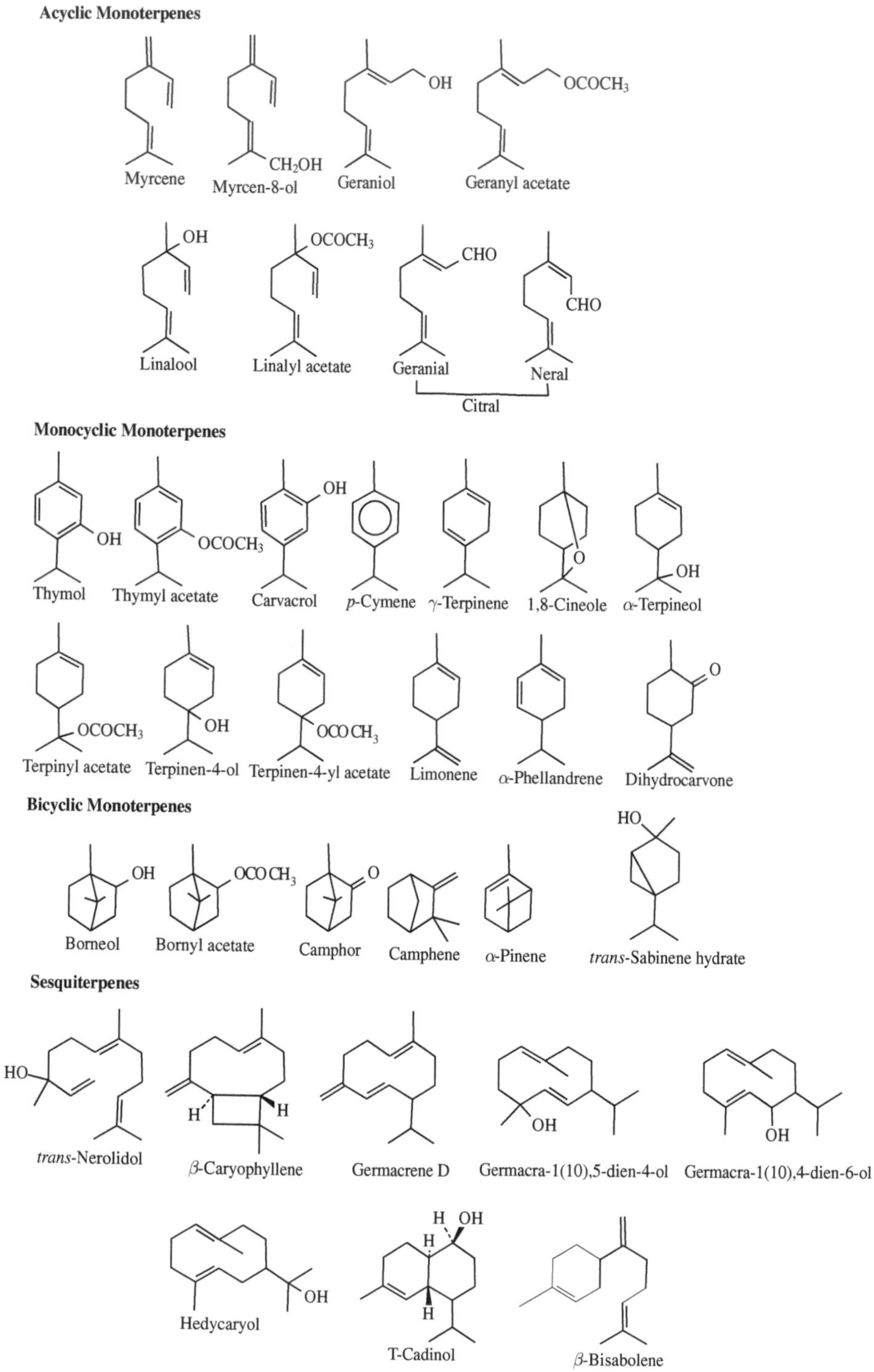

Acyclic Monoterpenes

Myrcene   Myrcen-8-ol   Geraniol   Geranyl acetate

Linalool   Linalyl acetate   Geranial   Neral

Citral

Monocyclic Monoterpenes

Thymol   Thymyl acetate   Carvacrol   *p*-Cymene   *γ*-Terpinene   1,8-Cineole   *α*-Terpineol

Terpinyl acetate   Terpinen-4-ol   Terpinen-4-yl acetate   Limonene   *α*-Phellandrene   Dihydrocarvone

Bicyclic Monoterpenes

Borneol   Bornyl acetate   Camphor   Camphene   *α*-Pinene   *trans*-Sabinene hydrate

Sesquiterpenes

trans-Nerolidol   *β*-Caryophyllene   Germacrene D   Germacra-1(10),5-dien-4-ol   Germacra-1(10),4-dien-6-ol

Hedycaryol   T-Cadinol   *β*-Bisabolene

Figure 3.2 Skeletons of the most important volatiles within the genus *Thymus*.

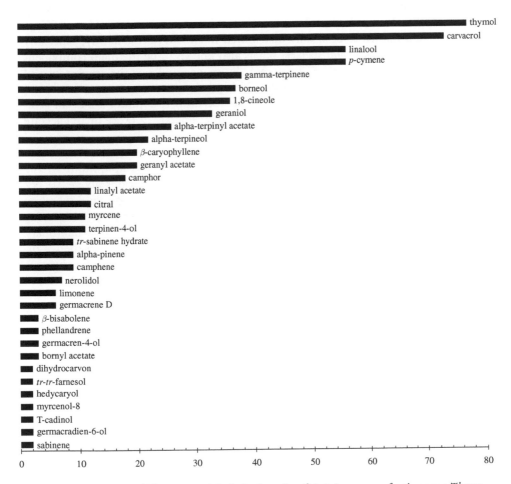

Figure 3.3 Components of *Thymus* essential oils in the order of their importance for the genus *Thymus*.

p-cymene or γ-terpinene, being the main component of the oil and going clearly beyond the phenols. It seems plausible that such oils are also treated as "phenolic" oils.

Linalool ranks third in importance for the genus *Thymus*. In the oils of 56 taxa it is found in percentages above 10 per cent. Its fine sweet smell is very opposed to that of the phenols and gives the plants quite a different character. The same is true for geraniol which occurs in 33 taxa. Linalool is widely distributed within essential oils in general and the genus *Thymus* has never been an important source of linalool. In this respect other Lamiaceae, namely lavender (*Lavandula angustifolia*) or clary sage (*Salvia sclarea*), and among the Apiaceae coriander (*Coriandrum sativum*), have always been more important. As will be shown later, within *Thymus* several taxa contain both, phenolic plants and plants containing linalool.

With borneol on the sixth place, a bicyclic monoterpene skeleton has got great importance. In 37 species it could be detected in concentrations above 10 per cent mostly accompanied by structurally-related monoterpenes such as camphor and camphene. 1,8-Cineole ranks seventh in frequency in *Thymus* essential oils and is represented by

36 taxa. In the essential oils it often occurs together with camphor and borneol, being responsible for the relatively high rank of camphor, which is also abundant in borneol-type oils. 1,8-cineole is known to be widely distributed in the Myrtaceae family, especially in the essential oils of *Eucalyptus* species, but is also known to be the main component of oil from *Rosmarinus officinalis* (Lamiaceae). From geraniol on, the importance of individual terpenes declines continuously. Nevertheless, some widely distributed monoterpenes can be met, such as α-terpinyl acetate (26 taxa), α-terpineol (22 taxa), geranyl acetate (20 taxa), camphor (18 taxa), citral (geranial + neral, 12 taxa), linalyl acetate (12 taxa), myrcene (11 taxa), and terpinen-4-ol (11 taxa).

As mentioned above, within the genus *Thymus* sesquiterpenes are not very important. The most frequently represented is β-caryophyllene, more or less ubiquitous in the essential oil kingdom. Nevertheless, it was detected in concentrations above 10 per cent in 20 *Thymus* taxa but never formed the main component of the oils and therefore hardly gives a special character to the oils. The same stands for germacrene D, ubiquitous in essential oils but hardly reaching more than 10 per cent in the oils. More extraordinary is the presence of the oxygenated germacranes, namely germacra-1(10), 5-dien-4-ol, germacra-1(10),4-dien-6-ol and hedycaryol. The latter two are known to be thermolabile and to decompose during GC, forming elemol and shyobunol respectively, documented by broad peaks in the gas chromatogram (Stahl, 1984b). Together with two other sesquiterpenes, namely T-cadinol and nerolidol, they are widely distributed within the *Thymus* species of northern Europe and therefore deserve to be mentioned.

Eighteen terpenes though present in concentrations above 10 per cent are not included in the diagram because they reach the 10 per cent limit only within one taxon. Compared with the high number of taxa investigated (162) their occurrence must be classified as sporadic. Nevertheless, they are listed here: α-cadinol, carveol, carvone, cinnamol, citronellol, elemol, fenchone, geranyl butyrate, germacrene B, intermedeol, isoborneol, isoeugenol, *cis*-myrcen-8-yl acetate, neryl acetate, spathulenol, α-terpinene, thymyl methyl ether, and thymyl acetate. A few of them give doubt of a correct identification.

Non-terpenoid aliphates

Non-terpenoid aliphates are present in many *Thymus* oils but only in minor percentages. Compounds with a chain length of 8 carbon atoms are the most frequent ones, e.g. octanol-3, octen-1-ol-3, octanone-3, octyl-3 acetate, octen-1-yl-3 acetate. The corresponding hexane derivatives rank second in frequency followed by nonane derivatives. In addition, these branched chains are also common, e.g. 6-methyl-5-heptanol, 5-methyl-3-heptanone, isoamyl acetate, methyl isovalerate, etc. However more than 62 different non-terpenoid aliphates could be detected in the oils, thus representing 17.2 per cent of the oil constituents within *Thymus*.

Non-terpenoid aromatics and phenylpropane derivatives

This group comprises C_6C_1-, C_6C_2-, and C_6C_3-derivatives, the latter better known as phenylpropanoids. Together they represent about 7.8 per cent of the oil constituents of *Thymus*. Among these compounds isoeugenol and cinnamol are the most prominent compounds, because they occur in at least one *Thymus* species in concentrations above 10 per cent and therefore they deserve to be mentioned here. All the others can more or less be found only in traces and sometimes their identification is doubtful.

Enantiomeric composition of essential oil compounds

Only few essential oils of the genus *Thymus* have been the object of enantioselective analyses. This is due to the fact that thymol and carvacrol, the most interesting compounds in *Thymus* oils, are both achiral terpenes because of their aromatic ring system. The same is true for 1,8-cineole with its symmetrical skeleton. Even the enantiomeric puritiy of linalool showing two enantiomers, R-(+)- and S-(−)-linalool, has never been studied in *Thymus* oils. Only two publications can be found focusing on some minor terpenes in oils of the phenolic group (Kreis *et al.*, 1990, 1991). Because of a very limited number of samples these studies cannot claim to reflect the true situation within the *Thymus* oils. Nevertheless, they can be considered as a first attempt to use enantioselective analysis as an analytical tool for quality control of thyme oils as it is common for the authenticity control e.g. of lavender oils (Kreis *et al.*, 1993) and Neroli oils (Juchelka *et al.*, 1996).

In the course of a screening of several medicinal essential oils, the oils of thyme (*T. vulgaris*) and wild thyme (*T. serpyllum*) were investigated concerning the enantiomeric proportions of α-pinene (1S- and 1R-), β-pinene (1S- and 1R-) and limonene (4S- and 4R-) (Kreis *et al.*, 1990). With percentages between 1 and 3 all three terpenes are minor constituents of these oils. Applying enantioselective GC on β-cyclodextrane phases β-pinene was shown to be the enantiomerically most pure compound with 4 per cent R- and 96 per cent S-β-pinene in thyme oil and 7 per cent R- and 93 per cent S-β-pinene in wild-thyme oil. The enantiomeric proportion of α-pinene in thyme oil was 89 per cent S- and 11 per cent R-, in wild-thyme oil 86 per cent S- and 14 per cent R-. For limonene the proportion 70 per cent S- and 30 per cent R- was found in thyme oil and 73 per cent S- and 27 per cent R- in wild-thyme oil.

In another publication the authors focused on the borneol and isoborneol contents in the essential oils of *T. vulgaris*, *T. serpyllum*, and *T. zygis* (Kreis *et al.*, 1991). The borneol content in these oils amounts to a maximum of 3 per cent, that of isoborneol clearly below that limit. Borneol as well as isoborneol show two enantiomers each, (−)-borneol (1S, 2R, 4S-), (+)-borneol (1R, 2S, 4R-) and (−)-isoborneol (1R, 2R, 4R-), (+)-isoborneol (1S, 2S, 4S-). For the enantioselective analysis a combination of thin-layer chromatography (TLC) with GC on permethylated β-cyclodextrin as chiral stationary phase was applied, the (−)-enantiomers eluting before the (+)-isomers. Analysing lab-distilled oils from dried herbs, it was found that the proportion of (−)-borneol to (+)-borneol was quite homogeneous with 98.1–99.6 : 0.4–1.5 in four *T. vulgaris* oils, and >99.9 : <0.1 in both, *T. zygis* and *T. serpyllum* oils. The enantiomeric proportion of isoborneol was investigated only in 7 samples of commercial thyme oils (*T. vulgaris*) showing extremely varied results with 21.2–57.0 per cent (−)-isoborneol and 43.0–78.8 per cent (+)-isoborneol.

BIOSYNTHESIS OF THE AROMATIC TERPENES

Volatile terpenoids in plants are usually of aliphatic character. Only few exceptions exist, e.g. *p*-cymene, thymol, carvacrol, *p*-cymen-8-ol, cuminalcohol, calamenene, and xanthorrhizol. Therefore the processes leading to the aromatization of the cyclohexane ring have always been of great interest. *T. vulgaris* served as the main object for the elucidation of the biogenetic pathway of the aromatic monoterpenes due to the fact that the volatile oil of thyme consists mainly of thymol, carvacrol, and *p*-cymene. Here special attention will be dedicated to the biosynthesis of these terpene phenols, whereas the biosynthesis of monoterpenes and sesquiterpenes in general will only be touched.

Terpenes contain a sequence of two or more isoprenoid units (C5H8) joined either head to tail (more common) or head to head (less common) followed by secondary chemical transformations. The early steps in terpenoid biosynthesis are the reactions resulting in the isoprenoid units, namely isopentenyl diphosphate (IPP) and dimethylallyl diphosphate (DMAPP). Nowadays two different pathways are known (Figure 3.4). The first defined and today called the "classical" pathway (Figure 3.4, left column) was the acetate/mevalonate pathway (Little and Croteau, 1999) in which three molecules of acetyl-Coenzyme A (acetyl-CoA) are fused by the enzymes acetyl-CoA acyltransferase and hydroxymethylglutaryl-CoA (HMG-CoA) synthase to produce the C6 compound HMG-CoA. A reduction of HMG-CoA forms mevalonic acid which is then converted to the C5 compound IPP, the central precursor of the terpenoid synthesis.

As a result of detailed feeding experiments it became obvious that mevalonic acid can not be the only intermediate in the terpenoid biosynthesis, but that IPP was derived by assembly of glycolytic intermediates. As a result of this finding the pyruvate/glyceraldehyde-3-phosphate pathway has been discovered (Figure 3.4, right column). The initial step of this pathway has been formulated as involving a transketolase reaction of glyceraldehyde-3-phosphate with carbons 2 and 3 of pyruvate to yield the C5 intermediate 1-deoxy-d-xylulose followed by a series of reduction and dehydration steps and a phosphorylation affording IPP or DMAPP as end products (Rohmer *et al.*, 1996). The classical acetate/mevalonate pathway seems to predominate in the cytosol leading to sesquiterpenes and triterpenes, whereas the pyruvate/glyceraldehyde-3-phosphate pathway (also called the deoxyxylulose pathway) is realized in the plastids to form monoterpenes, diterpenes, and tetraterpenes. Consequently, as could be proved the isoprenoic units of thymol in *T. vulgaris* are formed via the latter pathway (Eisenreich *et al.*, 1997). Labelling patterns of some monoterpenes and sesquiterpenes were found to bear two isoprene units derived from both pathways (Adam and Zapp, 1998).

The coupling of two isoprene units, IPP and DMAPP, yields a C10 molecule, namely geranyl pyrophosphate (GPP). It serves as a precursor to the monoterpenes. A head-to-tail coupling of a third isoprene unit provides farnesyl pyrophosphate (FPP), the precursor of the sesquiterpenes with 15 carbon atoms. Over twenty monoterpene synthases, mostly cyclases, convert the acyclic precursors (geranyl, neryl, linalyl pyrophosphates) to various cyclohexanoid monoterpenes (Charlwood and Banthorpe, 1978; Croteau, 1987).

Focusing on the phenolic monoterpenes, it was postulated in 1964 that γ-terpinene was the starting product for the biosynthesis of the aromatic monoterpenes, and *p*-cymene was considered to be formed as the first aromatic product via a non-enzymatic aromatization of γ-terpinene (Granger *et al.*, 1964). In principle this path was affirmed several years later and, indeed, *p*-cymene could be attributed the role as a key intermediate, but the non-enzymatic process was refuted and proved to be strongly enzymatic (Poulose and Croteau, 1978). Young, rapidly expanding thyme leaves (*T. vulgaris*) were utilized for these latter biosynthetic experiments. The time course of incorporation of $^{14}CO_2$ into the volatile terpenoids of thyme cuttings suggested a biosynthetic sequence by which γ-terpinene gave rise to *p*-cymene. Further incorporation experiments with exogeneous γ-(G-^{3}H)terpinene and *p*-(G-^{3}H)cymene gave strong evidence that thymol is biosynthesized by hydroxylation of *p*-cymene. Thus, *p*-cymene is the central precursor of the oxygenated compounds (Figure 3.5).

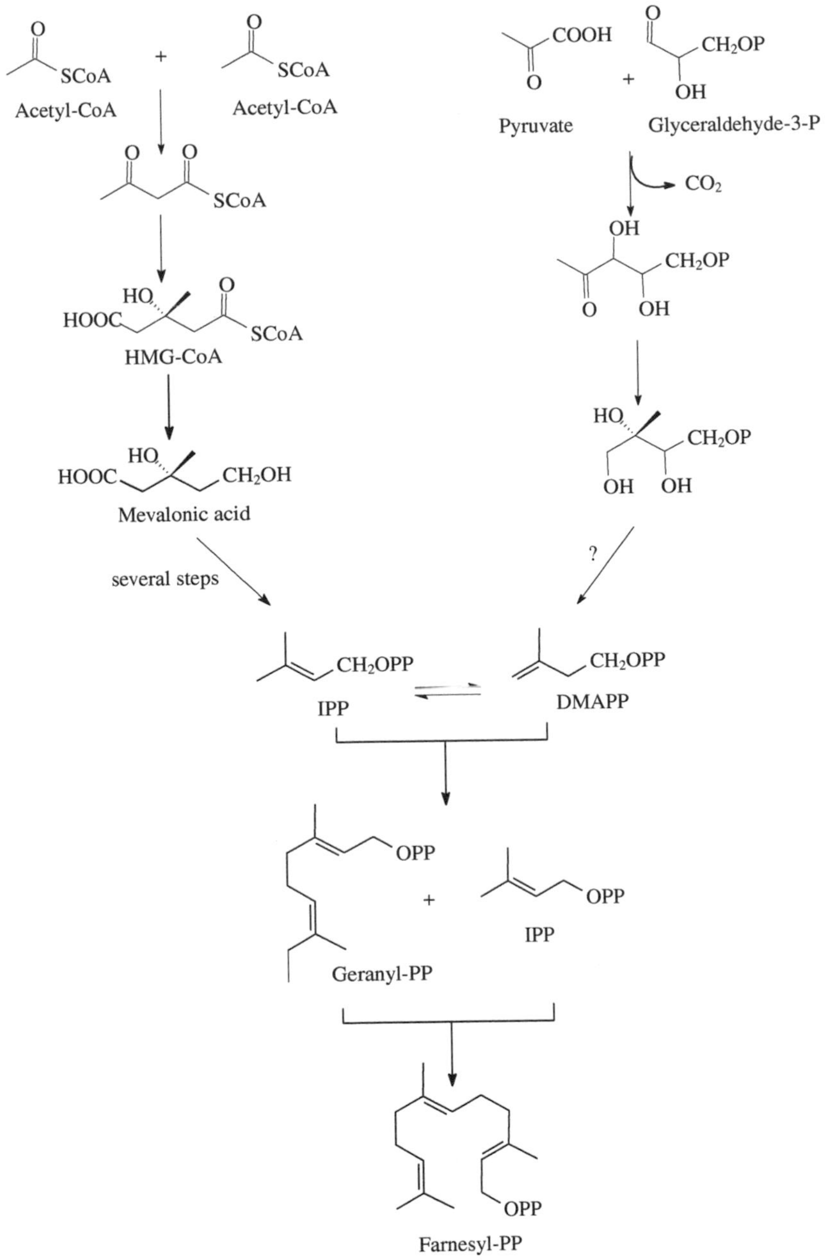

Figure 3.4 Terpenoid biosynthesis. Left column: classical pathway (acetate/mevalonate pathway); right column: the pyruvate/glyceraldehyde-3-phosphate pathway.

In the meantime the isolation of γ-terpinene synthase, a soluble enzyme which catalyses the cyclisation from neryl and geranyl pyrophosphate to γ-terpinene, was successful (Alonso and Croteau, 1991). It could be achieved with 21-day-old plants which had been subjected to an epidermal abrasion technique which selectively extract the contents

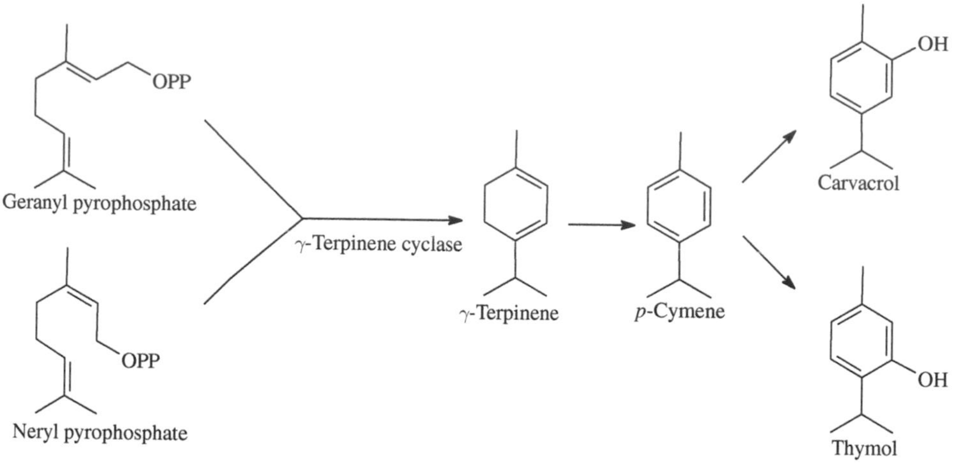

Figure 3.5 Biosynthesis of thymol and carvacrol.

of the epidermal oil glands (Gershenzon *et al.*, 1987). The phenolic and lipophilic materials present in the crude enzyme solution were removed by adsorption before the enzyme was purified by isoelectric focusing and dye-ligand anion-exchange. As a result of the production test with the γ-terpinene cyclase reacting with geranyl pyrophosphate it was observed that besides γ-terpinene (the major product set at 100 parts) a plot of side products were generated: α-thujene (16 parts), myrcene (6 parts), α-terpinene (7 parts), limonene (4 parts), linalool (5 parts), terpinene-4-ol (3 parts), α-terpineol (5 parts), and geraniol (8 parts). These results suggest that all of these monoterpenes are synthesised as co-products by the γ-terpinene synthase *in vitro* (Alonso and Croteau, 1992).

In Lamiaceae, the localisation of the biosynthesis and accumulation of monoterpenoids in the peltate glandular trichomes have never been doubted although they were experimentally proved by Croteau only in 1977 investigating *Majorana* (Croteau, 1977). He reported that the excised *Majorana* leaf epidermis with glandular trichomes incorporated the radioactivity of (U-^{14}C)-sucrose into monoterpenes. However, it was not clear whether the biosynthesis or accumulation of monoterpenes was restricted to peltate glandular trichomes alone or occurred in both peltate glandular trichomes and capitate glandular trichomes. This question was studied with *T. vulgaris* being the experimental object (Yamaura *et al.*, 1992). Quantitative analyses of the essential oils in intact glandular trichomes isolated from thyme cotyledons with the use of adhesive tape and a glass capillary tube showed that the content of thymol per cotyledon was approximately equal to the total sum of thymol in peltate and capitate glandular trichomes. The radioactivity of (U-^{14}C)-sucrose administered to cotyledonal segments was incorporated into γ-terpinene and thymol most actively by the peripheral part abundant in peltate glandular trichomes. This enzymatic reaction failed when peltate glandular trichomes were removed from cotyledons, indicating that the biosynthesis and accumulation of monoterpenes in thyme seedlings take place primarily in peltate glandular trichomes and only to a minor extent in the capitate glandular trichomes (Yamaura *et al.*, 1992). In previous experiments the light-dependent formation of peltate glandular trichomes and monoterpenes in thyme seedlings had already been demonstrated (Yamaura *et al.*,1989) as well as the participation of phytochrome in the photoregulation of monoterpene production (Tanaka *et al.*, 1989; Yamaura *et al.*,1991).

SUMMARY TABLE OF THE PRINCIPAL OIL COMPOUNDS OF ALL *THYMUS* SPECIES STUDIED

The presentation of information found in all the publications on *Thymus* essential oils adequately requires a special attention due to the high variability in techniques, sources, etc. The time prior to 1960 is covered by the publication of Gildemeister and Hoffmann (1961). From the analytical point of view, the publication of Gildemeister and Hoffmann represents a certain borderline because since that time analytical techniques, especially GC and later GC/MS, have developed considerably. With respect to these earlier publications, a list of all *Thymus* species described only in the publication of Gildemeister and Hoffmann (1961) is given in Table 3.1. All the results from chemical research work on *Thymus* species from 1960 to 2000 are summarised in Table 3.2. This Table 3.2 is intended to help the interested reader to consult original publications for further study. Table 3.2 is an expanded and updated version of Stahl-Biskup's 1991 table.

It must be stressed that questioning chemical data were adopted without filtering meaning, neither published nor checking although in some cases the correctness of the results is in doubt. Not all the authors present their results with sufficient accuracy. This may also concern the correct assignment of the plant material investigated. Nevertheless, Table 3.2 will give a very valuable basis for reflections on the chemical nature of volatiles in *Thymus*. In Table 3.2 for each species analysed the oil description is given presenting the five strongest constituents of the essential oils together with their percentages in the oils if beyond 10 per cent. Those whose concentrations lay below 10 per cent are listed in decreasing order.

The findings are arranged with respect to geographical aspects. Political names of the countries were chosen because in this way the plant source seemed to be associated sufficiently correctly. In the case of "Caucasia", the geographical term was preferred since the small states in the Caucasian region are hardly known and therefore do not allow correct location. Previously they were part of the former Soviet Union. The order of the countries follows two lines: it begins in the north with Greenland and Iceland going via Great Britain and Scandinavia to the east to Siberia. The second line begins with Morocco in the western Mediterranean region and goes via the Iberian Peninsula eastward to the Balkan Peninsula touching France, Austria, Germany, and Italy. Turkey and the

Table 3.1 Thymus species exclusively treated in the Gildemeister and Hoffmann (1961)

T. *brachyphyllus* Opiz
T. *cephalotus* L.
T. *cimicinus* Blum.
T. *clivorum* Lyka f. *borosianus*
T. *eltonicus* Klok. et Schost.
T. *odoratissimus* Bieb.
T. *serpyllum* L. ssp. *carniolicus*
T. *squarrosus* Fisch. et Mey.

Table 3.2 Summary table of principal results of oils analysed from all *Thymus* species studied

Thymus species	*Essential oil composition * (Main components only)*
Greenland (East Coast)	
T. praecox Opiz ssp. *arcticus* (E. Durand) Jalas (Stahl, 1984a)	in all the 4 types, linalyl acetate is the main component (61.4–73.1%), sesquiterpene alcohols are type characterising: (1) linalyl acetate, hedycaryol 12.7%, nerolidol, β-caryophyllene, oct-1-en-3-yl acetate, (2) linalyl acetate, nerolidol 11.4%, β-caryophyllene 10.6%, oct-1-en-3-yl acetate, (3) linalyl acetate, β-caryophyllene, germacrene D, β-sesquiphellandrene, and (4) linalyl acetate, hedycaryol, β-caryophyllene, oct-1-en-3-yl acetate
Iceland	
T. praecox Opiz ssp. *arcticus* (E. Durand) Jalas (Stahl, 1984b)	types 1 to 7 contain linalyl acetate as main component (about 70%), sesquiterpene alcohols are type characterizing, type 8: no linalyl acetate. (1) linalyl acetate, hedycaryol, nerolidol, T-cadinol, (2) linalyl acetate, hedycaryol, nerolidol, (3) linalyl acetate, no sesquiterpene alcohols, (4) linalyl acetate, hedycaryol, T-cadinol, (5) linalyl acetate, hedycaryol, (6) linalyl acetate, T-cadinol, (7) linalyl acetate, nerolidol, and (8) hedycaryol 45%, β-caryophyllene, germacrene D, β-bisabolene
Great Britain *(Scotland, Ireland, south of England)*	
T. praecox Opiz ssp. *arcticus* (E. Durand) Jalas (Bischof-Deichnik, 1987; Schmidt, 1998)	polymorphous, type characterizing compounds: hedycaryol 40.2%, linalool 25.5%/linalyl acetate 25.2%, germacra-1(10), 4-dien-6-ol 35.4%, *tr*-nerolidol 36.2%, T-cadinol 24.5%/hedycaryol 22.1%, β-caryophyllene 31.5%, linalool 61.2%, α-cadinol 23.6%/hedycaryol 19.3%, germacra-1(10),5-dien-4-ol 32.7%, *tr-tr*-farnesol 23.6%, *tr*-sabinene hydrate 21.5%/germacra-1(10), 4-dien-6-ol 19.9%, α-pinene 19.2%, geranyl acetate 27.7%, γ-terpinene 35.1%, α-terpineol 51.7%, α-terpinyl acetate 36.0%/α-terpineol 20%, *cis*-myrcenyl-8-acetate 29.6%
T. pulegioides L. (Schmidt, 1998)	(1) thymol 38.3%, γ-terpinene 12.2%, *p*-cymene, 3-octanone, β-bisabolene, (2) linalool 68.9%, thymol, geraniol, β-bisabolene, thymyl methyl ether, (3) geraniol 36.4%, neral, β-bisabolene, germacra-1(10),5-dien-4-ol, germacrene D, and (4) carvacrol 28.1%, γ-terpinene 21.4%, *p*-cymene 10.8%, linalool, germacra-1(10),4-dien-6-ol
Norway (West Coast)	
T. praecox Opiz ssp. *arcticus* (E. Durand) Jalas (Stahl-Biskup, 1986a)	in all types linalyl acetate is the main component (about 70%), sesquiterpene alcohols are type characterizing: (1) linalyl acetate, hedycaryol, nerolidol, T-cadinol, (2) linalyl acetate, hedycaryol, nerolidol, (3) linalyl acetate, no sesquiterpene alcohols, (4) linalyl acetate, hedycaryol, T-cadinol, (5) linalyl acetate, hedycaryol, and (6) linalyl acetate, T-cadinol
T. pulegioides L. (Stahl-Biskup, 1986b)	(1) carvacrol 35.2%, γ-terpinene 24.8%, *p*-cymene 10.2%, β-caryophyllene, β-bisabolene, and (2) thymol 37.2%, γ-terpinene 23.2%, *p*-cymene, β-caryophyllene, thymyl methyl ether
Finland	
T. serpyllum L. ssp. *serpyllum* (Stahl-Biskup and Laakso, 1990)	(1) monoterpene hydrocarbons 33%, 1,8-cineole 12.5–15.0%, germacra-1(10),5-dien-4-ol 3–12%, germacrene D 10.0–12.0%, germacra-1(10),4-dien-6-ol, (2) monoterpene hydrocarbons 30%, 1,8-cineole 26%, β-caryophyllene, germacrene D, hedycaryol, and (3) monoterpene hydrocarbons 27%, 1,8-cineole 19%, germacrene D, β-caryophyllene, camphor
T. serpyllum L. ssp. *tanaensis* (Hyl.) Jalas (Ivars, 1964; Von Schantz and Ivars, 1964)	(1) linalool 21.9–43.8%, linalyl acetate 8.9–17.6%, β-caryophyllene, 1,8-cineole, camphor, and (2) 1,8-cineole 17.2–27.6%, myrcene 15.4–22.4%, β-caryophyllene 6.8–19.1%, camphor, linalool

(Stahl-Biskup and Laakso, 1990)	(1) linalool 52.2%, monoterpene hydrocarbons 13%, germacrene D, germacra-1(10),4-dien-6-ol, (2) linalyl acetate 58.3%, monoterpene hydrocarbons 15%, germacrene D, germacra-1(10),4-dien-6-ol, and (3) monoterpene hydrocarbons 33%, 1,8-cineole 12.5–15.0%, germacra-1(10),5-dien-4-ol 3–12%, germacrene D 10.0–12.0%, germacra-1(10),4-dien-6-ol

Lithuania

T. pulegioides L. (Mockuté and Bernotiené, 1998, 1999, 2001)	(1) geraniol 16.3–29.2%, geranial 9.7–16.1%, linalool 0.4–13.7%, β-caryophyllene, neral, (2) carvacrol 16.0–22.2%, β-caryophyllene 11.4–15.9%, β-bisabolene 11.1–12.2%, γ-terpinene (5.9–14.5), and (3) α-terpinyl acetate 49.5–70.4%, β-caryophyllene 6.2–11.5%, geraniol, β-bisabolene, α-terpineol
T. serpyllum L. s.l. (Ložiené *et al.*, 1998)	1,8-cineole 16.3–19.0%, β-caryophyllene 9.6–11.3%, myrcene 9.7–10.7%, germacrene D, camphor

White Russia

T. serpyllum L. (Popov and Odynets, 1977)	γ-terpinene 21.4%, *p*-cymene 19.0%, thymol, α-terpineol, carvacrol

Kazakhstan

T. marschallianus Willd. (Dembitskii *et al.*, 1985)	*p*-cymene 22.4%, thymol 20.0%, γ-terpinene 19.3%, β-bisabolene, thymyl methyl ether

Siberia

T. krylovii Byczenn. (Tikhonov *et al.*, 1988)	γ-terpinene 26.2%, *p*-cymene 17.5%, myrcene, limonene, α-terpineol

Morocco

T. algeriensis Boiss. (Benjilali *et al.*, 1987a,b)	(1) thymol 14.4–65.1%, and (2) carvacrol 22.8–70.3%, both types additionally: *p*-cymene 0–36%, γ-terpinene 0–16%, linalool 0.7–11%, borneol 1–15.2%
T. broussonettii Boiss. (Benjilali *et al.*, 1987b)	(1) carvacrol 53–83.2%, (2) borneol 24–42.9%, *p*-cymene 19–24.3%, and (3) thymol 18.2–58.5%, additional intermediates
(Tantaoui-Elaraki *et al.*, 1993)	carvacrol 53.3%, *p*-cymene 13.5%, α-pinene, α-terpinene
T. ciliatus (Desf.) Benth. (Benjilali *et al.*, 1987b)	polymorphous, most important compounds: thymol 0.3–29.3%, carvacrol 0.4–21.7%, α-terpinyl acetate 16.4/42.9%, geranyl butyrate 14.6–26.7%, geranyl acetate 21.7%, camphor 0.4–28.4%, borneol 0.1–31.6%
T. hirtus Willd. (Benjilali *et al.*, 1987b)	(1) carvacrol 91.6%, and (2) thymol 19.2%, geraniol, camphor, caryophyllene epoxide, carvacryl acetate
T. maroccanus Ball (Richard *et al.*, 1985)	carvacrol 74.0/55.5%, thymol 0.4/18.4%, *p*-cymene 10.0/5.6%, linalool, γ-terpinene
T. pallidus Cosson ex Blatt (Richard *et al.*, 1985; Benjilali *et al.*, 1987a)	(1) thymol 20.6, borneol 12.7%, *p*-cymene 23.6%, γ-terpinene 14.3%, carvacrol, (2) carvacrol 24.2%, borneol 17.4%, thymol 17.7%, γ-terpinene 11.1%, *p*-cymene 10.3%, (3) *tr*-dihydrocarvone 39.9–61.1%, *cis*-dihydrocarvone 6.2–26%, and (4) camphor 54.8%, camphene 13.8%, borneol 11.0%, α-pinene, linalool
T. riatarum Humbert et Maire (Velasco Negueruela *et al.*, 1991a)	carvacrol 24.5%, *p*-cymene 17.7%, γ-terpinene 17.6%, borneol, thymol
(Iglesias *et al.*, 1991)	carvacrol 22.3%, *p*-cymene 17.5%, γ-terpinene 10.3%, borneol, β-bisabolene
T. satureioides Cosson (Benjilali *et al.*, 1987; Richard *et al.*, 1985)	(1) borneol 26–77.6%, phenols (thymol + carvacrol) 8.7–21.9%, α-terpineol 5.8–21%, camphene 0.1–11.2%, bornyl acetate, and (2) phenols (thymol + carvacrol) 34.7–50%, borneol 13.0–19.0%
(Tantaoui-Elaraki *et al.*, 1993)	borneol 31.2%, camphene 27.4%, α-pinene 17.5%, linalool, *p*-cymene

Table 3.2 (Continued)

Thymus species	*Essential oil composition* * *(Main components only)*
T. *zygis* L. (Richard *et al.*, 1985)	(1) thymol 30.7%, *p*-cymene. 23.3%, β-caryophyllene epoxide, carvacrol, thymyl methyl ether, and (2) carvacrol 42.9%, *p*-cymene 28.5%, γ-terpinene, linalool, borneol
(Tantaoui-Elaraki *et al.*, 1993)	*p*-cymene 50.6%, carvacrol, borneol, thymol, linalool
Spain	
T. *albicans* Hoffmanns. et Link (Morales, 1986)	(1) linalool 51.1%, 1,8-cineole 32.9%, α-terpineol, β-pinene, α-pinene, and (2) 1,8-cineole 70.5%, α-terpineol, β-pinene, α-pinene, alloaromadendrene
T. *antoninae* Rouy et Coincy (Velasco Negueruela and Pérez Alonso, 1986)	1,8-cineole 38.3%, camphor 17.5%, borneol 11.0%, β-caryophyllene
(Vila *et al.*, 1991a; Cañigueral *et al.*, 1994)	1,8-cineole 19.2/24.4%, camphor 15.6/14.9%, borneol, camphene, myrcene
T. *x arundanus* Willk. (Soriano Cano *et al.*, 1997)	1,8-cineole 46.4%, linalool 11.7%, limonene, *tr*-sabinene hydrate, *p*-cymene
T. *baeticus* Boiss. ex Lacaita (Morales, 1986)	(1) α-pinene 28.6%, *p*-cymene 13.9%, γ-terpinene 12.9%, terpinen-4-ol, linalool, (2) terpinen-4-ol 24.1%, α-terpineol 23.5%, *p*-cymene, α-pinene, borneol, and (3) linalool 35.4%, borneol, α-terpineol, *p*-cymene, α-pinene
(Adzet *et al.*, 1988)	1,8-cineole 14.4%, α-pinene 10.0%, *p*-cymene, terpinen-4-ol, borneol
(García Vallejo *et al.*, 1992a)	polymorphous; 1,8-cineole 0.5–15.1%, geraniol 0.2–14.9%, terpinen-4-ol 2.4–12.6%, borneol 0.9–10.0%, *p*-cymene, α-pinene
(Cruz *et al.*, 1993; Cabo *et al.*, 1990, 1992)	female: 1,8-cineole 20.9/10.3%, terpinen-4-ol 11.2/22.8%, borneol 9.8/12.7%, α-terpineol + α-terpinyl acetate, linalool + linalyl acetate, hermaphrodite: 1,8-cineole 13.8–21.4%, geraniol 20.7/7.3%, borneol 3.5/11.2%, terpinen-4-ol 8.1/10.3%, citral 10.8–8.9%
(Soriano Cano *et al.*, 1997)	*tr*-sabinene hydrate 18.2%, borneol 13.7%, α-pinene, verbenone, *p*-cymene
(Sáez, 1999)	polymorphous, type characterising components: terpinen-4-ol, *tr*-sabinene hydrate, *p*-cymene, 1,8-cineole, borneol, linalool, geranyl acetate, linalyl acetate, geranial
T. *borgiae* Rivas-Martínez, Molina et Navarro (Blázquez *et al.*, 1990)	carvacrol 59.7%, camphor, β-caryophyllene, germacrene D, germacrene B
T. *bracteatus* Lange ex Cutanda (Morales, 1986)	(1) linalool 74.3%, borneol, camphor, and (2) Linalool 17.2%, carvacrol 16.9%, borneol 12.8%, α-terpineol, terpinen-4-ol
T. *caespititius* Brot. (Seoane *et al.*, 1972)	α-terpineol 68%, borneol 11.0%, *p*-cymene, linalool
(Morales, 1986)	α-terpineol 26.2%, *p*-cymene 24.5%, camphene, α-pinene, thymol
T. *capitatus* Hoffmanns. et Link, today: *Thymbra capitata* (L.) Cav. (Mateo *et al.*, 1976; Morales, 1986; Velasco Negueruela and Pérez Alonso, 1986)	carvacrol 61.0%, *p*-cymene, γ-terpinene, β-caryophyllene, myrcene
T. *carnosus* Boiss. (Marhuenda and Alarcón de la Lastra, 1987; Marhuenda *et al.*, 1987, 1988)	borneol 41.2/43.7%, camphene 10.8/10.7%, bornyl acetate, terpinen-4-ol, camphor

T. x enicensis Blanca (Sáez, 1995a,b)	*p*-cymene 36.4%, thymol 29.9%, γ-terpinene, borneol, α-pinene
T. ericoides Rouy (Mateo *et al.*, 1978)	*p*-cymene 51.1%, γ-terpinene 10.2%, α-pinene, borneol, carvacrol
T. fontqueri (Jalas) Molero et Rovira (Molero and Rovira, 1983)	unknown 14.8%, myrcene 14.4%, β-citral 13.4%, 1,8-cineole 11.5%
T. funkii Coss. (Mateo *et al.*, 1978; Morales, 1986)	1,8-cineole 48.0%, camphor 11.2%, camphene, α-pinene, β-pinene
(Morales, 1986)	1,8-cineole 58.0/48.0/22.1%, camphor 6.9/11.2/18.8%, borneol 1.9/2.7/11.4%, myrcene, α-terpineol
(Velasco Negueruela and Pérez Alonso, 1986)	1,8-cineole 52.0/20.0%, camphor 6.5/11.0%, borneol, camphene, myrcene
(Adzet *et al.*, 1988)	1,8-cineole 47.6%, camphor 10.0%, β-pinene, camphene, α-pinene
(Vila *et al.*, 1995)	1,8-cineole 28.0/27.5%, camphor 17.7/14.2%, borneol, camphene, β-pinene
T. glandulosus Lag. ex H. del Villar (Adzet *et al.*, 1989b)	*p*-cymene 58.0%, borneol, α-pinene, camphene, γ-terpinene
T. godayanus (Blázquez and Zafra-Polo, 1989)	α-terpinyl acetate 23.6/6.1%, bornyl acetate 3.8/18.2%, carvacrol 4.4/17.3%, geranyl acetate 1.0/11.0%, 1,8-cineole
T. granatensis Boiss. (Cabo *et al.*, 1986a, b)	myrcene 18.6%, β-caryophyllene 14.0%, camphor 10.6%, borneol, limonene
T. hyemalis Lange (Adzet *et al.*, 1976)	polymorphous, type characterizing main components: (1) *p*-cymene/carvacrol, (2) *p*-cymene/thymol, (3) Carvacrol, and (4) Borneol/camphor/1,8-cineole
(Mateo *et al.*, 1978)	*p*-cymene 38.0%, thymol 26.3%, γ-terpinene, linalool, myrcene
(Morales, 1986)	*p*-cymene 51.1%, γ-terpinene 10.2%, α-pinene, borneol, carvacrol
(Cabo *et al.*, 1980, 1986c)	(1) *p*-cymene 52.7%, 1,8-cineole 15.5%, linalool, α-pinene, thymol, and (2) myrcene 16.9/31.5%, 1,8-cineole 17.1/13.6%, camphor 12.0/1.2%, terpinen-4-ol, 0.1/16.8%
(Cabo *et al.*, 1987)	myrcene 31.3%, 1,8-cineole 17.1%, camphor 12.5%, camphene, α-pinene, limonene
(Jiménez Martín *et al.*, 1989, 1992)	(1) *p*-cymene 31.5%, borneol 17.8%, carvacrol, linalool/linalyl acetate, camphene, (2) *p*-cymene 43.8–49.2%, thymol 5.4–13.9%, carvacrol 1.4–13.3%, γ-terpinene, borneol, and (3) thymol 52.3%, *p*-cymene 13.9%, carvacrol, borneol, linalool/linalyl acetate
(Sáez 1995a, 1998)	(1) thymol 36.7%, γ-terpinene 27.5%, *p*-cymene 17.5%, (2) carvacrol 39.9%, *p*-cymene 22.4%, γ-terpinene 10.6%, α-pinene, verbenone, (3) linalyl acetate 53.0%, linalool 20.2%, α-terpineol, (4) linalool 34.4%, *p*-cymene 16.4%, carvacrol 15.1%, α-pinene, borneol, (5) *p*-cymene 35.8%, carvacrol 24.7%, borneol, γ-terpinene, (6) *p*-cymene 34.3%, γ-terpinene 12.4%, borneol 12.4%, α-pinene 10.9%, thymol, and (7) *p*-cymene 45.3%, borneol 16.9%, camphene 11.1%, γ-terpinene
T. hyemalis Lange x *T. vulgaris* L. (Sáez, 1995a)	*p*-cymene 44.7%, γ-terpinene 12.3%, carvacrol 11.2%, borneol 10.4%, camphene
T. x indalicus Blanca (Sáez, 1995a)	1,8-cineole 25.7%, *p*-cymene 22.7%, γ-terpinene 12.3%, thymol, borneol
T. lacaitae Pau (García Martín and García Vallejo, 1984)	1,8-cineole 59.2–68.8%, β-pinene, limonene, sabinene, myrcene
(Velasco Negueruela and Pérez-Alonso, 1985a)	1,8-cineole 61.3%, limonene, β-pinene, sabinene, nerolidol
(Morales, 1986)	1,8-cineole 32.9%, myrcene, α-pinene, β-caryophyllene, carvacrol

Table 3.2 (Continued)

Thymus species	*Essential oil composition * (Main components only)*
T. leptophyllus ssp. *leptophyllus* Lange (Mateo *et al.*, 1978; Morales, 1986)	1,8-cineole 21.6%, linalool, thymol, *p*-cymene, myrcene
(Zafra-Polo *et al.*, 1988a; Blázquez *et al.*, 1989)	linalyl acetate 68.5%, linalool 17.0%, 1,8-cineole, α-terpineol
T. longiflorus Boiss. (Mateo *et al.*, 1978)	1,8-cineole 58.0%, camphor, α-pinene, camphene, β-pinene
(Morales, 1986)	(1) 1,8-cineole 24.8%, borneol 10.6%, myrcene, camphor, camphene, and (2) terpinen-4-ol 26.2%, γ-terpinene, myrcene, α-pinene, limonene
(Velasco Negueruela and Pérez-Alonso, 1986)	camphor 34.6%, 1,8-cineole 22.6%, borneol 13.0%, terpinen-4-ol, α-terpineol
(Cruz *et al.*, 1988)	1,8-cineole 40.6%, camphor, camphene, α-pinene, borneol
T. loscosii Willk. (Molero and Rovira, 1983)	citral 15.0%, unknown 12.8%, camphene, camphor
(Morales, 1986)	1,8-cineole 39.8/50.4%, camphor 5.5/14.0%, α-terpineol, borneol, *p*-cymene
T. mastigophorus Lacaita (Morales, 1986)	β-caryophyllene 14.2%, myrcene 13.8%, camphene, linalool, α-pinene
(Velasco Negueruela and Pérez-Alonso, 1986)	myrcene 25.0%, 1,8-cineole 19.2%, camphene 18.5%, β-caryophyllene, α-pinene
(García Vallejo and García Martín, 1986)	β-caryophyllene 7.9–27.0%, myrcene 0.1–23.1%, germacrene D 4.8–14.0%, elemol 0.7–10.9%
T. mastichina L. (García-Martín *et al.*, 1974)	linalool 82.5%, limonene + 1,8-cineole, camphor, *p*-cymene
(Adzet 1977a, b)	(1) linalool 74%, camphor, borneol, 1,8-cineole (Portugal), (2) 1,8-cineole 60–75%, linalool 8.5–20.5%, borneol, camphor (Spain), and (3) linalool 43%, 1,8-cineole 41%, borneol, camphor (Spain)
(García Vallejo *et al.*, 1984)	(1) 1,8-cineole 81.3%, (2) linalool 82.9%, and (3) 1,8-cineole 47.7%, linalool 29.0%
(Mateo *et al.*, 1978; Morales, 1986)	1,8-cineole 60.1%, linalool 14.5%, α-pinene, β-pinene, α-terpineol
(Velasco Negueruela and Pérez-Alonso, 1986)	1,8-cineole 74.8%, borneol + bornyl acetate, linalool, α-phellandrene
(Velasco Negueruela *et al.*, 1992)	(1) linalool, (2) 1,8-cineole 71%, and (3) 1,8-cineole/linalool
(Soriano Cano *et al.*, 1997)	1,8-cineole 42.6%, linalool 32.8%, β-pinene, sabinene, α-terpineol
T. mastichina L. ssp. *mastichina* (Morales, 1986)	1,8-cineole 66.5%, linalool, β-pinene, α-pinene, α-terpineol
T. membranaceus Boiss. incl. *T. murcicus* Porta (Mateo *et al.*, 1978; Morales, 1986)	1,8-cineole 35.9/62.4/55.1%, camphor 17.8/2.8/17.8%, camphene, myrcene, β-pinene
(Velasco Negueruela and Pérez-Alonso, 1986)	camphor 29.1%, 1,8-cineole 27.4%, borneol 10.4%, α-terpinene
(Zarzuelo *et al.*, 1987)	1,8-cineole 28.9%, camphor 16.7%, borneol, camphene, carvacrol
(Vila *et al.*, 1987)	1,8-cineole 41.2%, camphor 13.7%, camphene, α-pinene, β-pinene

T. membranaceus Boiss. x *T. moroderi* Pau ex R. Morales (Vila *et al.*, 1987)	1,8-cineole 32.9%, camphor 15.5%, camphene, α-pinene, borneol
T. x monrealensis Pau ex R. Morales (Soriano Cano *et al.*, 1992)	thymol 41.1–44.2%, γ-terpinene 13.9–15.4%, *p*-cymene 10.1–15.1%, 1,8-cineole, carvacrol
T. x monrealensis Pau ex R. Morales nothosubsp. *garcia-vallejoi* Sánchez-Gómez et Alcaraz (Sáez, 1995b)	geranyl acetate 24.3%, camphene 18.8%, camphor 14.5%, myrcene, α-pinene
T. moroderi Pau ex Martinez (Adzet *et al.*, 1989a)	1,8-cineole 24.5%, camphor 22.8%, camphene 10.6%, α-pinene, borneol
(Cañigueral *et al.*, 1994)	1,8-cineole 28.1/18.5%, camphor 19.8/20.3%, camphene, borneol, α-pinene, additional sesquiterpene alcohols are mentioned as type characterizing components
T. orospedanus H. del Villar (Crespo *et al.*, 1986)	*p*-cymene 22.5%, γ-terpinene 22.4%, carvacrol 15.6%, thymol, linalool
(Velasco Negueruela and Pérez-Alonso, 1985b)	1,8-cineole 25.2%, camphor 18.1%, borneol 12.3%, camphene, γ-terpinene
T. piperella L. (Adzet and Passet, 1976)	(1) *p*-cymene 26.3/50%, carvacrol 38.6/21.8%, γ-terpinene 14.5/2.1%, linalool, thymol, and (2) thymol 30.4–48.5%, *p*-cymene 21.2–31%, γ-terpinene, terpinen-4-ol, linalool
(Morales, 1986)	*p*-cymene 47.0%, carvacrol 11.5%, γ-terpinene, β-caryophyllene, borneol
(Blanquer *et al.*, 1998; Boira and Blanquer, 1998)	(1) *p*-cymene 43.3%, carvacrol 15.8%, γ-terpinene 14.0%, limonene, β-caryophyllene, (2) *p*-cymene 44.8%, thymol 23.0%, γ-terpinene, borneol, and (3) *p*-cymene 52.1%, carvacrol 18.1%, γ-terpinene, borneol
T. serpylloides Bory ssp. *gadorensis* (Pau) Jalas (Morales, 1986)	carvacrol 41.2%, *p*-cymene 24.9%, γ-terpinene 15.5%, borneol, myrcene
(Crespo *et al.*, 1988)	carvacrol 34.4%, γ-terpinene 25.6%, *p*-cymene 19.3%, α-phellandrene, linalool
(Arrebola *et al.*, 1995)	carvacrol 73.5/50.1/24.5/38.5%, γ-terpinene 0.73/11.31/26.7/12.1%, *p*-cymene 0.94/16.1/12.8/9.3%, thymol, borneol
(Arrebola *et al.*, 1997)	carvacrol 70.6%, *p*-cymene, γ-terpinene, α-pinene, thymol
(Sáez, 2001)	polymorphous (1) geraniol 80.0%, isobornyl acetate, borneol, citronellol, (2) *p*-cymene 30.9%, carvacrol 20.1%, thymol 18.5%, *tr*-ocimene 10.3%, γ-terpinene, (3) linalool 29.4%, γ-terpinene 11.7%, thymol 11.6%, p-cymene 11.3%, α-terpinene, (4) linalyl acetate 39.4%, linalool 24.5%, camphor, camphene, α-terpineol, (5) myrcene 17.1%, p-cymene 15.1%, 1,8-cineole 10.5%, terpinen-4-ol 10.8%, and (6) myrcene 30.4%, α-terpineol 27.0%, terpinen-4-ol 11.7%, 1,8-cineole 10.1%
T. serpylloides Bory ssp. *serpylloides* (Arrebola *et al.*, 1994)	carvacrol 45.0–62.5%, γ-terpinene 8.8–19.9%, *p*-cymene 8.3–16.0%, thymol, terpinen-4-ol, α-terpinene
T. villosus L. ssp. *lusitanicus* (Boiss.) Coutinho (Morales, 1986)	camphor 30.2%, borneol 15.5%, camphene, α-pinene, terpinen-4-ol
(Pérez Alonso and Velasco Negueruela 1984; Velasco Negueruela *et al.*, 1992)	camphor 37.0/25.7/14.2%, borneol 15.6/8.8/ 4.4%, linalool 1.4/12.8%,18.2%, 1,8-cineole 1.9/14.7/13.8%, camphene

Table 3.2 (Continued)

Thymus species	*Essential oil composition * (Main components only)*
T. vulgaris L. (only of wild origin) (García-Martín et al., 1974)	1,8-cineole 35.7–44.4%, camphor 11.6–16.3%, camphene 8.1–10.9%, linalool, borneol
(Elena-Rosselló, 1976)	(1) linalool + linalyl acetate 90–98%, (2) α-terpineol + terpinyl acetate 90–96%, and (3) thymol + p-cymene 50–65%
(Adzet et al., 1977b)	(1) 1,8-cineole, (2) thymol, (3) carvacrol, (4) tr-sabinene hydrate/terpinen-4-ol, (5) linalool, and (6) α-terpineol
(Mateo et al., 1978; Morales, 1986)	1,8-cineole 33.0%, camphor 14.5%, camphene 11.4%, p-cymene, γ-terpinene
(García Vallejo et al., 1992b)	1,8-cineole 6.0–38.9%, camphor 3.5–18.0%, camphene 2.9–13.2%, γ-terpinene 0–12.4%, myrcene 1.2–10.6%
T. vulgaris ssp. aestivus (Reuter ex Willk.) H. Bolós et O. Bolós, syn. T. aestivus Reut. ex Willk. (Mateo et al., 1978)	1,8-cineole 23.2%, thymol 17.2%, camphor 12.8%, p-cymene, borneol
(Adzet et al., 1988)	linalool 62.8%, geraniol, 1,8-cineole, borneol, camphor
(Morales, 1986; Elena-Rosselló, 1976; Adzet et al., 1988; Blázquez and Zafra-Polo, 1990)	(1) 1,8-cineole 22.5/55.1%, α-pinene 13,0%, β-pinene 12.6%, myrcene, γ-terpinene, (2) linalool 62.8%, geraniol, 1,8-cineole, borneol, camphor, (3) α-terpineol, and (4) 1,8-cineole 22.2%, geranyl acetate 20.0%, geraniol 17.4%, linalool, borneol
T. vulgaris L. ssp. vulgaris (Morales, 1986)	1,8-cineole 23.2%, thymol 17.2%, camphor 12.8%, p-cymene, borneol
T. webbianus Rouy (Zafra-Polo et al., 1988b)	germacrene B 18.8%, terpinen-4-ol, β-caryophyllene, 1,8-cineole, borneol
T. willkommii Ronn. (Adzet et al., 1991)	polymorphous, main types: (1) linalool 30–57%, (2) α-terpinyl acetate 36–69%, linalool 0.4–14%, (3) tr-sabinene hydrate, terpinen-4-ol, myrcen-8-ol, and (4) linalool, terpinyl acetate, 1,8-cineole
T. zygis L. (Adzet et al., 1977b)	(1) thymol, and (2) linalool
(Mateo et al., 1978)	(1) thymol 20.9–61.1%, p-cymene 9.1–18.0%, 1,8-cineole 0.2–14.4%, camphor 0–11.3%, γ-terpinene, and (2) p-cymene 22.4%, carvacrol 20.6%, γ-terpinene 13.0%, thymol 11.7%, borneol
(Cabo et al., 1981)	(1) p-cymene 30.3%, carvacrol 22.2%, borneol, 1,8-cineole, thymol, and (2) thymol 36.0%, p-cymene 19.8%, 1,8-cineole 15.3%, borneol, carvacrol
(García Martín and García Vallejo, 1983)	(1) thymol 49.8%, p-cymene 18.9%, γ-terpinene, linalool, α-pinene, (2) carvacrol 43.9%, p-cymene 20.8%, γ-terpinene 11.9%, linalool, α-pinene, (3) linalool 79.0%, linalyl acetate, carvacrol, myrcene, (4) geranyl acetate 68.6%, geraniol 16.1%, linalool, (5) α-terpinyl acetate 70.3%, α-terpineol + borneol 13.3%, limonene, and (6) myrcenol 28.6%, terpinen-4-ol 10.0%, myrcene, tr-thujanol, γ-terpinene,
(García Martín and García Vallejo, 1984)	thymol 21.3–38.1%, p-cymene 25.3–35.8%, γ-terpinene 6.5–11.6%, linalool, α-pinene, carvacrol
(Jiménez et al., 1993)	thymol 74.0%, p-cymene 10.3%, carvacrol, terpinen-4-ol
T. zygis L. ssp. gracilis (Boiss.) Morales (Morales, 1986)	(1) thymol 37.2%/61.1%, p-cymene 20.6/18.0%, terpinen-4-ol 11.3/0.8%, γ-terpinene, linalool, and (2) carvacrol 20.6%, p-cymene 22.4%, γ-terpinene 13.0%, thymol 11.7%, borneol
(Sánchez Gómez et al., 1995)	linalool 33.3%, myrcene, terpinen-4-ol, γ-terpinene, tr-sabinene hydrate

(Sáez, 1995b)	(1) tthymol 71.8%, linalool, carvacrol, *p*-cymene, (2) linalool 28.6%, α-terpineol 17.0%, *p*-cymene 13.4%, α-pinene, myrcene, (3) linalool 91.4%, and (4) thymol 25.5%, carvacrol 22.8%, *p*-cymene 18.8%, γ-terpinene, geranyl acetate
T. zygis L. ssp *sylvestris* (Hoffmanns. et Link) Brot. ex Coutinho (Velasco Negueruela and Pérez Alonso, 1984)	thymol 37.4–53.2%, *p*-cymene 10.1–17.3%, linalool 1.9–11.4%, borneol 1.0–10.2%, γ-terpinene
(Morales, 1986)	thymol 20.9–46.8%, *p*-cymene 9.1–15.2%, 1,8-cineole 0.5–14.4%, camphor 5.6–11.3%, γ-terpinene
(Sáez, 1995b)	(1) 1,8-cineole 34.5%, limonene 19.0%, thymol, linalool, *p*-cymene, (2) thymol 34.2%, *p*-cymene 27.6%, γ-terpinene 11.0%, linalool, α-pinene, (3) linalool 73.0%, 1,8-cineole 16.1%, borneol, and (4) *p*-cymene 28.2%, thymol 24.4%, carvacrol 18.2%, γ-terpinene 10.9%, linalool
T. zygis L. ssp. *zygis* (García Martín *et al.*, 1974)	linalool 32.8%, *p*-cymene 17.9%, thymol 15.1%, camphor, carvacrol
(Morales, 1986)	(1) terpinyl acetate 73.1/65.4%, α-terpineol, carvacrol, linalyl acetate, thymol, (2) terpinyl acetate 37.8%, linalool 37.5%, thymol, α-terpineol, (3) linalool 57.1%, borneol, camphene, terpinen-4-ol, and (4) carvacrol 54.5%, thymol 13.5%, *p*-cymene 12.3%, γ-terpinene, β-caryophyllene
T. x zygophorus R. Morales (García Martín and García Vallejo, 1984)	linalool + linalyl acetate 42.6%, 1,8-cineole, α-terpineol, camphene, α-pinene
Portugal	
T. albicans Hoffmanns. et Link (Miguel *et al.*, 1999)	1,8-cineole 50–65%, borneol, α-pinene, β-pinene, terpineol
(Salgueiro *et al.*, 1997c)	(1) 1,8-cineole 42.6%, α-terpineol, borneol, linalool, β-pinene, (2) linalool 44.5%, borneol, α-terpineol, 1,8-cineole, camphene, and (3) Linalool 40.8%, 1,8-cineole 25.8%, α-terpineol, borneol, β-caryophyllene
T. caespititius Brot. (Salgueiro *et al.*, 1997b)	(1) α-terpineol 30.6–40.5%, *p*-cymene, T-cadinol, γ-terpinene, γ-cadinene, and (2) carvacrol 36.3%, thymol 16.1%, carvacryl acetate, *p*-cymene, α-terpineol-type 2: on the Azores only
(Pereira *et al.*, 1999)	(1) sabinene 66.1/74.3%, (2) thymol 30.2/39.0%, sabinene 26.8/9.7%, carvacrol, γ-terpinene, (3) α-terpineol 34.3–55.9%, thymol 1.6–12.2%, γ-terpinene, and (4) Carvacrol 36.3/43.4, thymol 20.9/12.0%, α-terpineol
T. camphoratus Hoffmanns. et Link (Velasco Negueruela and Pérez-Alonso, 1987)	1,8-cineole 19.9%, borneol + α-terpineol + bornyl acetate 15.5%, terpinen-4-ol 10.2%, camphene 10.0%
(Adzet *et al.*, 1988)	terpinen-4-ol 29.3%, γ-terpinene 12.2%, *p*-cymene, α-terpinene, borneol
(Salgueiro, 1992)	(1) 1,8-cineole, (2) linalool/linalyl acetate, (3) camphene/borneol, (4) camphene/1,8-cineole/borneol, (5) α-pinene/linalool, and (6) α-pinene/1,8-cineole; mean percentages: borneol 1.2–35.0%, 1,8-cineole 0.6–35.5%, linalool 1.1–26.1%, camphene 0.2–13.5%, linalyl acetate 0.3–13.2%, α-pinene 0.9–10.5%
(Salgueiro *et al.*, 1997a)	(1) linalool 16.7%, linalyl acetate, T-cadinol, geranyl acetate, intermedeol, (2) borneol 29.5%, camphene 11.4%, camphor 10.7%, α-pinene, linalool, (3) 1,8-cineole 32.8%, T-cadinol, α-pinene, linalool, γ-cadinene, and (4) 1,8-cineole 22.9%, borneol 18.2%, α-pinene, camphene, camphor

Table 3.2 (Continued)

Thymus species	Essential oil composition * (Main components only)
T. capitatus Hoffmanns. et Link, today: *Thymbra capitata* (L.) Cav. (Proença da Cunha and Roque, 1986)	carvacrol 69%, γ-terpinene, *p*-cymene, α-pinene, myrcene
T. capitellatus Hoffmanns. et Link (Velasco Negueruela *et al.*, 1991b)	linalool 31.6%, linalyl acetate, α-terpineol, 1,8-cineole, borneol
(Salgueiro, 1992)	(1) 1,8-cineole, (2) Camphene/1,8-cineole/borneol, and (3) linalool/linalyl acetate; mean percentages: 1,8-cineole 25.1–59.1%, borneol 0.5–21.0%, linalool 0.8–13.5%, camphene 0.6–10.9%
(Figueiredo *et al.*, 1993)	1,8-cineole 50.4/55.5%, borneol, α-pinene, sabinene, linalool
T. carnosus Boiss. (Salgueiro *et al.*, 1995)	(1) borneol 25.2%, *cis*-sabinene hydrate 12.7%, terpinen-4-ol 10.1%, *tr*-sabinene hydrate, camphene, (2) borneol 30.0%, camphene 10.8%, bornyl acetate, terpinen-4-ol, *cis*-sabinene hydrate, and (3) linalool 26.9%, borneol 17.5%, *tr*-sabinene hydrate, *tr*-sabinene hydrate, terpinen-4-ol
T. lotocephalus G. López et R. Morales (Figueiredo *et al.*, 1993)	1,8-cineole 10.3/24.1%*, linalyl acetate 22.8/5.5%, linalool 11.3/6.0%, α-pinene, α-terpineol; * flowers/leaves
(Salgueiro *et al.*, 2000b)	(1) linalool 24.6%, β-caryophyllene 10.1%, camphor, intermedeol, viridiflorol, (2) 1,8-cineole 18.4%, camphor, α-pinene, viridiflorol, borneol, (3) linalool 13.9%, 1,8-cineole 11.7%, α-pinene, camphor, viridiflorol, (4) linalyl acetate 16.1%, linalool 11.5%, caryophyllene oxide 10.6%, 1,8-cineole, camphor, and (5) geranyl acetate 20.3%, intermedeol 10.9%, camphor, caryophyllene oxide, viridiflorol
T. mastichina (L.) L. ssp. *donyanae* R. Morales (Salgueiro *et al.*, 1997c)	1,8-cineole 38.4%, borneol 15.3%, camphene, α-terpineol, β-pinene
T. mastichina (L.) L. ssp. *mastichina* (Salgueiro *et al.*, 1997c)	(1) 1,8-cineole 60%, α-terpineol, β-pinene, α-pinene, sabinene, (2) linalool 58.5%, 1,8-cineole, camphor camphene, borneol, and (3) linalool 38.0%, 1,8-cineole 21.9%, β-pinene, α-terpineol, α-pinene
(Miguel *et al.*, 1999)	1,8-cineole 40–50%, camphor, camphene
T. x mourae (Salgueiro *et al.*, 2000 b)	1,8-cineole 23.5%, borneol 14.7%, camphor, intermedeol, β-caryophyllene
T. pulegioides L. (Salgueiro *et al.*, 1993)	thymol 40.0%, *p*-cymene 12.5%, γ-terpinene 12.2%, octan-3-one, carvacrol
T. x viciosoi (Pau) Morales (Salgueiro *et al.*, 1993)	carvacrol 30.0%, *p*-cymene 18.0%, thymol 17.3%, γ-terpinene, linalool
T. villosus L. ssp. *lusitanicus* (Boiss.) Coutinho (Salgueiro *et al.*, 2000a)	(1) linalool 41.0%, terpinen-4-ol 16.4%, *tr*-sabinene hydrate 11.2%, (2) linalool 31.5%, 1,8-cineole 22.3%, (3) linalool 69.7%, and (4) geranyl acetate 39.2%, geraniol 24.2%, 1,8-cineole 19.5%
T. villosus L. ssp. *villosus* (Salgueiro, 1992)	(1) *p*-cymene/borneol, and (2) *p*-cymene/camphor; mean percentages: *p*-cymene 22.5–39.8%, borneol 10.1–22.5%, camphor 2.5–13.9%, linalool, γ-terpinene
(Salgueiro *et al.*, 1997d; Salgueiro and Proença da Cunha, 1992)	(1) *p*-cymene 24.0%, camphor 11.0%, linalool γ-terpinene, borneol, (2) *p*-cymene 23.2%, borneol 19.6%, camphor, linalool, α-pinene, (3) linalool 20.9%, geraniol 12.8%, geranyl acetate 10.2%, camphor, α-pinene, and (4) α-terpineol 16.1%, camphor 13.2%, myrcene 10.6%, α-pinene, linalool

T. zygis L. ssp. *sylvestris* (Hoffmanns. et Link) Brot. ex Coutinho (Roque and Salgueiro, 1987)	thymol 14.8%, *p*-cymene 15.5%, geraniol 14.5%, geranyl acetate 12.0%, borneol + α-terpineol
(Salgueiro and Proença da Cunha, 1989)	(1) linalool 87.0%, thymol, β-caryophyllene, linalyl acetate, (2) thymol 49.2%, *p*-cymene 19.4%, γ-terpinene, borneol + α-terpineol, camphor, (3) geraniol 52.5%, geranyl acetate 38.0%, borneol + α-terpineol, (4) linalool 42.3%, 1,8-cineole 32.5%, borneol + α-terpineol, β-pinene, thymol and (5) 1,8-cineole 29.2%, thymol 25.6%, *p*-cymene 10.3%, γ-terpinene, borneol + α-terpineol
(Proença da Cunha and Salgueiro, 1991)	(1) linalool 68.0–92.1%, (2) thymol 25.0–62.1%, *p*-cymene 11.2–49.9%, γ-terpinene 3.5–28.5%, (3) carvacrol 32.5–65.2%, *p*-cymene 12.5–31.0%, γ-terpinene 3.9–15.5%, (4) geraniol 21.1–67.5%, geranyl acetate 13.1–59.1%, (5) 1,8-cineole 25.0–42.5%, linalool 24.5–49.1%, (6) 1,8-cineole 19.0–39.5%, thymol 18.5–37.1%, *p*-cymene 9.5–16.0%, (7) α-terpineol + α-terpinyl acetate 60.8–71.1%, (8) linalool 20.2–58.5%, thymol 16.0–33.5%, *p*-cymene 10.1–18.5%, and (9) 1,8-cineole 15.0–39.5%, linalool 28.2–39.9%, thymol 12.5–21.5%
T. zygis L. ssp. *zygis* (Salgueiro *et al.*, 1992)	geranyl acetate 0.1–59.9%, thymol 0.4–38.5%, carvacrol 0.5–28.6%, *p*-cymene 0.5–29.0%, γ-terpinene 0.5–19.5%
(Salgueiro *et al.*, 1993)	carvacrol 42.0%, *p*-cymene 19.9%, γ-terpinene 15.0%, linalool, β-caryophyllene

South of France

T. nitens Lamotte (Granger *et al.*, 1973)	(1) geraniol 90%, (2) α-terpineol 90%, (3) thymol 50%, *p*-cymene 30%, (4) carvacrol 15–40%, *p*-cymene 50–70%, and (5) *p*-cymene 80%
(Adzet *et al.*, 1977b)	(1) phenol type, (2) geraniol type, and (3) α-terpineol type; percentages not given
T. serpyllum praecox (Opiz) Wollm. (Vernin *et al.*, 1994)	geranyl acetate 25.0%, β-caryophyllene 13.8%, geraniol 11.8%, nerolidol, T-cadinol
T. vulgaris L. (only of wild origin) (Granger and Passet, 1971, 1973)	(1) geraniol + geranyl acetate up to 93%, (2) linalool + linalyl acetate 95%, (3) α-terpineol + α-terpinyl acetate 90–96%, (4) *trans*-sabinene hydrate up to 56%, terpinen-4-ol + terpinen-4-yl acetate up to 43%, *cis*-myrcen-8-ol 10–20%, (5) carvacrol up to 85%, and (6) thymol up to 65%

Sardinia

T. capitatus Hoffmanns. et Link, today: *Thymbra capitata* (L.) Cav. (Arras and Grella, 1992)	carvacrol 73.4%, *p*-cymene 10.5%, γ-terpinene, borneol, β-caryophyllene
T. herba-barona Loisel. (Falchi, 1967)	(1) carvacrol 43%, and (2) thymol 45%

Corsica

T. herba-barona Loisel. (Granger and Passet, 1974)	(1) carvacrol 59%; *p*-cymene, γ-terpinene, (2) carvacrol 44%, *p*-cymene 13%, γ-terpinene, (3) carvone 75%, dihydrocarvone 10%, limonene, (4) carvone 85%, limonene, and (5) carvacrol 50%, dihydrocarvone 20%
(Corticchiato *et al.*, 1995, 1998)	(1) carvone 74.6%, limonene, carvacrol, α-terpineol, *cis*-dihydrocarvone, (2) α-terpinyl acetate 55.9%, α-terpineol 15.7%, linalyl acetate, geranial, linalool, (3) carvacrol 59.4%, *p*-cymene, 2-heptanone, γ-terpinene, borneol, (4) thymol 47.9%, carvacrol, *p*-cymene, γ-terpinene, 2-heptanone, (5) linalool 31.8%, linalyl acetate 16.1%, α-terpineol, geranial, geraniol, (6) *cis*-dihydrocarvone 77.1%, *tr*-dihydrocarvone, borneol, 1,8-cineole, α-terpineol, and (7) geraniol 60.6%, geranyl acetate 19.1%, terpinen-4-ol, β-caryophyllene, α-terpineol

Table 3.2 (Continued)

Thymus species	*Essential oil composition* * *(Main components only)*
Italy	
T. alpigenus (Kerner) Ronn. (Tomei *et al.*, 1998)	β-caryophyllene 14.9%, germacrene D 13.9%, camphene, caryophyllene epoxide, α-pinene
T. capitatus Hoffmanns. et Link, today: *Thymbra capitata* (L.) Cav. (Ruberto *et al.*, 1992; Biondi *et al.*, 1993)	carvacrol 86.3%, β-caryophyllene, elemol, terpinen-4-ol
T. longicaulis C. Presl. (Valentini *et al.*, 1987)	(1) thymol 28.5/11.5%, γ-terpinene 28.6/0.2%, cinnamol 11.5/0.2%, *p*-cymene, α-terpineol, (2) α-terpineol 18.2%, α-phellandrene 11.9%, carvacrol, β-caryophyllene, and (3) *tr*-nerolidol 19.1%, thymol 10.2%, cinnamol, anisol, linalool
T. longicaulis C. Presl var. *longicaulis* (Bellomaria *et al.*, 1981)	(1) α-terpineol 24.9%, myrcene 12.0%, *p*-cymene 10.2%, nerolidol, thymol, (2) carvacrol 22.0%, thymol, bornyl acetate, α-pinene, isoborneol, and (3) myrcene 24.6%, bornyl acetate 23.8%, limonene 18.8%, thymol 13.9%, fenchone
T. longicaulis C. Presl var. *subisophyllus* (Bellomaria *et al.*, 1981)	(1) carvacrol 26.6%, thymol 18.2%, α-pinene, camphene, β-caryophyllene, (2) borneol 22.6%, bornyl acetate 12.3%, fenchone, myrcene, *p*-cymene, (3) borneol 29.1%, thymol 20.3%, nerolidol 13.9%, β-pinene, and (4) isoborneol 28.4%, borneol 22.0%, bornyl acetate 10.3%, geraniol, fenchone
T. praecox Opiz ssp. *polytrichus* (A. Kerner ex Borbas) Jalas (Valentini *et al.*, 1987)	(1) citronellol 29.9/45.4%, *cis*-nerolidol, piperitone, isoborneol, cinnamol, (2) isoeugenol 18.0%, α-terpineol, anisol, β-caryophyllene, *cis*-nerolidol, and (3) *cis*-nerolidol 15. 2%, terpinen-4-ol, linalyl acetate, *p*-cymene, carvacrol
T. pulegioides L. (Senatore, 1996)	thymol 33.8%, γ-terpinene 13.5%, *p*-cymene, β-caryophyllene, carvacrol
T. striatus Vahl. (Valentini *et al.*, 1987)	(1) carvacrol 11.7%, thymol 11.7%, linalyl acetate, camphor, β-caryophyllene, and (2) α-terpineol 16.4/14.9%, β-caryophyllene 14.5/17.7%, α-phellandrene 8.0/12.3%, *cis*-nerolidol 2.4/15.9%
T. vulgaris L. (only of wild origin) (Cioni *et al.*, 1990)	carvacrol 58.7–78.6%, thymol 0.4–21.8%, borneol, *p*-cymene, β-caryophyllene
(Maccioni *et al.*, 1992)	(1) carvacrol 23.3/28.0%, *p*-cymene 26.3/22.0%, thymol 11.0/11.3%, γ-terpinene 10.5–11.0%, borneol, and (2) thymol 31.4/36.7%, *p*-cymene 17.0/17.5%, γ-terpinene 11.1/13.1%, carvacrol 12.4/5.8%, borneol
(Piccaglia and Marotti 1991, 1993)	thymol 16.6/38.2%, *p*-cymene 25.3/18.6%, γ-terpinene 12.3/12.0%, β-caryophyllene, linalool
Germany	
T. x citriodorus (Pers.) Schreb. (Stahl-Biskup and Holthuijzen, 1995)	geraniol 61.3%, geranial, neral, β-caryophyllene, nerol
(Mikus and Zobel, 1996)	geraniol 56.6%, citral (geranial + neral)12.6%, γ-terpinene, citronellal, nerol
T. pulegioides L. (Messerschmidt, 1965)	linalool + linalyl acetate 22.3–45%, carvacrol 4.9–32.5%, geraniol, 1,8-cineole, geranyl acetate
(Sievers, 1971)	polymorphous, percentages not given, main compounds: (1) citral, (2) thymol, (3) carvacrol, (4) thymol + carvacrol, and (5) citral + phenols

Austria

T. praecox Opiz ssp. *polytrichus* (Kerner ex Borbas) Ronn. emend. Jalas (Bischof-Deichnik *et al.*, 2000)

very polymorphous, type characterizing components: thymol, geraniol/geranyl acetate, *tr*-sabinene hydrate/terpinen-4-ol, α-terpineol, linalool, linalool/linalyl acetate, *tr*-nerolidol, hedycaryol, T-cadinol, germacra-1(10),5-dien-4-ol, germacra-1(10),4-dien-6-ol, borneol.

Hungary

T. serpyllum L. (Oszagyán *et al.*, 1996)

carvacrol 39.5/45.9%, thymol, *p*-cymene, linalool, nerol

Slovakia

T. alpestris Tausch (Mártonfi, 1992a)

(1) thymol 41.0%, β-caryophyllene 10.5%, *p*-cymene, γ-terpinene, carvacrol, and (2) carvacrol 47.0%, β-caryophyllene, γ-terpinene, *p*-cymene, thymol

T. kosteleckyanus Opiz, syn. *T. pannonicus* All. (Mechtler *et al.*, 1994a)

(1) α-terpinyl acetate 74.9–84.5%, limonene 3.3–13.0%, α-terpineol, (2) thymol 41.0–50.5%, *p*-cymene 16.7–25.5%, γ-terpinene 7.2–14.8%, geraniol, (3) linalool 70.0–72.0%, (4) thymol 12.3–43.1%, *p*-cymene 6.5–36.7%, geraniol 1.3–29.6%, γ-terpinene 1.6–27.3%, thymyl methyl ether, and (5) *p*-cymene 32.6–66.2%, thymol 11.7–29.9%, thymyl methyl ether 6.7–10.3%, γ-terpinene, geraniol

T. praecox Opiz (Mechtler *et al.*, 1994b)

carvacrol 23–52%, *p*-cymene 15–19%, β-caryophyllene 16–21%

T. pulcherrimus Schur (Mechtler *et al.*, 1994b)

percentages not given, main components: β-caryophyllene, β-bisabolene

T. pulegioides L. (Mártonfi *et al.*, 1994)

(1) thymol 17.9–49.6%, β-caryophyllene 14.8–29.6%, γ-terpinene 7.5–19.2%, carvacrol 0.7–12.9%, citral 1.3–11.3%, (2) carvacrol 24.0–58.1%, β-caryophyllene 11.0–34.9%, γ-terpinene 1.9–25.6%, *p*-cymene 0–17.7%, citral 0–10.2%, (3) linalool 33.4–92.3%, β-caryophyllene 0–32.0%, citral 0–23.8%, geraniol 0–15.8%, carvacrol 0–7.9%, (4) citral 24.7–65.5%, geraniol 14.3–57.0%, β-caryophyllene 10.2–30.4%, linalool 1.1–22.9%, carvacrol, and (5) fenchone 18.5–46.3%, β-caryophyllene 9.7–18.5%, citral, γ-terpinene, carvacrol

(Mechtler *et al.*, 1994b)

p-cymene 18–27%, thymol 18–25%, geraniol 15–19%, β-caryophyllene 8–12%

T. pulegioides L. ssp. *chamaedrys* (Fries) Gusul. (Mártonfi, 1992b)

(1) thymol 20.8%, β-caryophyllene 15.2%, γ-terpinene 14.9%, carvacrol 10.8%, *p*-cymene, (2) carvacrol 32.9%, γ-terpinene 17.2%, β-caryophyllene 16.6%, citral, *p*-cymene, (3) linalool 54.8%, citral, β-caryophyllene, geraniol, (4) citral 29.1%, geraniol 22.4%, β-caryophyllene 14.6%, linalool, carvacrol, and (5) fenchone 33.9%, β-caryophyllene 10.4%, citral, γ-terpinene, carvacrol

Ukraine

T. borysthenicum (Sur *et al.*, 1988)

(1) thymol 33.6%, borneol + α-terpineol 24%, limonene 18.3%, unknown 14.5%, *p*-cymene, and (2) borneol + α-terpineol 45.9%, limonene, 1,8-cineole, linalool, thymol

T. dimorphus Klok. et Shost. (Prikhod'ko *et al.*, 1999)

carvacrol 13.1%, γ-terpinene 12.7%, 1,8-cineole, *p*-cymene, β-myrcene

T. marschallianus Willd. (Sur *et al.*, 1988)

(1) thymol 29.0–59.3%, γ-terpinene, *p*-cymene, carvacrol, (2) carvacrol 39.1–61.4%, thymol, γ-terpinene, *p*-cymene, (3) thymol 26.4/19.7%, carvacrol 28.1/17.1%, γ-terpinene, *p*-cymene, (4) geraniol 33.7–70.2%, geranyl acetate 5.4–11.6%, geranial 1.7–10.1%, and (5) carveol 65.8%, borneol, α-terpineol

T. serpyllum L. (Sur *et al.*, 1988)

(1) thymol 50.0/35.1%, γ-terpinene 12.7/18.0%, *p*-cymene 8.6/14.1%, and (2) carvacrol 48.4/55.2%, γ-terpinene 10.1/27.1%, *p*-cymene 8.0/7.1%

Table 3.2 (Continued)

Thymus species	*Essential oil composition * (Main components only)*

Romania

T. balcanus Borb.
(Kisgyörgy *et al.*, 1983)
α-terpineol 21.1%, linalool 14.1%, linalyl acetate 10.3%, α-terpinyl acetate, neryl acetate

T. comosus Heuff.
(Kisgyörgy *et al.*, 1983)
neryl acetate 24.4%, carvacrol 12.5%, thymol 10.5%, camphene

T. dacicus Borb.
(Kisgyörgy *et al.*, 1983)
carvacrol 30.0%, thymol 16.8%, nerol, α-terpineol, linalyl acetate

T. glabrescens Willd.
(Kisgyörgy *et al.*, 1983)
unknown compound 21.3%, geraniol + geranyl acetate 15.5%, linalool, nerol, neryl acetate

T. pulegioides L.
(Kisgyörgy *et al.*, 1983)
carvacrol 33.6%, thymol 31.2%, *p*-cymene 12.9%, neryl acetate, nerol

Croatia

T. capitatus Hoffmanns.
et Link, today: *Thymbra capitata* (L.) Cav.
(Kuštrak *et al.*, 1990)
carvacrol 75.9/82.6%, 1,8-cineole, limonene, bornyl acetate, linalyl acetate

T. glabrescens Willd.
(Kuštrak *et al.*, 1990)
(1) 1,8-cineole 29.4%, myrcene, camphene, α-pinene, β-pinene, and (2) thymyl acetate 14.3%, carvacrol 10.7%, *p*-cymene 10.0%, thymol, bornyl acetate

T. longicaulis C. Presl
(Kuštrak *et al.*, 1990)
thymol 40.1%, *p*-cymene 26.3%, carvacrol, α-terpineol, γ-terpinene

T. pulegioides L.
(Mastelic *et al.*, 1992)
geraniol 38.3%/4.7%, linalool 28.2/28.6%, β-caryophyllene, thymol, geranyl acetate

(Kuštrak *et al.*, 1990)
(1) carvacrol 29.8%, *p*-cymene 15.1%, γ-terpinene 11.8%, β-caryophyllene 11.5%, α-pinene, (2) thymol 21.7%, carvacrol, *p*-cymene, thymyl acetate, borneol, and (3) linalool 49.4%, carvacrol 13.2%, *p*-cymene, thymol, β-caryophyllene

Bosnia-Herzegowina

T. glabrescens Willd.
(Karuza-Stojakovic *et al.*, 1989)
α-terpinyl acetate 32.0%, terpinen-4-ol, thymol, myrcene, α-pinene

T. jankae Celak.
(Karuza-Stojakovic *et al.*, 1989)
p-cymene 30.3/15.5%, carvacrol 21.2/6.5%, geraniol 2.3/26.3%, geranyl acetate 0.5/12.0%, α-terpinyl acetate

T. maly Ronn. in Hayek
(Karuza-Stojakovic *et al.*, 1989)
terpinyl acetate 4.8/13.8%, α-pinene 4.3/13.5%, geraniol, camphene, myrcene

T. marschallianus Willd.
(Karuza-Stojakovic *et al.*, 1989)
thymol 13.7%, *p*-cymene, carvacrol, terpinen-4-ol, α-pinene

T. pannonicus All. (syn.
T. kosteleckyanus Opiz)
(Karuza-Stojakovic *et al.*, 1989)
α-terpinyl acetate 31.3%, terpinen-4-ol, carvacrol, thymol, geranyl acetate

T. pulegioides L. (Karuza-Stojakovic *et al.*, 1989)
geraniol 29.8/7.1%, linalool 20.0/5.4%, thymol 6.3/14.2%, carvacrol 4.3/11.1%, α-terpinyl acetate 8.1/16.7%

T. striatus Vahl. (Karuza-Stojakovic *et al.*, 1989)
terpinen-4-ol 23.3/11.0%, α-terpinyl acetate 8.1/11.2%, linalool 6.6/10.6%, myrcene, limonene

Albania

T. serpyllum L. (Asllani, 1973) — phenols 47–74%, *p*-cymene 8.5–36.5%

Macedonia

T. albanus ssp. *albanus* H. Braun (Kulevanova et al., 1998c) — (E)-nerolidol 20.3/48.4%, β-caryophyllene 18.0/14.8%, β-pinene, geranyl acetate, linalool

T. jankae Celak. var. *jankae* (Kulevanova et al., 1998a) — linalool 28.1/35.6%, geranial 15.3/20.2%, α-terpinyl acetate 11.3/0.4%, β-caryophyllene, thymol

T. jankae Celak. var. *pantotrichus* Ronn. (Kulevanova et al., 1998a) — linalool 30.4/30.9%, geranial 16.0/22.6%, β-pinene, α-terpineol, geraniol

T. jankae Celak. var. *patentipilus* Lyka (Kulevanova et al., 1998a) — linalool 31.2%, geranial 24.9%, β-pinene, borneol, (E)-nerolidol

T. longidens Velen. var. *dassareticus* Ronn. (Kulevanova et al., 1996a) — α-terpinyl acetate 16.2%, α-terpineol 15.6%, linalool 14.8%, geraniol 14.5%, geranyl acetate

T. longidens Velen. var. *lanicaulis* Ronn. (Kulevanova et al., 1996a) — (1) carvacrol 33.6%, geraniol 31.5%, *p*-cymene, γ-terpinene, geranyl acetate, and (2) thymol 41.5/35.7%, geraniol 12.0/18.2%, α-terpinyl acetate, *p*-cymene, γ-terpinene

T. macedonicus (Degen et Urum.) Ronn. ssp. *macedonicus* (Kulevanova et al., 1995) — geraniol 18.5%, *tr*-sabinene hydrate 14.3%, linalool 11.7%, α-terpinyl acetate 11.3%, β-pinene

(Kulevanova et al., 1999) — (1) linalool 46.6%, (2) geraniol 43.3%, geranyl acetate 37.6%, linalool, and (3) α-terpinyl acetate 44.1–60.6%, linalool 11.7–13.6%, α-terpineol

T. moesiacus Velen. (Kulevanova et al., 1996b) — geraniol 14.8–33.3%, geranyl acetate 4.11–16.6%, linalool 8.1–25.0%, carvacrol 12.3–13.3%, thymol

T. rohlenae Velen. (Kulevanova et al., 1998b) — *p*-cymene 33.3%, γ-terpinene 11.8%, thymol, geraniol, linalool

T. tosevii Velen. ssp. *tosevii* (Kulevanova et al., 1995) — (1) geraniol 21.8%, *tr*-sabinene hydrate 12.7%, α-terpinyl acetate 13.6%, α-terpineol, linalool, and (2) α-terpinyl acetate 19.7%, thymol 19.4%, carvacrol, linalool, neryl acetate

(Kulevanova et al., 1997) — polymorphous, various combinations, main components: thymol, carvacrol, α-terpinyl acetate, geraniol, linalool, *tr*-sabinene hydrate

T. tosevii Velen. ssp. *tosevii* var. *longifrons* Ronn. (Kulevanova et al., 1996c) — (1) thymol 33.4%, *p*-cymene 11.5%, γ-terpinene, linalool, β-pinene, (2) carvacrol 45.3%, *p*-cymene, γ-terpinene, thymol, geraniol, (3) carvacrol 33.1%, thymol 21.9%, *p*-cymene 13.0%, γ-terpinene 11.1%, (4) carvacrol 17.4%, thymol 17.3%, α-terpinyl acetate 15.1%, γ-terpinene, (5) α-terpinyl acetate 22.3%, carvacrol 21.1%, γ-terpinene, thymol, *p*-cymene, and (6) geraniol 37.8%, linalool 25.2%, geranyl acetate 12.0%, carvacrol 10.0%

T. tosevii Velen. ssp. *substriatus* (Kulevanova et al., 1997) — (1) thymol 24.5%, carvacrol 16.4%, *p*-cymene, γ-terpinene, β-pinene, (2) thymol 45.6%, linalool 13.1%, *p*-cymene, γ-terpinene, and (3) thymol 35.8%, linalool 32.9%, γ-terpinene, *p*-cymene

Greece

T. capitatus Hoffmanns. et Link, today: *Thymbra capitata* (L.) Cav. (Skrubis, 1972) — carvacrol 67%, thymol, borneol, α-pinene, linalool

Table 3.2 (Continued)

Thymus species	*Essential oil composition * (Main components only)*
(Philianos *et al.*, 1982)	carvacrol 55.5–81.0%, *p*-cymene 5.8–15.1%, γ-terpinene 2.0–15.5%, 1,8-cineole, β-caryophyllene
T. longicaulis C. Presl ssp. *chaubardii* (Reichb.fil.) Jalas (Tzakou *et al.*, 1998)	(1) geraniol 56.8%, geranyl acetate, nerol, β-caryophyllene, α-terpinyl acetate, (2) linalool 63.1%, α-terpinyl acetate 20.4%, α-terpineol, limonene, and (3) thymol 19.4%, limonene 18.7%, borneol, carvacrol, terpinen-4-ol
T. parnassicus Halácsy (Tzakou and Constantinidis, 1998)	β-caryophyllene 4.8–14.0%, spathulenol 1.0–20.9%, 1,8-cineole 1.7–11.0%, caryophyllene epoxide, carvacrol, thymol
T. sibthorpii Benth. (Katsiotis *et al.*, 1990)	(1) geraniol 31.6%, thymol 22.1%, geranyl acetate, *p*-cymene, β-caryophyllene, and (2) geraniol 30.1%, linalool 23.0%, citronellyl acetate, β-caryophyllene, geranyl acetate
T. tosevii Velen. (Katsiotis and Iconomou, 1986)	linalool 36.5%, geraniol 27.5%, thymol, borneol, citronellol

Cyprus

T. integer Grieseb. (Bellomaria *et al.*, 1994)	borneol 18.7–23.0%, *p*-cymene 15.7–25.0%, γ-terpinene 9.2–12.3%, thymol, linalool

Turkey

T. argaeus Boiss. et Bal. (Sezik and Başaran, 1986)	linalool 26.6%, linalyl acetate 19.5%, borneol 15.0%, geraniol, nerol
T. aznavourii Velen. (Tümen *et al.*, 1998b)	germacrene D 22.8%, (E)-β-farnesene 16.1%, α-pinene 11.1%, β-caryophyllene, limonene
T. atticus Celak. (Tümen *et al.*, 1997a)	thymol 37.2%, *p*-cymene, carvacrol, β-bisabolene, borneol
T. bornmuelleri Velen. (Başer *et al.*, 1993a)	thymol 45.0%, *p*-cymene 12.6%, carvacrol, γ-terpinene, α-pinene,
T. canoviridis Jalas (Başer *et al.*, 1998)	carvacrol 29.5%, geraniol 13.3%, thymol, β-caryophyllene, geranyl acetate
T. capitatus Hoffmanns. et Link, today: *Thymbra capitata* (L.) Cav. (Tanker and Meriçli, 1988)	carvacrol 49.8–60.8%, thymol 1.3–18.8%, *p*-cymene, γ-terpinene, terpinen-4-ol
(Özek *et al.*, 1995)	carvacrol 68.7–77.9%, *p*-cymene 6.1–10.5%, γ-terpinene, terpinen-4-ol, myrcene
T. cariensis Hub.-Mor. et Jalas (Başer *et al.*, 1992a)	borneol 13.4%, 1,8-cineole 12.7%, α-pinene 12.2%, camphor, camphene
T. cilicicus Boiss. et Bal. (Meriçli and Tanker, 1986)	α-terpineol 33.4%, camphor, citronellol, camphene, 1,8-cineole
(Tümen *et al.*, 1994)	α-pinene 16.7%, 1,8-cineole 10.4%, *cis*-verbenol, camphor, *tr*-verbenol
(Akgül *et al.*, 1999)	α-terpineol 16.4%, camphor, 1,8-cineole, α-pinene, linalool
T. eigii (M. Zohary et P.H. Davis) Jalas (Sezik and Saracoğlu, 1987)	carvacrol 75.1%, γ-terpinene, β-caryophyllene, *p*-cymene, α-thujone
(Başer *et al.*, 1996a)	carvacrol 64.6%, *p*-cymene, β-caryophyllene, isoborneol, γ-terpinene
T. fallax Fisch. and Mey. (Tümen *et al.*, 1999)	carvacrol 68.1%, thymol, *p*-cymene, β-caryophyllene, γ-terpinene

T. fedtschenkoi var. *handelii* (Ronn.) Jalas (Meriçli, 1986b)	linalool 17.2%, borneol 10.4%, thymol, carvacrol, bornyl acetate
(Başer *et al.*, 2002)	linalool 12.9%, α-terpineol, 1,8-cineole, *tr*-sabinene hydrate, camphor
T. haussknechtii Velen. (Başer *et al.*, 1992a)	linalool 19.9%, borneol 10.4%, 1,8-cineole, camphor, β-caryophyllene
T. kotschyanus Boiss. et Hohen. var. *glabrescens* Boiss. (Meriçli, 1986b)	carvacrol 44.2%, linalool, camphene, limonene
T. leucostomus Hausskn. et Velen. var. *argillaceus* Jalas (Başer *et al.*, 1992b)	thymol 27.0%, carvacrol 22.0%, linalool, *p*-cymene, borneol
T. leucostomus Hausskn. et Velen. var. *gypsaceus* Jalas (Başer *et al.*, 1999c)	thymol 33.2%, borneol 22.2%, *p*-cymene, carvacrol, camphene
T. leucostomus Hausskn. et Velen. var. *leucostomus* (Tümen *et al.*, 1997b)	(1) carvacrol 21.6%, *p*-cymene 17.8%, thymol 14.1%, borneol, γ-terpinene, and (2) α-terpinyl acetate 23.8%, linalool 13.7%, borneol 12.9%, thymol 11.3%, bornyl acetate
T. longicaulis C. Presl ssp. *chaubardii* (Boiss. et Heldr. ex Reichb.fil.) Jalas (Başer and Koyuncu, 1994)	thymol 66.5/69.7%*, carvacrol, *p*-cymene, γ-terpinene, β-bisabolene, * var. *chaubardii*/var. *alternatus*
T. longicaulis C. Presl ssp. *longicaulis* (Başer *et al.*, 1993b)	(1) thymol 52.9%, *p*-cymene 18.3%, γ-terpinene, β-bisabolene, carvacrol, (2) α-terpinyl acetate 82.1%, limonene, α-pinene, β-bisabolene, sabinene, and (3) geraniol 68.8%, geranyl acetate 16.4%, β-bisabolene, nerol
T. longicaulis C. Presl var. *subisophyllus* (Borbas) Jalas (Başer *et al.*, 1992b)	thymol 21.7%, borneol 15.8%, *p*-cymene 15.4%, camphene, α-pinene
T. migricus Klok. et Des.-Shost. (Başer *et al.*, 2002)	(1) carvacrol 36.3%, thymol, (2) Thymol 44.2%, carvacrol, and (3) carvacrol 36.5%, thymol 36.3%
T. pectinatus Fisch. et Mey. var. *pectinatus* (Başer *et al.*, 1992c)	thymol 35.0%, borneol 17.7%, *p*-cymene 11.1%, carvacrol, camphene
(Tümen *et al.*, 1998a; Başer *et al.*, 1999d)	thymol 52.5%, *p*-cymene 14.6%, γ-terpinene 12.1%, carvacrol, borneol
T. praecox Opiz ssp. *grossheimii* (Ronn.) Jalas var. *grossheimii* (Başer *et al.*, 1996c)	thymol 26.6%, *p*-cymene 24.9%, α-pinene, α-terpinyl acetate, β-caryophyllene
T. praecox Opiz ssp. *skorpilii* (Velen.) Jalas var. *laniger* (Borb.) Jalas (Başer *et al.*, 1996c)	thymol 17.8/41.4%, carvacrol 10.5/7.6%, 1,8-cineole, β-caryophyllene, *p*-cymene
T. praecox Opiz ssp. *skorpilii* (Velen.) Jalas var. *skorpilii* (Başer *et al.*, 1996c)	geraniol 24.2%, α-terpinyl acetate 22.7%, geranyl acetate, linalool, linalyl acetate

Table 3.2 (Continued)

Thymus species	*Essential oil composition * (Main components only)*
T. pseudopulegioides Klokov et Des.-Shost. (Başer et al., 1999a)	(1) thymol 23.1%, linalool 20.1%, γ-terpinene, p-cymene, carvacrol, (2) thymol 50.1%, carvacrol 10.7%, p-cymene 10.7%, γ-terpinene, methyl carvacrol, and (3) linalool 21.6%, α-terpinyl acetate 16.7%, geraniol 11.2%, p-cymene, thymol
T. pubescens Boiss. et Kotschy var. cratericola Jalas (Başer et al., 1999c)	carvacrol 17.5%, p-cymene 16.4%, thymol 10.8%, α-pinene, borneol
T. revolutus Celak. (Meriçli and Tanker, 1986)	α-terpineol 30.5%, linalool 22.5%, p-cymene, β-terpineol, α-terpinene
T. roegneri C. Koch (Tümen et al., 1997a)	thymol 58.2%, p-cymene 12.9%, carvacrol, β-bisabolene, borneol
T. sibthorpii Benth. (Başer et al., 1992d)	thymol 34.8%, γ-terpinene, p-cymene, borneol, myrcene
T. sipyleus ssp. sipyleus var. davisianus (Meriçli and Tanker, 1986)	geranial 32.8%, α-terpineol, neral, isoborneol, 1,8-cineole
T. sipyleus Boiss. ssp. sipyleus var. sipyleus (Başer et al., 1995a)	(1) geranial 25.5–37.0%, neral 13.6–25.6%, 1,8-cineole, α-terpineol, camphor, (2) linalool 21.8%, α-terpineol, geranial, linalyl acetate, neral, and (3) α-terpineol + isoborneol 25.5%, geranial 12.8%, neral, torreyol, camphor
T. spathulifolius Hausskn. et Velen. (Meriçli, 1986a)	carvacrol 47.5%, p-cymene, γ-terpinene, linalool, linalyl acetate
T. striatus Vahl var. interruptus Jalas (Başer et al., 1999e)	(1) thymol 10.5%, borneol, carvacrol, tr-sabinene hydrate, β-bisabolene, (2) β-caryophyllene 29.6%, carvacrol 20.6%, caryophyllene epoxide, (E)-β-farnesene, germacere D, (3) thymol 34.7%, β-caryophyllene 12,7%, β-bisabolene, caryophyllene epoxide, carvacrol, and (4) β-caryophyllene 56.5%, germacere D 11.1%, carvacrol, linalool, caryophyllene epoxide
T. subcollinus Klok. (Başer et al., 1997)	germacrene D 31.9%, β-caryophyllene 17.6%, α-pinene, δ-cadinene, limonene
T. syriacus Boiss. (Tümen and Başer, 1994)	thymol 49.0%, carvacrol 15.9%, p-cymene, borneol, γ-terpinene
T. thracicus Velen. var. longidens (Velen.) Jalas (Başer et al., 1995b)	(1) geraniol 15.7%, thymol 12.3%, p-cymene 12.2%, γ-terpinene 10.7%, carvacrol 10.5%, and (2) geraniol 47.3%, geranyl acetate 18.3%, camphor, 1,8-cineole, β-bisabolene
T. zygioides Griseb. var. lycaonicus (Celak.) Ronn. (Meriçli and Tanker, 1986)	thymol 24.6%, linalool 12.2%, borneol, carvacrol, 1,8-cineole
(Başer et al., 1996b)	(1) carvacrol 48.1%, γ-terpinene 12.0%, thymol, p-cymene, β-bisabolene, (2) geraniol 68.6%, geranyl acetate, β-bisabolene, nerol, β-caryophyllene, (3) α-terpinyl acetate 36.2%, α-terpineol 19.5%, β-bisabolene, bornyl acetate, limonene, and (4) thymol 41.8–57.2%, γ-terpinene 1.3–19.5%, β-bisabolene 1.2–15.9%, p-cymene 4.1–12.0%, carvacrol
T. zygioides Griseb. var. zygioides (Başer et al., 1999b)	linalool 33.7%, (E)-nerolidol 12.5%, neral, geranial, camphor
Caucasia T. collinus Bieb. (Kasumov, 1988)	thymol 57.0%, p-cymene, 1,8-cineole, carvacrol, terpinolene

T. coriifolius Ronn. (Kasumov, 1987)	α-pinene 11.1%, limonene, γ-terpinene, linalool, *p*-cymene
T. dagestanicus Klok. et Shost. (Kasumov and Davidenko, 1985)	thymol 32.7%, 1,8-cineole 12.6%, *p*-cymene, carvacrol, γ-terpinene
T. eriophorus Ronn. (Kasumov, 1981)	thymol 22.0%, geraniol 11.5%, linalool 11.5%, borneol 10.7%, linalyl acetate
(Kasumov and Komarova, 1983)	linalool 12.2%, borneol, thymol, carvacryl acetate, limonene
T. fedtschenkoi Ronn. (Kasumov, 1988)	thymol 63.4%, γ-terpinene, carvacrol, α-pinene, β-caryophyllene, terpinolene
T. fominii (Kasumov, 1981)	carvacrol 15.8%, thymol 15.8%, *p*-cymene, γ-terpinene, terpinen-4-ol
T. karamarianicus Klok. et Shost. (Kasumov and Farkhadova, 1986)	thymol 9.1/26.3/26.3%, carvacrol 9.1/10.0/14.8%, citral 10.3/28.6/20.4%, α-terpineol 12.4/15.0/ 18.2%, linalool
T. kjapazii G. Grossh. (Novruzova and Kasumov, 1987)	thymol 24.3%
T. kotschyanus Boiss. et Hohen. (Kulieva *et al.*, 1979; Kasumov and Gadzhieva 1980; Guseinov *et al.*, 1987; Kasumov, 1980)	carvacrol 13.7/14.7%, thymol 10.1/11.1%, γ-terpinene, terpinolene, β-caryophyllene
(Kasumov, 1988)	thymol 35.5%, *p*-cymene 17.7%, carvacrol 11.7%, α-pinene, α-terpineol
T. marschallianus Willd. (Kasumov, 1987)	geraniol 30.3%, limonene 10.1%, α-pinene, α-terpineol, β-caryophyllene
T. migricus (Kasumov, 1981)	carvacrol 35.7%, γ-terpinene 13.3%, thymol 13.3%, *p*-cymene, β-caryophyllene
T. nummularius M.B. (Kasumov and Ismailov, 1975)	1,8-cineole 29.4%, thymol 28.3%, borneol 13.3%, carvacrol, limonene
(Kasumov and Gavrenkova, 1982)	β-caryophyllene 10.5%, thymol, linalool, borneol, terpinen-4-ol
T. pastoralis Iljin (Kasumov and Davidenko, 1985)	thymol 32.6%, carvacrol 21.1%, terpinen-4-yl acetate 19.5%, *p*-cymene, γ-terpinene
T. rariflorus C. Koch (Kasumov, 1979)	23 compounds, no percentages given; main compounds thymol and carvacrol
T. serpyllum L. (Abetisjan *et al.*, 1988)	(1) thymol 81.5/76.1%, *p*-cymene, carvacrol, β-caryophyllene, α-terpineol, and (2) carvacrol 49.0–62.0%, thymol 21.5–29.7%, *p*-cymene, β-caryophyllene, α-terpineol
T. tiflisiensis Klok. et Shost. (Kasumov, 1987)	carvacrol 20.1%, β-caryophyllene 18.2%, γ-terpinene, linalool, geraniol
T. transcaucasicus Ronn. (Kasumov, 1981; Kasumov and Gavrenkova, 1985)	thymol 36.6%, *p*-cymene 15.7%, carvacrol, γ-terpinene, borneol
(Kasumov and Komarova, 1983)	thymol 46.8%, carvacrol 10.8%, γ-terpinene, β-caryophyllene, α-pinene

Table 3.2 (Continued)

Thymus species	*Essential oil composition* * *(Main components only)*
(Kasumov, 1987)	β-caryophyllene 14.3%, geraniol, geranyl acetate, carvacrol, linalool
(Kasumov, 1988)	thymol 33.6%, linalool 12.8%, carvacrol 11.7%, terpinolene, α-pinene
T. trautvetteri Klok. (Kasumov *et al.*, 1979)	geraniol 9.8/12.7%, linalool, borneol, geranyl acetate, β-caryophyllene
(Ismailov *et al.*, 1981)	geraniol 10.6%, linalool, borneol, geranyl acetate

Libya

T. algeriensis Boiss. (Aboutabl and El-Dahmy, 1995)	carvacrol 36.8%, myrcene 20.2%, thymol 12.5%, α-terpinene 10.7%, α-thujene

Egypt

T. bovei Benth. (Aboutabl *et al.*, 1986a,b)	thymol 68.4%, *p*-cymene, thymyl acetate, carvacrol, γ-terpinene
T. decussatus L. (Khodair *et al.*, 1993)	thymol 69.7%, phellandrene, carvacrol, α-terpinene, γ-terpinene

Israel

T. capitatus Hoffmanns. et Link, today: *Thymbra capitata* (L.) Cav. (Ravid and Putievsky, 1985, 1986)	(1) thymol 39.3/52.6%, carvacrol 12.7/18.1%, γ-terpinene 19.4/6.8%, *p*-cymene, β-caryophyllene, and (2) carvacrol 34.8/39.8%, γ-terpinene 14.7/12.2%, *p*-cymene 13.6/11.3%, thymol, β-caryophyllene
(Fleisher *et al.*, 1984)	(1) thymol 50.8–67.0%, carvacrol, (2) carvacrol 60.3–76.1%, thymol, and (3) carvacrol 34.5–43.7%, thymol 14.4–29.6%

Ethiopia/Erithree

T. schimperi Ronn. (Nigist Asfaw *et al.*, 2000)	(1) thymol 36.8%, *p*-cymene 20.5%, γ-terpinene 12.9%, carvacrol, α-terpinene, and (2) carvacrol 63.3%, γ-terpinene, thymol, linalool, *p*-cymene
T. serrulatus Hochst. ex Benth. (Nigist Asfaw *et al.*, 2000)	thymol 48.5%, γ-terpinene 12.9%, *p*-cymene 12.8%, carvacrol, linalool

Saudi Arabia

T. vulgaris L. (Mossa *et al.*, 1987)	thymol 62.2%, carvacrol, *p*-cymene, linalool, Δ-3-carene

Iran

T. carmanicus Jalas (Rustaiyan *et al.*, 2000)	thymol 40.8%, carvacrol 24.8%, γ-terpinene, *p*-cymene, borneol
T. kotschyanus Boiss. et Hohen. (Rustaiyan *et al.*, 2000)	thymol 38.0%, carvacrol 14.2%, 1,8-cineole 13.2%, linalool, *p*-cymene
(Sefidkon *et al.*, 1999; Sefidkon and Dabiri, 1999)	carvacrol 40.7–61.2%, thymol 7.5–26.9%, γ-terpinene, *p*-cymene, borneol
T. pubescens Boiss. et Kotschy ex Celak. (Rustaiyan *et al.*, 2000)	thymol 37.9%, carvacrol 14.1%, *p*-cymene 13.1%, γ-terpinene, linalool

Pakistan

T. serpyllum L. (Sattar *et al.*, 1991)	thymol 42.6%, *p*-cymene, carvacrol, borneol, terpinen-4-ol

India
T. serpyllum L. (Razdan and Koul, 1975)

carvacrol 49.4%, *p*-cymene, thymol, zingiberene, eugenol

(Gulati and Gupta, 1977)

thymol 57.6%, *p*-cymene 20.0%, γ-terpinene, zingiberene, borneol

Mongolia
T. dahuricus Serg. (Shavarda *et al.*, 1980)

α-terpineol 29%, linalool 14%, camphor, *p*-cymene, terpinen-4-ol

T. gobicus Tschern. (Shavarda *et al.*, 1980)

thymol 38%, *p*-cymene 20%, γ-terpinene 16%, carvacrol, borneol

China
T. magnus Nakai (Han and Kim, 1980)

thymol, carvacrol, *p*-cymene, γ-terpinene, α-pinene; percentages not given

T. mongolicus Ronn. syn. **T. serpyllum** L. var. *mongolicus* Ronn. (Fang Hong-ju *et al.*, 1988)

carvacrol 51.2%, *p*-cymene 11.7%, borneol, 1,8-cineole, thymol

(Luo and Song, 1989)

p-cymene 30.3%, thymol + carvacrol 20.0%, β-phellandrene 14.0%

(Zhang Hongli *et al.*, 1992)

thymol 23.9%, 2,4,5-trimethyl benzyl alcohol 16.9%, *p*-cymene 16.3%, carvacrol 10.6%, o-tert. butylphenol

T. quinquecostatus Celak. (Shyuan Qi *et al.*, 1987)

(1) linalool 45.2%, borneol 13.5%, γ-cadinene, *tr*-menthen-1-ol, camphene, and (2) thymol 21.9, carvacrol 20.3%, sabinene 11.0%, borneol, *p*-cymene

(Fang Hong-ju *et al.*, 1988)

(1) phenol type: see var. prazewalskii, (2) linalool 72.9/29.7%, borneol 7.8/13.1%, *p*-cymene 0.3/16.5, terpinen-4-ol, camphene, and (3) ester type: see var. *asiaticus*

T. quinquecostatus Celak. var. *asiaticus* (Fang Hong-ju *et al.*, 1988)

geranyl acetate 18.4%, carvacrol 11.2%, geraniol, borneol, *p*-cymene, myrcene

T. quinquecostatus Celak. var. *prazewalskii* (Fang Hong-ju *et al.*, 1988)

carvacrol 21.2%, γ-terpinene 13.7%, *p*-cymene 10.6%, thymol, camphene

Japan
T. quinquecostatus Celak. (Kameoka *et al.*, 1973)

thymol 56.1%, carvacrol, *p*-cymene, oct-1-en-3-ol, camphor

New Zealand
T. vulgaris L. (Morgan, 1989)

(1) thymol 49.7%, and (2) carvacrol 48.8%, further components in both types: *p*-cymene 8.8–41.4%, γ-terpinene 0.6–15.6%, linalool

Chile
T. vulgaris L. (Montes Guyot *et al.*, 1981)

carvacrol 43%, *p*-cymene 41%, α-pinene, limonene, borneol

Cameroon
T. vulgaris L. (Amvam Zollo *et al.*, 1998)

thymol 27.2%, *p*-cymene 23.6%, γ-terpinene 22.7%, linalool, β-caryophyllene

Note
*Main components only: first five components; components >10% with percentages, components <10% without percentages, in numerical order.

adjoining Caucasian region follow. At the end, some countries of the eastern Mediterranean area in Africa, the Near East, the Middle East and the Asiatic countries of the Far East are listed, ending with New Zealand; at the very end Cameroon and the only plant source of South America in Chile is treated. Specialities of the evaluation will be referred to as footnotes within Table 3.2 itself.

VARIABILITY IN ESSENTIAL OIL COMPOSITION

In plants the essential oil yield and chemical composition vary considerably due to different factors. Both intrinsic and extrinsic factors have to be considered. As intrinsic factors we encounter genetic and sexual variations as well as seasonal and ontogenetic variations. Extrinsic factors are described by ecological and environmental aspects such as altitude, soil, climate, light, etc. There are no systematic investigations of all these circumstances influencing the oil composition for the genus *Thymus*. Therefore only results gathered from individual analyses, arising from different interests of the researchers in both applied science and basic science can be summarised. It must be stressed that there are a few early reports dealing with oil variations which erroneously interpret them as being caused by extrinsic factors. Nowadays they are more correctly described as being a result of chemical polymorphism. Since this phenomenon is widespread in the genus *Thymus*, Chapter 4 is dedicated to this important topic.

Infraspecific variations

In Lamiaceae the phenomenon of infraspecific variability concerning the essential oil composition, meaning chemical differences that exist in morphologically identical species, was examined systematically by Lawrence (1980). His concept guaranteed that only taxonomically authenticated plants harvested at approximately the same stage of growth were analysed, an indispensable prerequisite which has often been neglected by other scientists. He found five species of the genus *Mentha*, namely *Mentha arvensis*, *M. longifolia*, *M. pulegium*, *M. spicata*, and *M. suaveolens*, to be polymorphous as well as three species of the genus *Monarda*, and 9 species of the genus *Pycnanthemum*.

Regarding the genus *Thymus* he referred to the noteworthy publication by Adzet *et al.* (1977a) who had presented an examination of the polychemism in Mediterranean *Thymus* species some years before. The authors had found some species to show a distinct tendency towards chemical differentiation, especially *T. aestivus*, *T. herba-barona*, *T. hyemalis*, *T. mastichina*, *T. nitens*, *T. vulgaris*, and *T. zygis*. Other southern species, such as *T. antoninae*, *T. longiflorus*, *T. membranaceus*, and *T. piperella*, only showed minor chemical variations. Regarding the geographical distribution of the investigated species and chemotypes, Adzet (1977a) and Passet (1979) interpreted the polychemism within the genus *Thymus* as a result of a dynamic evolution, which does not only preserve the species, but also ensures a territorial advantage by a process of adaptation to the environmental conditions. This very far-reaching interpretation has not been substantiated.

Within the genus *Thymus* the phenomenon of polychemism, an expression coined by Tétényi (1970), was described for the first time when Granger and Passet (1971, 1973) published the results of their studies of the essential oil chemistry in *T. vulgaris* collected in the south of France. Detailed analyses of populations as well as of a multitude of individuals proved *T. vulgaris* to be polymorphous, showing six different chemotypes characterised by the following main oil constituents: thymol, carvacrol, *tr*-sabinene hydrate/terpinen-4-ol, α-terpineol, linalool, and geraniol. A correlation between climate and distribution could be established because the phenolic types were growing in hot and dry regions whereas the linalool and the α-terpineol types preferred a humid climate, the geraniol type humid and cold regions.

The six chemotypes of *T. vulgaris* in the south of France fulfill the requirements for "chemical races" as defined by Hegnauer (1978) who described them as growing in populations which are geographically separated and presenting hereditary chemical characteristics. Contrary to that, other *Thymus* species such as *T. baeticus*, *T. camphoratus*, *T. hyemalis*, *T. praecox* ssp. *arcticus*, *T. praecox* ssp. *polytrichus*, *T. tosevii*, *T. willkommii*, and *T. zygis* ssp. *sylvestris* proved to be more polymorphous, showing a higher or even an uncertain number of chemotypes. Moreover, no restricted occurrence of a certain chemotype can be observed and different chemotypes grow side by side within one population. Such a non-homogeneity of populations can only be reliably proven when a multitude of individual plants of a species are analysed. This should always be considered in the experimental concept.

Nowadays within the genus *Thymus* the flood of publications reporting on infraspecific variations concerning the essential oil composition does not allow a scientifically profound compilation of such data. When available they have been included in Table 3.2. The column "Essential Oil Composition" provides a rough idea on the variability of the oil composition by indicating there the chemotypes with (1), (2), (3)... This enumeration of chemotypes does not include any information on the quality of the experimental concepts, which unfortunately differs considerably, nor on the authors' interpretations if any. Further details can easily be gathered from the original publications cited in Table 3.2. In order to establish the concept of polychemism within the genus *Thymus* more thoroughly, Chapter 4 of this book is dedicated to this phenomenon and presents representative studies in this field, in particular the results elaborated by Salgueiro *et al.* from species growing in the western part of the Iberian Peninsula, by García-Vallejo *et al.*, Arrebola, Cañigueral *et al.*, Blanquer *et al.*, and Sáez from species growing in the central, eastern and southern parts of the Iberian Peninsula, as well as by Stahl-Biskup focusing on the species of northern European latitudes. Mártonfi contributed greatly to the knowledge of the polychemism in the Slovenian *Thymus pulegioides* and Kulevanova discussed the polychemism of the Macedonian *Thymus tosevii*.

Sexual variations

Within the genus *Thymus* gynodioecy is widespread, revealing female and hermaphrodite plants. Two subspecies of *T. serpylloides*, ssp. *gadorensis* and ssp. *serpylloides*, were investigated with regard to the differences in oil yield and oil composition between female and hermaphrodite plants. Both subspecies belong to the phenolic group containing carvacrol as the main constituent of their essential oils. Concerning the variations of the carvacrol content in two different phenological stages (full flowering, fruiting) of *T. serpylloides* ssp. *serpylloides*, no significant differences in the percentages of carvacrol between hermaphrodite and female plants were found when the mean of three years was evaluated, although in one case a concentration of this component higher by 17 per cent with respect to the hermaphrodites was striking (Arrebola *et al.*, 1994). The oil content changed reciprocally.

Differences between female and hermaphrodite plants were more distinct in *T. serpylloides* ssp. *gadorensis*. There the highest oil content was always obtained from female individuals in both years, as well as at all three phenological stages investigated (full flowering, fruiting, post-fruiting). In contrast to that the carvacrol content was always found to be higher in the hermaphrodite individuals (Arrebola *et al.*, 1995). Sexual differences were also discovered in *T. baeticus* (Cabo *et al.*, 1990). There the contents of 1,8-cineole and terpinen-4-ol were found to be higher in the oils of female plants than in hermaphrodite plants (21 per cent and 11 per cent resp. versus 14 per cent and 8 per cent), whereas the contents of citral and geraniol were lower (7 per cent and 5 per cent versus 11 per cent and 21 per cent, respectively). However, the results give rise to some doubts because of the fact that *T. baeticus* later turned out to be a highly polymorphous plant (Sáez, 1999), and the differences then found may have been caused by having investigated different oil types.

Seasonal and ontogenetical variations

Results published on the chemical composition of *Thymus* oils reveal that most of the oils were produced from flowering plants. This period in the plants' life cycle is chosen because the oil yield usually peaks at that time. At least this was found for *T. vulgaris* and *T. pulegioides* in the early 1960s (Messerschmidt, 1964; Tucakov, 1964), and later for *T. herba-barona* (Falchi-Delitala *et al.*, 1983), *T. granatensis* (Cabo *et al.*, 1986a), *T. pectinatus* var. *pectinatus* (Başer *et al.*, 1999d), *T. zygis* ssp. *sylvestris* (Moldão-Martins *et al.*, 1999). But this fact does not always seem to be valid because in a different year the oil content was found to be peak at different times for *T. vulgaris* (Messerschmidt, 1964; Weiss and Flück, 1970). A detailed study of *T. hyemalis* during its complete vegetative cycle revealed that the oil yield varying between 0.15 and 0.58 per cent peaks twice, the first time in April (0.52 per cent) at the flowering stage and a second time in July (0.58 per cent) (Cabo *et al.*, 1987). The same phenomenon was said to be found for *T. pulegioides* collected from mid-April to mid-September in Campania, Italy (Senatore, 1996). The oil increased from 0.38 per cent in April to 1.11 per cent in May when the plant was in full flower. But studying the presented data in detail, the second peak found in June (0.93 per cent) might be an error in the method, at least too few collections were examined. Egyptian *T. vulgaris* was reported to show only minor seasonal variations in the oil content (Karawya and Hifnawy, 1974).

With respect to the oil composition most attention was focused on the phenol containing species due to the economic interest in these substances. Detailed results are documented for *T. vulgaris*, which shall be summarized shortly. In the two older papers it was stated that the composition of the oils was very constant over the whole season (Messerschmidt, 1964; Weiss and Flück, 1970), whereas other authors could observe considerable variations. For example Egyptian *T. vulgaris* showed an increase of the total phenolic portion before the full flower with thymol and carvacrol temporarily forming contrarotating curves (Karawya and Hifnawy, 1974). Holthuijzen (1994) observed a clear decrease in the thymol content from the beginning of June (60.7 per cent) to the end of July (25.5 per cent) increasing again until October (39.0 per cent). The sum of the biogenetically related terpenes (thymol, carvacrol, *p*-cymene and γ-terpinene) however varied only within limited borders (74.3–82.1 per cent). In *T. pulegioides* collected in Campania (Italy) the highest phenol content was observed at full flower at the end of May (43.3 per cent) increasing from 19.2 per cent in April and decreasing to 22.4 per cent in September (Senatore, 1996). The thymol content of *T. pectinatus* var. *pectinatus* was found to peak in the pre-flowering stage (Başer *et al.*, 1999d).

In order to determine the optimal harvest time for *T. vulgaris* in New Zealand naturalized plants were studied intensively during 13 months (McGimpsey and Douglas, 1994). It could be shown that seasonal variation has a significant effect on the yield and composition of thyme oil. Oil yield and phenol content peaked after flowering had finished (December). The maximum total phenol content peaked at a total of 37 per cent after flowering in summer (December/January). *p*-Cymene, which was an important component of Central Otago thyme oils, ranged from 40 to 50 per cent in winter and early spring (May to October), declining to 21 per cent in January. The authors conclude that harvest after flowering in December (summer time) delivers the highest yield and the best quality. Aiming the use of the essential oil from *T. zygis* ssp. *sylvestris* in Portugal as a food ingredient, the most interesting stage is the post-flowering period, the essential oil at this time being rich in thymol, geranyl acetate and geraniol, with *p*-cymene presenting lower levels (Moldão-Martins *et al.*, 1999).

Seasonal variations of compounds other than phenols were studied in various other *Thymus* species. Both, the linalyl acetate content of *T. praecox* ssp. *arcticus* (Holthuijzen, 1994) and the geraniol content of *T. x citriodorus* (Stahl-Biskup and Holthuijzen, 1995) varied to different extents during the period from June to October, the first within the limits 59–68 per cent the latter between 62 and 79 per cent. A low fluctuation was established for linalool within the oil of a linalool-chemotype of *T. vulgaris* in the south of France (Granger *et al.*, 1965). Contrary results were gained from *T. leptophyllus* whose linalyl acetate content varied considerably between 35 per cent in March and 73 per cent in September (Zafra-Polo *et al.*, 1988a).

Marhuenda *et al.* (1987) focused on the borneol content in *T. carnosus* ranging from 38.6–50.6 per cent with one peak in April and another in August. The high content of borneol in April coincided with a minimum of terpinen-4-ol and with a general decrease of hydrocarbons. Very high seasonal variations are reported for *T. granatensis* (Cabo *et al.*, 1986a), *T. marschallianus* (Schratz and Hörster, 1970), and *T. hyemalis* (Cabo *et al.*, 1987). Referring to the latter, throughout the growth cycle of the plants, the essential oil was rich in hydrocarbons varying from a minimum of 31 per cent in November to a maximum of 53 per cent in July, with a secondary peak in April and May. Among these compounds myrcene exhibits the widest fluctuations (between 9 and 31 per cent). Variations in the alcohol–acetate fraction covered a much narrower range from 10 per cent in August to a maximum of 20 per cent in October; 1,8-cineole peaked at 27 per cent in August and fell to 13 per cent in October, camphor from 21 per cent in September to 11 per cent in July.

These few examples make clear that seasonal variations in the oil compositions are encountered and have to be investigated individually. General predictions cannot be made. Rather we learn that the oil compositions given (e.g. in Table 3.2) are not more than momentary snaps that must be treated with caution. Especially if oils or even the herbs are considered for commercial purposes, the decision for the time of harvest must be made individually.

With regard to ontological variations only one study is dedicated to the essential oil of *T. marschallianus* (Schratz and Hörster, 1970). Considerable differences in the oils comparing young leaves and one-year-old leaves within one plant were documented, the more if the whole season is considered. These findings reveal different biosynthetic capacities of oil glands *in statu nascendi* or adult oil glands.

GLYCOSIDICALLY BOUND VOLATILES

The existence of glycosidically bound monoterpenes became evident for the first time, when Francis and Allcock (1969) reported on the detection of geranyl, neryl, and citronellyl glucosides in rose petals. At that time the discovery of monoterpene glycosides focused attention on a new field of research, especially in Lamiaceae, and has led to speculations about their role. Glycosidically bound volatiles were assumed to be transport forms of monoterpenes or involved in monoterpene catabolism (Skopp and Hörster, 1976; Croteau and Martinkus, 1979). *T. vulgaris* was one of the early objects which gave rise to such hypotheses.

After acid and enzymatic treatment of fresh leaves of thyme (*T. vulgaris*), thymol and carvacrol could be detected as main hydrolysis products besides minor amounts of linalool and geraniol (Skopp and Hörster, 1976). Glucose and galactose were found to be the sugar moieties. The same hydrolysis products were found when fresh plant material of *T. vulgaris* was exclusively subjected to an enzymatic hydrolysis, besides further aglycones, namely hexan-1-ol, *cis*-hexen-3-ol-1, octanol-3, octen-1-ol-3, benzyl alcohol, phenethyl alcohol and eugenol (Van den Dries and Baerheim Svendsen, 1989). In a geraniol type of *T. pulegioides* geraniol was the main glycoside apart from smaller amounts of eugenol, linalool, and 1-octen-3-ol (Mastelic *et al.*, 1992).

These findings gave reason for a more intensive study of the composition of the glycosidic fraction of four *Thymus* taxa, *T. vulgaris*, *T. pulegioides*, *T. x citriodorus*, *T. praecox* ssp. *polytrichus* and *T. praecox* ssp. *arcticus* (Holthuijzen, 1994; Stahl-Biskup and Holthuijzen, 1995). On balance the result of these investigations can be summarised as follows (Holthuijzen, 1994): a) the content of glycosidically bound volatiles usually is much lower than the essential oil content with proportions of 1:60–100 (*T. vulgaris*), 1:30–120 (*T. x citriodorus*), and 1:400 (*T. praecox* ssp. *arcticus*). b) After enzymatic hydrolysis of the glycosidic fraction a multitude of compounds can be detected, a considerable number of them occurring in all the five taxa investigated, namely *cis*-hexen-3-ol-1, *tr*-hexen-3-ol, octanol-3, octen-1-ol-3, linalool, terpinen-4-ol, α-terpineol, geraniol, benzyl alcohol, phenyl ethyl alcohol, and eugenol. c) Nevertheless, besides these ubiquitous compounds each species had its individual pattern correlated to the composition of its free volatiles, at least in *T. vulgaris* (thymol and carvacrol), *T. pulegioides* (thymol and linalool), and *T. x citriodorus* (geraniol). In *T. praecox* ssp. *polytrichus* and ssp. *arcticus* a structural equivalence between glycosidically bound volatiles and free volatiles was much less distinct.

Comparing the variation (development, course) of the compositions of the glycosidically bound volatiles and the free volatiles during one vegetation period, a difference between *T. vulgaris* and *T. x citriodorus* on the one hand and *T. arcticus* on the other hand could be observed. In *T. vulgaris* and *T. x citriodorus* the curves of main components of the aglycone fraction run parallel to the free volatiles, thymol or geraniol, respectively, always being the main component within both fractions (Stahl-Biskup and Holthuijzen, 1995). In contrast to that, the composition of the aglycone fraction of *T. praecox* ssp. *arcticus* varied irregularly without any structural correlation to the essential oil constituents (Holthuijzen, 1994).

Glycosylation in vegetable tissue is quite common and in essential oil-bearing plants it might be a protective mechanism to prevent the lipophilic volatiles such as phenols or alcohols from destroying membranes. Therefore it is of little wonder that in essential oil plants the key intermediates of the terpene biosynthesis such as geraniol, nerol, linalool, terpineol, terpinen-4-ol or intermediates of other pathways, such as aliphatic alcohols, benzyl alcohol, phenylethyl alcohol are widespread in their glycosidic forms. From observations in *Thymus* (Skopp and Hörster, 1976; Holthuijzen, 1994; Stahl-Biskup and Holthuijzen, 1995) or in *Mentha* (Stengele and Stahl-Biskup, 1993, 1994), it can be derived that if the essential oils mainly consist of hydroxylated terpenes, e.g. thymol in *T. vulgaris*, geraniol in *T. x citriodorus*, or menthol in *M. x piperita*, the corresponding glycosides are present in the same plant. This is the case although in Lamiaceaes special accumulation sites exist in the form of glandular trichomes where such membrane destroying compounds are assumed to be stored safely. Electron microscopic observations of the secretory cells of glandular trichomes of young *Mentha piperita* leaves make it plausible that glycosylation takes place when the subcuticular space is full (Stahl-Biskup *et al.*, 1993).

REFERENCES

Abetisjan, R.G., Aslanjaic, C.K., Arutjunjan, E.G. and Akopjan, C.W. (1988) The essential oils of *Mentha longifolia* (L.) Huds. and *Thymus serpyllum* L. (Armenian SSR). *Rast. Res.*, 24, 605–610.

Aboutabl, E.A., Soliman, F.M., El-Zalabani, S.M., Brunke, E.J. and El-Kersh, T.A. (1986a) Essential oil of *Thymus bovei* Benth. *Egypt. J. Pharm. Sci.*, 27, 209–214.

Aboutabl, E.A., Soliman, F.M., El-Zalabani, S.M., Brunke, E.J. and El-Kersh, T.A. (1986b) Essential oil of *Thymus bovei* Benth. *Sci. Pharm.*, 54, 43–48.

Aboutabl, E.A. and El-Dahmy, S.I. (1995) Chemical composition and antimicrobial activity of essential oil of *Thymus algeriensis* Boiss. *Bull. Fac. Pharm. Cairo Univ.*, 33, 87–90.

Adam, K.-P. and Zapp, J. (1998) Biosynthesis of the isoprene units of chamomile sesquiterpenes. *Phytochemistry*, 48, 953–959.

Adzet, T. and Passet, J. (1976) Estudio quimotaxonómico de *Thymus piperella* L. *Collect. Bot. (Barcelona)*, 10, 1–6.

Adzet, T., Granger, R., Passet, J. and San Martín, R. (1976) Chimiotypes de *Thymus hiemalis* Lange. *Plant. Méd. Phytothér.*, 10, 6–15.

Adzet, T., Granger, R., Passet, J. and San Martín, R. (1977a) Le polymorphisme chimique dans le genre *Thymus*: sa signification taxonomique. *Biochem. Syst. Ecol.*, 5, 269–272.

Adzet, T., Granger, R., Passet, J. and San Martín, R. (1977b) Chimiotypes de *Thymus mastichina* L. *Plant. Méd. Phytothér.*, 11, 275–280.

Adzet, T., Vila, R., Ibáñez, C. and Cañigueral, S. (1988) Essential oils of some Iberian *Thymus*. *Planta Med.*, 54, 369–370.

Adzet, T., Vila, R., Batllori, X. and Ibáñez, C. (1989a) The essential oil of *Thymus moroderi* Pau ex Martínez (Labiatae). *Flavour Fragr. J.*, 4, 63–66.

Adzet, T., Vila, R., Cañigueral, S. and Ibáñez, C. (1989b) The herb essential oil of *Thymus glandulosus* Lag. ex H. del Villar. *Flavour Fragr. J.*, 4, 133–134.

Adzet, T., Cañigueral, S., Gabalda, N., Ibáñez, C., Tomàs, X. and Vila, R. (1991) Composition and variability of the essential oil of *Thymus willkommii*. *Phytochemistry*, 30, 2289–2293.

Akgül, A., Özcan, M., Chialva, F. and Monguzzi, F. (1999) Essential oils of four wild-growing Labiatae Herbs: *Salvia cryptantha* Montbr. et Auch., *Satureja cuneifolia* Ten., *Thymbra spicata* L. and *Thymus cilicicus* Boiss. et Bal. *J. Essent. Oil Res.*, 11, 209–214.

Alonso, W.R. and Croteau, R. (1991) Purification and characterization of the monoterpene cyclase gamma-terpinene synthase from *Thymus vulgaris. Arch. Biochem. Biophys.*, **286**, 511–517.

Alonso, W.R. and Croteau, R. (1992) Comparison of two monoterpene cyclases isolated from higher plants; γ-terpinene synthase from *Thymus vulgaris* and limonene synthase from *Mentha piperita*. In R.J. Petroski and S.P. McCormick (eds), *Secondary-metabolite Biosynthesis and Metabolism*. Plenum Press, New York, London, pp. 239–251.

Amvam Zollo, P.H., Biyiti, L., Tchoumbougnang, F., Menut, C., Lamaty, G. and Bouchet, Ph. (1998) Aromatic plants of tropical Central Africa, Part XXXII, Chemical composition and antifungal activity of thirteen essential oils from aromatic plants of Cameroon. *Flavour Fragr. J.*, **13**, 107–114.

Arras, G. and Grella, G.E. (1992) Wild thyme, *Thymus capitatus*, essential oil seasonal changes and antimycotic activity. *J. Hort. Sci.*, **67**, 197–202.

Arrebola, M.L., Navarro, M.C., Jiménez, J. and Ocaña, F.A. (1994) Yield and composition of the essential oil of *Thymus serpylloides* subsp. *serpylloides. Phytochemistry*, **36**, 67–72.

Arrebola, M.L., Navarro, M.C. and Jiménez, J. (1995) Variations in yield and composition of the essential oil of *Thymus serpylloides* Bory ssp. *gadorensis* (Pau) Jalas. *J. Essent. Oil Res.*, **7**, 369–374.

Arrebola, M.L., Navarro, M.C. and Jiménez, J. (1997) Essential oils from *Satureja obovata*, *Thymus serpylloides* ssp. *serpylloides* and T. *serpylloides* ssp. *gadorensis* micropropagated plants. *J. Essent. Oil Res.*, **9**, 533–536.

Asllani, U. (1973) Albanian Thyme varieties and their essential oils. *Bulletin I Shkencave te Natyres*, **27**, 111–127.

Başer, K.H.C., Özek, T. and Tümen, G.H. (1992a) Essential oils of *Thymus cariensis* and *Thymus haussknechtii*, two endemic species in Turkey. *J. Essent. Oil Res.*, **4**, 659–661.

Başer, K.H.C., Kirimer, N., Özek, T., Kürkçüoglu, M. and Tümen, G. (1992b) The essential oil of *Thymus leucostomus* var. *argillaceus. J. Essent Oil Res.*, **4**, 421–422.

Başer, K.H.C., Özek, T. and Kürkçüoglu, M. (1992b) Composition of the essential oil of *Thymus longicaulis* C. Presl. var. *subisophyllus* (Borbas) Jalas from Turkey. *J. Essent. Oil Res.*, **4**, 311–312.

Başer, K.H.C., Kirimer, N., Özek, T., Kürkçüoglu, M. and Tümen, G. (1992c) The essential oil of *Thymus pectinatus* Fisch. et Mey. var. *pectinatus. J. Essent. Oil Res.*, **4**, 523–524.

Başer, K.H.C., Özek, T. and Kürkçüoglu, M. (1992d) Characterization of the essential oil of *Thymus sibthorpii* Bentham. *J. Essent. Oil Res.*, **4**, 303–304.

Başer, K.H.C., Özek, T., Kirimer, N. and Malyer, H. (1993a) The essential oil of *Thymus bornmuelleri* Velen. *J. Essent. Oil Res.*, **5**, 691–692.

Başer, K.H.C., Özek, T., Kirimer, N. and Tümen, G. (1993b) The occurrence of three chemotypes of *Thymus longicaulis* C. Presl. subsp. *longicaulis* in the same population. *J. Essent. Oil Res.*, **5**, 291–295.

Başer, K.H.C. and Koyuncu, M. (1994) Composition of the essential oils of two varieties of *Thymus longicaulis* C. Presl. subsp. *chaubardii* (Boiss. et Heldr. ex Reichb. fil.) Jalas. *J. Essent. Oil Res.*, **6**, 207–209.

Başer, K.H.C., Kürkçüoglu, M., Özek, T. and Akgül, A. (1995a) Essential oil of *Thymus sipyleus* Boiss. subsp. *sipyleus* var. *sipyleus. J. Essent. Oil Res.*, **7**, 411–413.

Başer, K.H.C., Özek, T., Kürkçüoglu, M. and Tümen, G. (1995b) Essential oil of *Thymus thracicus* Velen. var. *longidens* (Velen.) Jalas. *J. Essent. Oil Res.*, **7**, 661–662.

Başer, K.H.C., Kürkçüoglu, M., Tümen, G. and Sezik, E. (1996a) Composition of the essential oil of *Thymus eigii* (M. Zohary et. P.H. Davis) Jalas from Turkey. *J. Essent. Oil Res.*, **8**, 85–86.

Başer, K.H.C., Kirimer, N., Ermin, N. and Kürkçüoglu, M. (1996b) Essential oils from four chemotypes of *Thymus zygioides* Griseb. var. *lycaonicus* (Celak) Ronniger. *J. Essent. Oil Res.*, **8**, 615–618.

Başer, K.H.C., Kirimer, N., Ermin, N., Özek, T. and Tümen G. (1996c) Composition of essential oils from three varieties of *Thymus praecox* Opiz growing in Turkey. *J. Essent. Oil Res.* **8**, 319–321.

Başer, K.H.C., Özek, T., Kürkçüoglu, M. and Duman, H. (1997) Composition of the essential oil of *Thymus subcollinus* Klokov from Turkey. *J. Essent. Oil Res.*, **9**, 105–106.

Başer, K.H.C., Kirimer, N., Tümen, G. and Duman, H. (1998) Composition of the essential oil of *Thymus canoviridis* Jalas. *J. Essent. Oil Res.*, 10, 199–200.

Başer, K.H.C., Kürkçüoglu, M., Ermin, N., Tümen, G. and Malyer, H. (1999a) Composition of the essential oil of *Thymus pseudopulegioides* Klokov et Des.-Shost. from Turkey. *J. Essent. Oil Res.*, 11, 86–88.

Başer, K.H.C., Demirci, B., Kürkçüoglu, M. and Tümen, G. (1999b) Essential oil of *Thymus zygioides* Griseb. var. *zygioides* from Turkey. *J. Essent. Oil Res.*, 11, 409–410.

Başer, K.H.C., Özek, T., Kürkçüoglu, M., Tümen, G. and Yildiz, B. (1999c) Composition of the essential oils of *Thymus leucostomus* Hausskn. et Velen. var. *gypsaceus* Jalas and *Thymus pubescens* Boiss. et Kotschy ex Celak. var. *cratericola* Jalas. *J. Essent. Oil Res.*, 11, 776–778.

Başer, K.H.C., Demirci, B., Kürkçüoglu, M. and Tümen, G. (1999d) Composition of the essential oils of *Thymus pectinatus* Fisch. et May. var. *pectinatus* at different stages of vegetation. *J. Essent. Oil Res.*, 11, 333–334.

Başer, K.H.C., Kürkçüoglu, M. and Tümen, G. (1999e) Essential oils of *Thymus striatus* Vahl var. *interruptus* Jalas from Turkey. *J. Essent. Oil Res.*, 11, 253–256.

Başer, K.H.C., Demirci, B., Kirimer, N., Satil, F. and Tümen, G. (2002) The essential oil of two *Thymus* species of Turkey: *T. migricus* and *T. fedtschenkoi* var. *handelii*. *Flavour Fragr. J.*, 17, 41–45.

Bellomaria, B., Hruska, K. and Valentini, G. (1981) Composizione degli olii essenziali di *Thymus longicaulis* C. Presl in varie località dell'Italia Centrale. *Giorn. Bot. Ital.*, 115, 17–27.

Bellomaria, B., Valentini, G., Arnold, N. and Arnold, H.J. (1994) Composition and variation of essential oil of *Thymus integer* Griseb. of Cyprus. *Pharmazie*, 49, 684–688.

Benjilali, B., Hammoumi, M., M'Hamedi, A. and Richard, H. (1987a) Composition chimique des huiles essentielles de diverses variétés de thym du Maroc. II. Analyse en composantes principales (ACP). *Sci. Aliment.*, 7, 275–299.

Benjilali, B., Hammoumi, M. and Richard, H. (1987b) Polymorphisme chimique des huiles essentielles de Thym du Maroc. 1. Caracterisation des composants. *Sci. Aliment.*, 7, 77–91.

Biondi, D., Cianci, P., Geraci, C., Ruberto, G. and Piattelli, M. (1993) Antimicrobial activity and chemical composition of essential oils from Sicilian aromatic plants. *Flavour Fragr. J.*, 8, 331–337.

Bischof-Deichnik, C. (1997) *Das ätherische Öl der schottischen Population von* Thymus praecox ssp. arcticus *(E. Durand) Jalas (Lamiaceae)*. Doctoral thesis, University of Hamburg.

Bischof-Deichnik, C., Stahl-Biskup, E. and Holthuijzen, J. (2000) Multivariate statistical analysis of the essential oil data from *T. praecox* ssp. *polytrichus* of the Tyrolean Alps. *Flavour Fragr. J.*, 15, 1–6.

Blanquer, A., Boira, H., Soler, V. and Pérez, I. (1998) Variability of the essential oil of *Thymus piperella*. *Phytochemistry*, 47, 1271–1276.

Blázquez, M.A. and Zafra-Polo, C. (1989) Essential oil analysis of *Thymus godayanus*, an endemic species growing in northeastern Spain. *Pharmazie*, 44, 651.

Blázquez, M.A., Zafra-Polo, M.C. and Villar, A. (1989) The volatile oil of *Thymus leptophyllus* growing in Spain. *Planta Med.*, 55, 198.

Blázquez, M.A. and Zafra-Polo, M.C. (1990) A new chemotype of *Thymus vulgaris* ssp. *aestivus* Reuter ex Willk. A. Bolos and O. Bolos. *Pharmazie*, 45, 802–803.

Blázquez, M.A., Bono, A. and Zafra-Polo, M.C. (1990) Essential oil from *Thymus borgiae*, a new Iberian species of the Hyphodromi section. *J. Chromatogr.*, 518, 230–233.

Boira, H. and Blanquer, A. (1998) Environmental factors affecting chemical variability of essential oils in *Thymus piperella* L. *Biochem. Syst. Ecol.*, 26, 811–822.

Bruni, A. and Modenesi, P. (1983) Development, oil storage and dehiscence of peltate trichomes in *Thymus vulgaris* (Lamiaceae). *Nord. J. Bot.*, 3, 245–251.

Cabo, J., Bravo, L., Jiménez, J. and Navarro, C. (1980) *Thymus hiemalis*. II. Etude quali-et quantitative de son huile essentielle par C.G. *Planta Med.*, 39, 39–40.

Cabo, J., Jiménez, J., Revert, A. and Bravo, L. (1981) Effect of ecological factors (altitude) on the content and composition of essential oils from the sample of *Thymus zygis* L. collected in different areas. *Ars. Pharm.*, 22, 187–194.

Cabo, J., Cabo, M.M., Crespo, M.E., Jiménez, J. and Navarro, C. (1986a) *Thymus granatensis* Boiss.: II. Etude de son cycle evolutif. *Plant. Méd. Phytothér.*, 20, 129–134.

Cabo, J., Cabo, M.M., Crespo, M.E., Jiménez, J. and Navarro, C. (1986b) *Thymus granatensis* Boiss.: I. Etude qualitative et quantitative de son huile essentielle. *Plant. Méd. Phytothér.*, 20, 18–24.

Cabo, J., Crespo, M.E., Jiménez, J. and Navarro, C. (1986c) A study of the essences from *Thymus hyemalis* collected in three different localities. *Fitoterapia*, 57, 117–119.

Cabo, J., Crespo, M.E., Jiménez, J., Navarro, C. and Risco, S. (1987) Seasonal variation of essential oil yield and composition of *Thymus hyemalis*. *Planta Med.*, 53, 380–383.

Cabo, M.M., Cabo, J., Castillo, M.J., Cruz, T. and Jiménez, J. (1990) Study of the essential oil of *Thymus baeticus* Boiss. *Plant. Méd. Phytothér.*, 24, 197–202.

Cabo, M.M., Cabo, J., Castillo, M.J., Cruz, T. and Jiménez, J. (1992) Estudio de la esencia de *Thymus baeticus* Boiss. In Pescay Alimentacion Ministerio de Agricultura, (eds), *I Jornadas Ibericas de Plantas Medicinales, Aromaticas y de Aceites Esenciales*. Instituto Nacional de Investigación y Tecnología Agraria y Alimentaria, Madrid, pp. 269–275.

Cañigueral, S., Vila, R., Vicario, G., Tomàs, X. and Adzet, T. (1994) Chemometrics and essential oil analysis: chemical polymorphism in two *Thymus* species. *Biochem. Syst. Ecol.*, 22, 307–315.

Charlwood, B.V. and Banthorpe, D.V. (1978) The biosynthesis of monoterpenes. In L. Reinhold, J.B. Harborne, and T. Swain (eds), *Progress in Phytochemistry*. Pergamon Press, Oxford, New York, Toronto, Sydney, Paris, Frankfurt, pp. 65–125.

Cioni, P.L., Tomei, P.E., Catalano, S. and Morelli, I. (1990) Studio sulla variabilità delle essenze individuali di una micropopolazione di piante di *Thymus vulgaris* L. *Riv. Ital.*, 55, 3–6.

Corticchiato, M., Bernardini, A., Costa, J., Bayet, C., Saunois, A. and Voirin, B. (1995) Free flavonoid aglycones from *Thymus herba-barona* and its monoterpenoid chemotypes. *Phytochemistry*, 40, 115–120.

Corticchiato, M., Tomi, F., Bernardini, A.F. and Casanova, J. (1998) Composition and infraspecific variability of essential oil from *Thymus herba barona* Lois. *Biochem. Syst. Ecol.*, 26, 915–932.

Crespo, M.E., Cabo, J., Jiménez, J., Navarro, C. and Zarzuelo, A. (1986) Composition of the essential oil in *Thymus orospedanus*. *J. Nat. Prod. (Lloydia)*, 49, 558–560.

Crespo, M.E., Gomis, E., Jiménez, J. and Navarro, C. (1988) The essential oil of *Thymus serpylloides* ssp. *gadorensis*. *Planta Med.*, 54, 161–162.

Croteau, R. (1977) Site of monoterpene biosynthesis in *Majorana hortensis* leaves. *Plant. Physiol.*, 59, 519–520.

Croteau, R. and Martinkus, C. (1979) Metabolism of monoterpenes: demonstration of (–)-neomenthyl-β-D-glucoside as a major metabolite of (–)-menthone in Peppermint (*Mentha piperita*). *Plant Physiol.*, 64, 169–175.

Croteau, R. (1987) Biosynthesis and catabolism of monoterpenes. *Chem. Rev.*, 87, 929–954.

Cruz, G.T., Jiménez, M.J., Navarro, M.C., Cabo, T.J. and Cabo, C.M.M. (1988) Sur l'huile essentielle de *Thymus longiflorus* Boiss.. *Plant. Méd. Phytothér.*, 22, 225–230.

Cruz, T., Cabo, M.M., Castillo, M.J., Jiménez, J., Ruiz, C. and Ramos-Cormenzana, A. (1993) Chemical composition and antimicrobial activity of the essential oils of different samples of *Thymus baeticus* Boiss. *Phytotherapy Res.*, 7, 92–94.

Dembitskii, A.D., Yurina, R.A. and Krotova, G.I. (1985) The composition of the essential oils of *Thymus marschallianus*. *Khim. Prir. Soed.*, pp. 510–514.

Eisenreich, W., Sagner, S., Zenk, M.H. and Bacher, A.(1997) Monoterpenoid essential oils are not of mevalonoid origin. *Tetrahedron Lett.*, 38, 3889–3892.

Elena-Rosselló, J.A. (1976) *Projet d'une étude de taxonomie expérimentale du genre Thymus*. Thèse doct., Univ.Sc. et Tech. du Languedoc. Montpellier.

Falchi, L. (1967) Ricerche sugli olii essenziali di *Thymus herba-barona* Lois. di Sardegna. *Riv. Ital.*, 49, 336–340.

Falchi-Delitala, L., Solinas, V. and Gessa, C. (1983) Variazioni stagonali quantitative e qualitative di olio essenziale e dei suoi fenoli in *Thymus capitatus* Hofmgg. et Lk. ed in *Thymus herba-barona* Loisel. *Fitoterapia*, 54, 87–96.

Fang Hong-ju, Ni Jing-hua, Lin Shou-quan and Feng Yu-xiu (1988) The chemical composition of *Thymus mongolicus* Ronn. and *T. quinquecostatus* Celak oils of Chinese origin. *Flavour Fragr. J.*, 3, 73–77.

Figueiredo, A.C., Barroso, J.G., Pedro, L.G., Pais, M.S.M. and Scheffer, J.J.C. (1993) The essential oils of two endemic Portuguese Thyme species: *Thymus capitellatus* Hoffmanns. & Link and *T. lotocephalus* G. López & R. Morales. *Flavour Fragr. J.*, 8, 53–57.

Fleisher, A., Fleisher, Z. and Abu-Rukung, S. (1984) Chemovarieties of *Coridothymus capitatus* L. Rchb. growing in Israel. *J. Sci. Food Agric.*, 35, 495–499.

Francis, M.J.O. and Allcock, C. (1969) Geraniol-β-D-glucoside; occurrence and synthesis in rose flowers. *Phytochemistry*, 8, 1339–1347.

García Martín, D., Fernández Vega, F.I., López de Bustamante, F.M. and García Vallejo, C. (1974). Aceites esenciales de la provincia de Guadalajara. In M. Gaviña Mugica and J. Torner Ochoa (eds), *Contribution al estudio de los aceites esenciales españoles*. Ministry Agriculture, Instituto Nacional de Investigaciones Agrarias, Madrid, pp. 405–420.

García Martín, D. and García Vallejo, M.C. (1983) Chemotypes of *Thymus zygis* (Löfl.) L. of Guadarramma Sierra and other places in Castile (Spain). *9th International Essential Oil Congress, Singapore 1983*, 134–140.

García Martín, D. and García Vallejo, M.C. (1984) Evidencia química del origen no hibridógeno de *Thymus lacaitae* Pau (*Thymus gypsicola* Riv.-Mart. *Thymus aranjuezii* Jalas). *An. Inst. Nac. Inv. Agr. (Madrid), Ser. Forestal*, 219–229.

García Vallejo, M.C., García Martín, D. and Muñoz, F. (1984) Avance de un estudio sobre las esencias de *Thymus mastichina* L. español ("Mejorana de espana"). *An. Inst. Nac. Inv. Agr. (Madrid), Ser. Forestal*, 8, 201–218.

García Vallejo, M.C. and García Martín, D. (1986) Aceites esenciales de *Thymus mastigophorus* Lacaita. *Jornadas Nacionales de Plantas Aromáticas*, Medicinales y Condimentarias, León (Spain).

García Vallejo, M.C., García Martín, D. and Carrasco García, J. (1992a) Aceite esencial de "tomillo basto" (*Thymus baeticus* Boiss. ex Lacaita). In Ministerio de Agricultura, Pesca y Alimentación (eds), *I Jornadas Ibéricas de Plantas Medicinales, Aromáticas y de Aceites Esenciales*. Instituto Nacional de Investigación y Tecnología Agraria y Alimentaria, Madrid, pp. 181–193.

García Vallejo, M.C., Rebollar Reier, M.P. and García Martín, D. (1992b) Composición química del aceite esencial de *Thymus vulgaris* L., en la Comunidad de Madrid. In Ministerio de Agricultura, Pesca y Alimentación (eds), *I Jornadas Ibéricas de Plantas Medicinales, Aromáticas y de Aceites Esenciales*. Instituto Nacional de Investigación y Tecnología Agraria y Alimentaria, Madrid, pp. 221–232.

Gershenzon, J., Duffy, M.A., Karp, F. and Croteau R. (1987) Mechanized techniques for the selective extraction of enzymes from plant epidermal glands. *Anal. Biochem.*, 163, 159–164.

Gildemeister, E. and Hoffmann, F. (1961) *Die ätherischen Öle*, Vol. VII, Akademie Verlag, Berlin.

Granger, R., Passet, J. and Verdier, R. (1964) Le γ-terpinene, precursor du *p*-cymene dans *Thymus vulgaris* L. *C. R. Acad. Sci.*, Ser. D, 258, 5539–5541.

Granger, R., Passet, J. and Verdier, R. (1965) Linalool in the essential oil of *Thymus vulgaris*. *C. R. Acad. Sci., Ser. D*, 260, 2619–2621.

Granger, R. and Passet, J. (1971) Types chimiques (chémotypes) de l 'espèce *Thymus vulgaris* L. *C. R. Acad. Sci. Paris*, 273, 2350–2353.

Granger, R. and Passet, J. (1973) *Thymus vulgaris* spontane de France: Races chimiques et chemotaxonomie. *Phytochemistry*, 12, 1683–1691.

Granger, R., Passet, J., Teulade-Arbousset, G. and Auriol, P. (1973) Types chimiques de *Thymus nitens* Lamotte, endémique cévénol. *Plant. Méd. Phytothér.*, 7, 225–233.

Granger, R. and Passet, J. (1974) Type chimique de *Thymus herba-barona* Loiseleur et Deslong-champs de Corse. *Riv. ital.*, 56, 622–628.

Gulati, B.C. and Gupta, R. (1977) Essential oil from *Thymus serpyllum. Indian Perfumer*, 21, 162–163.

Guseinov, S.Ya., Kagramanova, K.M., Kasumov, F.Yu. and Akhundov, R.A. (1987) Studies on the chemical composition and on some aspects of the pharmacological action of the essential oil of *Thymus kotschyanus* Boiss. *Farmakol. Toksikol. (Moscow)*, 50, 73–74.

Han, D.S. and Kim, K.W. (1980) Studies of the essential oil components of *Thymus magnus* Nakai. *Saengyakhak-hoeji (Kor. J. Pharmacogn.)*, 11, 1–6.

Hegnauer, R. (1978) Die systematische Bedeutung der ätherischen Öle (Chemotaxonomie der ätherischen Öle). *Dragoco Rep.*, pp. 204–230.

Hegnauer, R. (1966) *Chemotaxonomie der Pflanzen.* Vol. IV. Birkhäuser Verlag, Basel, pp. 289–316.

Holthuijzen, J. (1994) *Ätherische Öle und glykosidisch gebundene flüchtige Inhaltsstoffe in ausgewählten Arten der Gattung Thymus L.* Doctoral thesis, University of Hamburg.

Iglesias, J., Vila, R., Cañigueral, S., Bellakhdar, J. and Il Idrissi, A. (1991) Analysis of the essential oil of *Thymus riatarum. J. Essent. Oil Res.*, 3, 43–44.

Ismailov, N.M., Kasumov, F.Yu. and Akhmedova, Sh.A. (1981) Essental oil of *Thymus trautvetteri. Dokl. Akad. Nauk Az. SSR*, 37, 64–67.

Ivars, L. (1964) Kemotaxonomiska undersökningar av *Thymus serpyllum. Farm. Aikak.- Farmaceutisk Notisblad*, 73, 324–332.

Jiménez Martín, J., Navarro M.C., Arrebola, M.L. and Socorro A.O. (1989) Botanical and pharmacochemical study of *Thymus hyemalis* Lange. *Bol. Soc. Brot., Sér. 2*, 62, 249–261.

Jiménez Martín, J., Navarro Moll, C. and Arrebola M.L. (1992) Estudio botánico-farmaco-químico de *Thymus hyemalis* Lange. In Ministerio de Agricultura, Pesca y Alimentación (eds), *I Jornadas Ibéricas de Plantas Medicinales, Aromáticas y de Aceites Esenciales.* Instituto Nacional de Investigación y Tecnología Agraria y Alimentaria, Madrid, pp. 149–161.

Jiménez, J., Navarro M. C., Montilla, M.P. and Martín A. (1993) *Thymus zygis* oil: its effect on CCl_4-induced hepatotoxicity and free radical scavenger activity. *J. Essent. Oil Res.*, 5, 153–158.

Juchelka, D., Steil, A., Witt, A. and Mosandl, A. (1996) Chiral compounds of essential oils. XX Chirality evaluation and authenticity profiles of neroli and petitgrain oils. *J. Essent. Oil Res.*, 8, 487–497.

Kameoka, H., Miyake, A. and Hirao, N. (1973) Öl von *Thymus quinquecostatus. Nippon Kagaku Kaishi*, 775.

Karawya, M.S. and Hifnawy, M.S. (1974) Flavors and nonalcoholic beverages: analytical study of the volatile oil of *Thymus vulgaris* L. growing in Egypt. *J. Assoc. Off. Anal. Chem.*, 57, 997–1001.

Karuza-Stojakovic, L., Pavlovic, S., Zivanovic, P. and Todorovic, B. (1989) Composition and yield of essential oils of various species of the genus *Thymus* L. *Arh. Farm.*, 39, 105–111.

Kasumov, F.Yu. (1979) Essential oil of *Thymus rariflorus. Khim. Prir. Soed.*, p. 863, p. 770.

Kasumov, F.Y. (1980) Essential oils of thyme. *Maslo-Zhir. Prom.-st.*, p. 31–32.

Kasumov, F.Yu. (1981) Components of thyme essential oils. *Khim. Prir. Soed.*, p. 522.

Kasumov, F.Yu. (1987) Component compositions of the essential oils of some species of the genus *Thymus. Khim. Prir. Soed.*, pp. 761–762.

Kasumov, F.Yu. (1988) Chemical composition of essential oils of Thyme species in the flora of Armenia. *Khim. Prir. Soed.*, 134–136.

Kasumov, F.Yu. and Ismailov, N.M. (1975) Essential oil of coin thyme. *Maslo-Zhir. Prom.-st.*, pp. 34–35.

Kasumov, F.Yu. and Gadzhieva, T.G. (1980) Components of *Thymus kotschyanus. Khim. Prir. Soed.*, p. 728.

Kasumov, F.Yu. and Davidenko, S.E. (1985) Chemical composition of the essential oil of *Thymus pastoralis* and *Thymus dagestanicus*. *Khim. Prir. Soed.*, p. 840.

Kasumov, F.Yu. and Gavrenkova, S.I. (1982) Components of the essential oil of *Thymus nummularius*. *Khim. Prir. Soed.*, pp. 654–655.

Kasumov, F.Yu. and Gavrenkova, S.I. (1985) *Thymus transcaucasicus* Ronn. – promising essential oil containing plant of Azerbaijan flora. *Dokl. Akad. Nauk Az. SSR*, 41, 56–59.

Kasumov, F.Yu. and Farkhadova, M.T. (1986) The composition of *Thymus karamarianicus* essential oil. *Khim. Prir. Soed.*, pp. 642–643.

Kasumov, F.Yu., Akhmedzade, F.A. and Akhmedova, Sh.A. (1979) Infraspecific variation in *Thymus trautvetteri* in relation to the chemical composition of essential oil. *Izv. Akad. Nauk. Ser. Biol. Nauk*, pp. 23–28.

Kasumov, F.Yu. and Gavrenkova, S.I. (1982) Components of the essential oil of *Thymus nummularius*. *Khim. Prir. Soed.*, pp. 654–655.

Kasumov, F.Yu. and Komarova, V.L. (1983) Essential oils of *Thymus transcaucasicus* Ronn. and *Thymus eriophorus* Ronn. *Maslo-Zhir. Prom.-st.*, p. 29.

Katsiotis, S. and Iconomou, N. (1986) Contribution to the study of the essential oil from *Thymus tosevii* growing wild in Greece. *Planta Med.*, 52, 334–335.

Katsiotis, S.T., Chatzopoulou, P. and Baerheim Svendsen, A. (1990) The essential oil of *Thymus sibthorpii* Benth. growing wild in Greece. *Sci. Pharm.*, 58, 303–306.

Khodair, A.I., Hammouda, F.M., Ismail, S.I., El-Missiry, M.M., Shahed, F.A. and Abdel-Azim, H. (1993) Phytochemical investigation of *Thymus decussatus* L. 1. Flavonoids and volatile oils. *Qatar Univ. Sci. J.*, 13, 211–213.

Kisgyörgy, Z., Csedö, K., Hörster, H., Gergely, J. and Racz, G. (1983) The volatile oil of the more important indigenous *Thymus* species occurring in the composition of Serpylli herba. *Rev. Med. (Tirgu-Mures, Rom.)*, 124–130.

Kreis, P., Dietrich, A., Juchelka, D. and Mosandl, A. (1993) Methodenvergleich zur Stereodifferenzierung von Linalool und Linalylacetat in ätherischen Ölen von *Lavandula angustifolia* Miller. *Pharm. Ztg. Wiss.*, 6, 149–155.

Kreis, P., Hener, U. and Mosandl, A. (1990) Chirale Inhaltsstoffe ätherischer Öle, III. Stereodifferenzierung von alpha-Pinen und Limonen in ätherischen Ölen, Drogen und Fertigarzneimitteln. *Dtsch. Apoth. Ztg.*, 130, 985–988.

Kreis, P., Juchelka, D., Motz, C. and Mosandl, A. (1991) Chirale Inhaltsstoffe ätherischer Öle, IX. Stereodifferenzierung von Borneol, Isoborneol und Bornylacetat. *Dtsch. Apoth. Ztg.*, 1984–1987.

Kulevanova, S., Ristic, M. and Stafilov, T. (1995) The composition of the essential oils from *Thymus macedonicus* (Degen et Urumov) Ronn. subsp. *macedonicus* and *Thymus tosevii* Velen. subsp. *tosevii* growing in Macedonia. *Farmacija*, 43, 13–14.

Kulevanova, S., Ristic, M. and Stafilof, T. (1996a) Comparative essential oils study of *Thymus longidens* Velen. var. *lanicaulis* Ronn. and *Thymus longidens* var. *dassareticus* Ronn. *Boll. Chim. Farm.*, 135, 239–243.

Kulevanova, S., Ristic, M. and Stafilov, T. (1996b) Composition of the essential oil from *Thymus moesiacus* from Macedonia. *Planta Med.*, 62, 78–79.

Kulevanova, S., Ristic, M., Stafilov, T. and Dorevski, K. (1996c) Essential oil composition of *Thymus tosevii* ssp. *tosevii* var. *longifrons*. *Acta Pharm. (Zagreb)*, 46, 303–308.

Kulevanova, S., Ristic, M., Stafilov, T., Dorevski, K. and Ristov, T. (1997) Composition of essential oils of *Thymus tosevii* ssp. *tosevii* and *Thymus tosevii* ssp. *substriatus* from Macedonia. *Pharmazie*, 52, 382–386.

Kulevanova, S., Ristic, M., Stafilov, T. and Matevski, V. (1998a) Composition of the essential oils of *Thymus jankae* Chel. var. *jankae*, *T. jankae* var. *pantotrichus* Ronn. and *T. jankae* var. *patentipilus* Lyka from Macedonia. *J. Essent. Oil Res.*, 10, 191–194.

Kulevanova, S., Ristic, M., Stafilov, T. and Matevski (1998b) Composition of the essential oil of *Thymus rohlenae* Velen. from Macedonia. *J. Essent. Oil Res.*, 10, 537–538.

Kulevanova, S., Ristic, M., Stafilov, T. and Matevski (1998c) Composition of the essential oil of *Thymus albanus* ssp. *albanus* H. Braun from Macedonia. *J. Essent. Oil Res.*, **10**, 335–336.

Kulevanova, S., Ristic, M. and Stafilov, T. (1999) Composition of the essential oil of *Thymus macedonicus* subsp. *macedonicus* (Degen et Urum.) Ronn. from Macedonia. *Herba Pol.*, **45**, 80–86.

Kulieva, Z.T., Guseinov, D. Ya., Kasumov, F.Yu. and Akhundov, R.A. (1979) Investigations of the chemical composition and some pharmacology and toxicological properties of the *Thymus kotschyanus* essential oil. *Dokl. Akad. Nauk, Az. SSR*, **35**, 87–91.

Kuštrak, D., Martinis, Z., Kuftinec, J. and Blazevic, N. (1990) Composition of the essential oils of some *Thymus* and *Thymbra* species. *Flavour Fragr. J.*, **5**, 227–231.

Lawrence, B.M. (1980) The existence of infraspecific differences in specific genera in the labiatae family. *Annales Techniques, 8 ème Congrès International des Huiles Essentielles, Octobre 1980*, Fedarome, Grasse, 118–131.

Little, D.B. and Croteau, R.B. (1999) Biochemistry of essential oil terpenes – A thirty year overview. In R. Teranishi, E.L. Wick and I. Hornstein (eds), *Flavor Chemistry: 30 Years of Progress*. Kluwer Academic/Plenum Publishers, New York, pp. 239–253.

Ložiene, K., Vaiciuniene, J. and Venskutonis, P.R. (1998) Chemical composition of the essential oil of creeping thyme (*Thymus serpyllum* L. s.l.) growing wild in Lithuania. *Planta Med.*, **64**, 772–773.

Luo, J. and Song, Y. (1989) Components of essential oil from *Thymus mongolicus* Ronn. *Linchan Huaxue Yu Gongye*, **9**, 53–58.

Maccioni, S., Flamini, G., Cioni, P.L. and Tomei, P.E. (1992) Phytochemical typology in some *Thymus* populations of *Thymus vulgaris* growing on Caprione's promontory (East Liguria). *Riv. Ital.*, 13–18.

Marhuenda, E. and Alarcón de la Lastra, C.A. (1987) Composition of essential oil of *Thymus carnosus* and its variation. *Fitoterapia*, **57**, 448–450.

Marhuenda, R.E., Menéndez, M. and Alarcón de la Lastra, C. (1987) Trace constituents in the essential oil of *Thymus carnosus* Boiss. *Plant. Med. Phytother.*, **21**, 43–46.

Marhuenda, E., Menéndez, M. and Alarcón de la Lastra, C. (1988) Constituents of essential oil of *Thymus* carnosus Boiss. *J. Chromatogr.*, **436**, 103–106.

Mártonfi, P. (1992a) Essential oil content in *Thymus alpestris* in Slovakia. *Thaiszia, Kosice*, **2**, 75–78.

Mártonfi, P. (1992b) Polymorphism of essential oils in *Thymus pulegioides* subsp. *chamaedrys* in Slovakia. *J. Essent. Oil Res.*, **4**, 173–179.

Mártonfi, P., Greijtovsky, A. and Repcak, M. (1994) Chemotype pattern differentiation of *Thymus pulegioides* on different substrates. *Biochem. Syst. Ecol.*, **22**, 819–825.

Mastelic, J., Grzunov, K. and Kravar, A. (1992) The chemical composition of terpene alcohols and phenols from the essential oil and terpene glycosides isolated from *Thymus pulegioides* L. grown wild in Dalmatia. *Riv. Ital.*, **3**, 19–22.

Mateo, C., Morera, M.P., Sanz, J., Caldéron, J. and Hernández, A. (1978) Estudio analítico de aceites esenciales procedentes de plantas españolas. 1. Especies del género *Thymus*. *Riv. Ital.*, **60**, 621–627.

McGimpsey, J.A., Douglas, M.H., van Klink, J.W., Beauregard, D.A. and Perry, N.B. (1994) Seasonal variation in essential oil yield and composition from naturalized *Thymus vulgaris* L. in New Zealand. *Flavour Fragr. J.*, **9**, 347–352.

Mechtler, C., Strauß, G., Länger, R. and Jurenitsch, J. (1994a) Variability of the composition of the essential oil of *Thymus kosteleckyanus* Opiz. *Eur. J. Pharm. Sciences, Abstract: Meeting Eufebs Berlin, 1994*, **2**, 122.

Mechtler, C., Schneider, A., Länger, R. and Jurenitsch, J. (1994b) Intraindividuelle Variabilität der Zusammensetzung des ätherischen Öles von Quendel-Arten. *Sci. Pharm.*, **62**, 117.

Meriçli, F.I. (1986a) Evaluation of thymol contents of endemic *Thymus* species growing in Turkey. *Doga Tr-Tipve Ecz. D.*, **10**, 187–200.

Meriçli, F. (1986b) Volatile oils of *Thymus kotschyanus* var. *glabrescens* and *Thymus fedtschenkoi* var. *handelii*. *J. Nat. Prod. (Lloydia)*, **49**, 942.

Meriçli, F.I. and Tanker, M. (1986) The volatile oils of some endemic *Thymus* species growing in Southern Anatolia. *Planta Med.*, 52, 340–341.

Messerschmidt, W. (1964) Gas- und dünnschichtchromatographische Untersuchungen der ätherisches Öle einiger Thymusarten 1. Untersuchung über den Einfluß verschiedener Faktoren auf die Bildung und Veränderung des ätherischen Öls. *Planta Med.*, 12, 501–512.

Messerschmidt, W. (1965) Gas- und dünnschichtchromatographische Untersuchungen der ätherischen Öle einiger Thymusarten. 2. Einfluß verschiedener Herkünfte auf die Zusammensetzung des ätherischen Öls von Herba Thymi, Serpylli und Vorschläge für eine chromatographische Beurteilung. *Planta Med.*, 13, 56–72.

Miguel, M.G., Guerrero, C.A.C., Brito, J.M.C., Venancio, F., Tavares, R., Martins, A. and Duarte, F. (1999) Essential oils from *Thymus mastichina* (L.) L. ssp. *mastichina* and *Thymus albicans* Hoffmanns & Link. *Acta Hortic.*, pp. 500.

Mikus, B. and Zobel, I. (1996) A comparative study of lemon scented thyme species. *Drogenreport*, 9, 10–15.

Mockuté, D. and Bernotiené, G. (1998) Essential oil of lemon-scented *Thymus pulegioides* L. grown wild in Vilnius vicinity. *Rast. Res.*, 34, 131–134.

Mockuté, D. and Bernotiené, G. (1999) The main citral-geraniol and cavacrol chemotype of the essential oil of *Thymus pulegioides* L. growing wild in Vilnius district. *J. Agric. Food Chem.*, 47, 3787–3790.

Mockuté, D. and Bernotiené, G. (2001) The α-terpenyl acetate chemotype of essential oil of *Thymus pulegioides* L. *Biochem. Syst. Ecol.*, 29, 69–76.

Moldão-Martins, M., Bernardo-Gil, M.G., Beirao da Costa, M.L. and Rouzet, M. (1999) Seasonal variation in yield and composition of *Thymus zygis* L. subsp. *sylvestris* essential oil. *Flavour Fragr. J.*, 14, 177–182.

Molero, J. and Rovira, A. (1983) Contribución al estudio biotaxonómico de *Thymus loscosii* Willk. y *Thymus fontqueri* (Jalas) Molero & Rovira, Stat. Nov. *Anales Jard. Bot. Madrid*, 39, 279–296.

Montes Guyot, M.A., Valenzuela L. and Wilkomirsky F. (1981) Aceite esencial de tomillo (*Thymus vulgaris* L.). *An. R. Acad. Farm.*, 47, 285–292.

Morales, R. (1986) Taxonomía de los géneros *Thymus* (excluida la sección *Serpyllum*) y *Thymbra* en la península Ibérica. *Ruizia*, 3, 5–324.

Morgan, R.K. (1989) Chemotypic characteristics of *Thymus vulgaris* L. in Central Otago, New Zealand. *J. Biogeography*, 16, 483–491.

Mossa, J.S., Al-Yahya, M.A. and Hassan, M.M.A. (1987) Physicochemical characteristics and spectroscopy of the volatile oil of *Thymus vulgaris* growing in Saudi Arabia. *Int. J. Crude Drug Res.*, 25, 26–34.

Nigist Asfaw, Storesund, II.J., Skattebøl, L., Tønnesen, F. and Aasen, A.J. (2000) Volatile constituents of two *Thymus* species from Ethiopia. *Flavour Fragr. J.*, 15, 123–125.

Novruzova, Z.A. and Kasumov, F.Yu. (1987) Anatomical analysis of the Caucasian species of the genus *Thymus* L. (Lamiaceae) in connection with the component composition of essential oils. *Izv. Akad. Nauk. Azerbaidz. SSR Ser. Biol. Nauk*, 18–24.

Özek, T., Demirci, F., Başer, K.H.C. and Tümen, G. (1995) Composition of the essential oil of *Coridothymus capitatus* (L.) Reichb. fil. from Turkey. *J. Essent. Oil Res.*, 7, 309–312.

Oszagyan, M., Simandi, B., Sawinsky, J. and Kery, A. (1996) A comparison between the oil and supercritical carbon dioxide extract of Hungarian Wild Thyme (*Thymus serpyllum* L.). *J. Essent. Oil Res.*, 8, 333–335.

Passet, J. (1979) Chemische Differenzierung beim Thymianöl, seine Eigenschaften und seine Bedeutung. *Dragoco Rep.*, 234–242.

Pereira, S.I., Santos, P.A.G., Barroso, J.G., Pedro, L.G., Figueiredo, A.C., Salgueiro, L.R., Deans, S.G. and Scheffer, J.J.C. (1999) Composition of the essential oils from thirteen populations of *Thymus caespititius* Brot. grown on the island S. Jorge (Azores). *Abstract ISEO Leipzig*.

Pérez Alonso, M.J. and Velasco Negueruela, A. (1984) Essential oil analysis of *Thymus villosus* subsp. *lusitanicus*. *Phytochemistry*, 23, 581–582.

Philianos, S.M., Andriopoulou-Athanassoula, T. and Loukis, A. (1982) Constituents of thyme oil from *Thymus capitatus* and *Coridothymus capitatus* from various regions of Greece. *Biologia Gallo-Helenica*, 9, 285–290.

Piccaglia, R. and Marotti, M. (1991) Composition of the essential oil of an Italian *Thymus vulgaris* L. ecotype. *Flavour Fragr. J.*, 6, 241–244.

Piccaglia, R. and Marotti, M. (1993) Characterization of several aromatic plants grown in Northern Italy. *Flavour Fragr. J.*, 8, 115–122.

Popov, V.I. and Odynets, A.I. (1977) Study on the chemical composition of the essential oil of Ukrainian thyme grown in Belorussia (*Thymus serpyllum*). *Mater. S' ezda Farm. B. SSR 3rd*, pp. 166–168.

Poulose, A.J. and Croteau, R. (1978) Biosynthesis of aromatic monoterpenes. *Arch. Biochem. Biophys.*, 187, 307–314.

Prikhod'ko, A.B., Klyuev, N.A., Volkovich, S.V., Emets, T.I., Petrenko, V.V. and Dolya, E.V. (1999) Component composition of the essential oil of *Thymus dimorphus*. *Chem. Nat. Compd.*, 35, 46–51.

Proença da Cunha, A. and Roque, O.R. (1986) Contribuiçao para o estudo analitico do oleo essencial de *Thymus capitatus*. *Bol. Fac. Farm. Coimbra*, 10, 31–41.

Proença da Cunha, A. and Salgueiro, L.R. (1991) The chemical polymorphism of *Thymus zygis* ssp. *sylvestris* from Central Portugal. *J. Essent. Oil Res.*, 3, 409–412.

Ravid, U. and Putievsky, E. (1985) Essential oils of Israeli wild species of Labiatae. In A. Baerheim Svendsen and J.J.C. Scheffer (eds), *Essential Oils and Aromatic Plants*. Martinus Nijhoff Publishers, Dordrecht, pp. 155–161.

Ravid, U. and Putievsky, E. (1986) Carvacrol and thymol chemotypes of East Mediterranean wild Labiatae herbs. In E.-J. Brunke (ed.), *Progress in Essential Oil Research*, de Gruyter, Berlin New York, pp. 163–167.

Razdan, T.K. and Koul, G.L. (1975) Zur Zusammensetzung des Quendelöles (Thymian). *Riechstoffe, Aromen, Körperpflegemittel*, 25, 166–168.

Richard, H., Benjilali, B., Banquour, N. and Baritaux, O. (1985) Étude de diverses huiles essentielles de Thym du Maroc. *Lebensm.-Wiss. u. -Technol.*, 18, 105–110.

Richardson, P.M. (1992) The chemistry of the Labiatae: an introduction and overview. In R.M. Harley and T. Reynolds (eds), *Advances in Labiate Science*. Royal Botanic Gardens, Kew, pp. 291–297.

Rohmer, M., Seemann, M., Horbach, S., Bringer-Meyer and S. Sahm, H. (1996) Glyceraldehyde 3-phosphate and pyruvate as precursors of isoprenic units in an alternative non-mevalonate pathway for terpenoid biosynthesis. *J. Am. Chem. Soc.*, 118, 2564–2566.

Roque, O.R. and Salgueiro, L.R. (1987) Composição do óleo essencial de *Thymus zygis* L. subsp. *sylvestris* (Hoffmanns. & Link) Brot. ex Coutinho da região de Souselas – Coimbra. *Bol. Fac. Farm. Coimbra*, 11, 41–50.

Ruberto, G., Biondi, D. and Piatelli, M. (1992) The essential oil of Sicilian *Thymus capitatus* (L.) Hoffmanns. et Link. *J. Essent. Oil Res.*, 4, 417–418.

Rustaiyan, A., Masoudi, S., Monfared, A., Kamalinejad, M., Lajevardi, T., Sedaghat, S. and Yari, M. (2000) Volatile constituents of three *Thymus* species grown wild in Iran. *Planta Med.*, 66, 197–198.

Sáez, F. (1995a) Essential oil variability of *Thymus hyemalis* growing wild in southeastern Spain. *Biochem. Syst. Ecol.*, 23, 431–438.

Sáez, F. (1995b) Essential oil variability of *Thymus zygis* growing wild in southeastern Spain. *Phytochemistry*, 40, 819–825.

Sáez, F. (1998) Variability in essential oils from populations of *Thymus hyemalis* Lange in southeastern Spain. *J. Herbs, Spices & Med. Plants*, 5, 65–76.

Sáez, F. (1999) Essential oil variability of *Thymus baeticus* growing wild in southeastern Spain. *Biochem. Syst. Ecol.*, 27, 269–276.

Sáez, F. (2001) Volatile oil variability in *Thymus serpylloides* ssp. *gadorensis* growing wild in Southeastern Spain. *Biochem. Syst. Ecol.*, 29, 189–198.

Salgueiro, L.R. (1992) Essential oils of endemic *Thymus* species from Portugal. *Flavour Fragr. J.*, 7, 159–162.

Salgueiro, L.R. and Proença da Cunha, A. (1989) Determinaçao de quimiotipos no *Thymus zygis* L. subsp. *sylvestris* (Hoffmanns. et Link) Brot. ex Coutinho da regiao de Eiras – Coimbra. *Rev. Port. Farm.*, 39, 19–27.

Salgueiro, L.R., Neto, F.C. and Proença da Cunha, A. (1992) Les huiles essentielles de *Thymus* spontanes de Tras-os-Montes (Portugal). *Riv. Ital. (special issue)*, 3, 468–490.

Salgueiro, L.R. and Proença da Cunha, A. (1992) Composiçao quimica do oleo essencial de *Thymus zygis* L. subsp. *sylvestris* da regiao centro de Portugal. I. Distrito de Coimbra. In Ministerio de Agricultura, Pesca y Alimentación (eds), *I Jornadas Ibéricas de Plantas Medicinales, Aromáticas y de Aceites Esenciales*. Instituto Nacional de Investigación y Tecnología Agraria y Alimentaria, Madrid, pp. 203–220.

Salgueiro, L.R., Proença da Cunha, A. and Paiva, J. (1993) Chemotaxonomic characterization of a *Thymus* hybrid from Portugal. *Flavour Fragr. J.*, 8, 325–330.

Salgueiro, L.R., Vila, R., Tomàs, X., Tomi, F., Cañigueral, S., Casanova, J., Proença da Cunha, A. and Adzet, T. (1995) Chemical polymorphism of the essential oil of *Thymus carnosus* from Portugal. *Phytochemistry*, 38, 391–396.

Salgueiro, L. R., Vila, R., Tomi, F., Tomàs, X., Cañigueral, S., Casanova, J., Proença da Cunha, A. and Adzet, T. (1997a) Composition and infraspecific variability of essential oil from *Thymus camphoratus*. *Phytochemistry*, 45, 1177–1183.

Salgueiro, L.R., Vila, R., Tomi, F., Figueiredo, A.C., Barroso, J.G., Cañigueral, S., Casanova, J., Proença da Cunha, A. and Adzet, T. (1997b) Variability of essential oils of *Thymus caespititius* from Portugal. *Phytochemistry*, 45, 307–311.

Salgueiro, L.R., Vila, R., Tomàs, X., Cañigueral, S., Proença da Cunha, A. and Adzet, T. (1997c) Composition and variability of the essential oils of *Thymus* species from section *Mastichina* from Portugal. *Biochem. Syst. Ecol.*, 25, 659–672.

Salgueiro, L. R., Proença da Cunha, A., Tomàs, X., Cañigueral, S., Adzet, T. and Vila, R. (1997d) Essential oil of *Thymus villosus* ssp. *villosus*: composition and chemical polymorphism. *Flavour Fragr. J.*, 12, 117–122.

Salgueiro, L.R., Vila, R., Tomàs, X., Cañigueral, S., Paiva, J., Proença da Cunha, A. and Adzet, T. (2000a) Chemotaxonomic study on *Thymus villosus* from Portugal. *Biochem. Syst. Ecol.*, 28, 471–482.

Salgueiro, L.R., Vila, R., Tomàs, X., Cañigueral, S., Paiva, J., Proença da Cunha, A. and Adzet, T. (2000b) Essential oil composition and variability of *Thymus lotocephalus* and *Thymus* x *mourae*. *Biochem. Syst. Ecol.*, 28, 457–470.

Sánchez Gómez, P., Sotomayor Sánchez, J.A., Soriano Cano, M.C., Correal Castellanos, E. and García Vallejo, M.C. (1995) Chemical composition of the essential oil of *Thymus zygis* ssp. *gracilis* c.v. "Linalool type", and its performance under cultivation. *J. Essent. Oil Res.*, 7, 399–402.

Sattar, A., Malik, M.S. and Khan, S.A. (1991) Essential oils of the species of Labiatae part IV. Composition of the essential oil of *Thymus serpyllum*. *Pakist. J. Sci. Ind. Res.*, 34, 119–120.

Schmidt, A. (1998) *Polychemismus bei den ätherisches Öl führenden Arten Thymus pulegioides L. und* Thymus praecox *Opiz ssp. arcticus (E. Durand) Jalas (Lamiaceae) im nordatlantischen Europa*. Doctoral thesis, University of Hamburg.

Schratz, E. and Hörster, H. (1970) Zusammensetzung des ätherischen Öls von *Thymus vulgaris* und *Thymus marschallianus* in Abhängigkeit von Blattalter und Jahreszeit. *Planta Med.*, 19, 160–176.

Sefidkon, F. and Dabiri, M. (1999) The effect of distillation methods and stage of plant growth on the essential oil content and composition of *Thymus kotschyanus* Boiss. et Hohen.. *Flavour Fragr. J.*, 14, 405–408.

Sefidkon, F., Jamzad, Z., Yavari-Behrouz, R., Nouri Sharg, D. and Dabiri, M. (1999) Essential oil composition of *Thymus kotschyanus* Boiss. et Hohen. from Iran. *J. Essent. Oil Res.*, 11, 459–460.

Senatore, F. (1996) Influence of harvesting time on yield and composition of the essential oil of a thyme (*Thymus pulegioides* L.) growing wild in Campania (Southern Italy). *J. Agric. Food Chem.*, 44, 1327–1332.

Seoane, E., Francia, E. and Reñé, E. (1972) Estudio del *"Thymus caespititius"*. II. Componentes volátiles. *An. Quim.*, **68**, 951–954.

Sezik, E. and Bašaran, A. (1986) The volatile oil of *Thymus argaeus* Boiss. et Bal. *Acta Pharm. Turc.*, **28**, 93–98.

Sezik, E. and Saracoğlu, I. (1988) Morphological and anatomical investigations on the plants used as folk medicine and herbal tea in Turkey. V. *Thymus eigii. Doga Tu Tip. ve Ecz. D.*, 12, 32–37.

Shavarda, A.L., Markova, L.P., Nadezhina, T.P., Sinitskii, V.S., Belenovskaya, L.M., Fokina, G.A., Ligaa, U. and Tumbaa, K.H. (1980) Essential oil plants of Mongolia – Terpenoid composition of the essential oils of some species of Labiatae. *Rast. Res.*, 16, 286–292.

Shyuan Qi, Changkai Li, Dexiu Zhao and Chenshun Wu (1987) Two chemotypes of *Thymus quinquecostatus* and the chemical component of essential oil from the cultivated plant. *Abstract Botaniker-Tagung, Berlin 1987*.

Sievers, E. (1971) *Beiträge zum Polymorphismus und zur Polytypie verschiedener Populationen von Thymus pulegioides L.* Doctoral thesis, University of Münster.

Simeon de Bouchberg, M., Allegrini, J., Bessiere, C., Attisso, M., Passet, J. and Granger, R. (1976) Propriétés microbiologiques des huiles essentielles de chimiotypes de *Thymus vulgaris* Linnaeus. *Riv. Ital.*, **58**, 527–536.

Skopp, K. and Hörster, H. (1976) An Zucker gebundene reguläre Monoterpene. Teil I. Thymol- und Carvacrolglykoside in *Thymus vulgaris. Planta Med.*, **29**, 208–215.

Skrubis, B.G. (1972) Seven wild aromatic plants growing in Greece and their essential oils. *Flavour Ind.*, **3**, 566–568, 571.

Soriano Cano, C., Sánchez Gómez, P. and Correal Castellanos, E. (1992) Estudio del aceite esencial de *Thymus* x *monrealensis* Pau ex R. Morales nothosubsp. *garcía-vallejoi* Sánchez-Gómez & Alcaraz. In Ministerio de Agricultura, Pesca y Alimentación (eds), *I Jornadas Ibéricas de Plantas Medicinales, Aromáticas y de Aceites Esenciale*s. Instituto Nacional de Investigación y Tecnología Agraria y Alimentaria, Madrid, pp. 261–268.

Soriano Cano, M.C., Sotomayor, J.A., Correal Castellanos, E., Sánchez Gómez, P. and García Vallejo, M.C. (1997) Chemical composition of the essential oil of *Thymus* x *arundanus* Wilk. and its parents T. *mastichina* L. and T. *baeticus* Boiss. ex Lacaita. *J. Essent. Oil Res.*, 9, 593–594.

Stahl, E. (1984a) Chemical polymorphism of essential oil in *Thymus praecox* ssp. *arcticus* (Lamiaceae) from Greenland. *Nord. J. Bot.*, 4, 597–600.

Stahl, E. (1984b) Das ätherische Öl aus *Thymus praecox* ssp. *arcticus* isländischer Herkunft. *Planta Med.*, 50, 157–160.

Stahl-Biskup, E. (1986a) Das ätherische Öl norwegischer Thymusarten. I. *Thymus praecox* ssp. *arcticus. Planta Med.*, 52, 36–38.

Stahl-Biskup, E. (1986b) Das ätherische Öl norwegischer Thymianarten II. *Thymus pulegioides. Planta Med.*, 52, 223–235.

Stahl-Biskup, E. (1991) The chemical composition of *Thymus* oils: a review of the literature 1960–1989. *J. Essent. Oil Res.*, 3, 61–82.

Stahl-Biskup, E. and Laakso, I. (1990) Essential oil polymorphism in Finnish *Thymus* species. *Planta Med.*, 56, 464–468.

Stahl-Biskup, E., Intert, F., Holthuijzen, J., Stengele, M. and Schulz, G. (1993) Glycosidically bound volatiles – a review 1986–1991. *Flavour Fragr. J.*, 8, 61–80.

Stahl-Biskup, E. and Holthuijzen, J. (1995) Essential oil and glycosidically bound volatiles of lemon-scented Thyme, *Thymus* x *citriodorus* (Pers.) Schreb. *Flavour Fragr. J.*, 10, 225–229.

Stengele, M. and Stahl-Biskup, E. (1993) Glycosidically bound volatiles in Peppermint *(Mentha piperita* L.). *J. Essent. Oil Res.*, 5, 13–19.

Stengele, M. and Stahl-Biskup, E. (1994) Influencing the level of glycosidically bound volatiles by feeding experiments with a *Mentha* x *piperita* L. cultivar. *Flavour Fragr. J.*, 9, 261–263.

Sur, S.V., Tulyupa, F.M., Tolok, A.Ya. and Peresypkina, T.N. (1988) Composition of essential oils from the aboveground part of the Thyme. *Khim. Farm. Zh.*, 22, 1361–1366.

Tanaka, S., Yamaura, T., Shigemoto, R. and Tabata, M. (1989) Phytochrome-mediated production of monoterpenes in thyme seedlings. *Phytochemistry*, 28, 2955–2957.

Tanker, M. and Meriçli, F. (1988) Pharmacognostic researches on *Thymus capitatus* (L.) Hoffm. et Link. *J. Pharm. Univ. Mar.*, 4, 45–52.

Tantaoui-Elaraki, A., Lattaoui, N., Errifi, A. and Benjilali, B. (1993) Composition and antimicrobial activity of the essential oils of *Thymus broussonettii*, T. *zygis* and T. *satureioides*. *J. Essent. Oil Res.*, 5, 45–53.

Tétényi, P. (1970) *Infraspecific chemical taxa of medicinal plants*. Académiai Kiadó, Budapest.

Tikhonov, V.N., Khan, V.A. and Kalinkina, G.I. (1988) Composition of the essential oil of *Thymus krylovii*. *Khim. Prir. Soedin.*, pp. 886–887.

Tomei, P.E., Bertoli, A., Cioni, P.L., Flamini, G. and Spinelli, G. (1998) Composition of the essential oil of *Thymus alpigenus*. *J. Essent. Oil Res.*, 10, 667–669.

Tucakov, J. (1964) Influenza dei fattori esogeni sul rendimento e la qualita' dell'olio essenziale di *Thymus vulgaris* L. *Riv. Ital.*, 46, 376–380.

Tümen, G. and Başer, K.H.C. (1994) Essential oil of *Thymus syriacus* Boiss. *J. Essent. Oil Res.*, 6, 663–664.

Tümen, G., Koyuncu, M., Kirimer, N. and Başer, K.H.C. (1994) Composition of the essential oil of *Thymus cilicicus* Boiss. & Bal. *J. Essent. Oil Res.*, 6, 97–98.

Tümen, G., Kirimer, N., Kürkçüoglu, M. and Başer, K.H.C. (1997a) Composition of the essential oils of *Thymus atticus* and *Thymus roegneri* from Turkey. *J. Essent. Oil Res.*, 9, 473–474.

Tümen, G., Ermin, N., Kürkçüoglu, M. and Başer, K.H.C. (1997b) Essential oil of *Thymus leucostomus* Hausskn. et Velen. var. *leucostomus*. *J. Essent. Oil Res.*, 9, 229–230.

Tümen, G., Başer, K.H.C., Kürkçüoglu, M. and Kirimer, N. (1998a) Composition of the essential oils of *Thymus pectinatus* var. *pectinatus* from Turkey during different states of development. *Abstract 29th ISEO, Frankfurt*.

Tümen, G., Başer, K.H.C., Demirci, B. and Ermin, N. (1998b) The essential oils of *Satureja coerulea* Janka and *Thymus aznavourii* Velen. *Flavour Fragr. J.*, 13, 65–67.

Tümen, G., Yildiz, B., Kirimer N., Kürkcüoglu, M. and Başer, K.H.C. (1999) Composition of the essential oil of *Thymus fallax* Fisch. et Mey. from Turkey. *J. Essent. Oil Res.*, 11, 489–490.

Tzakou, O. and Constantinidis, Th. (1998) Essential oil of *Thymus parnassicus* Halacsy. *46th Annual Congress of the Society for Medicinal Plant Research*, Vienna.

Tzakou, O., Verykokidou, E., Roussis, V. and Chinou, I. (1998) Chemical composition and antibacterial properties of *Thymus longicaulis* subsp. *chaubardii* oils: three chemotypes in the same population. *J. Essent. Oil Res.*, 10, 97–99.

Valentini G., Hruska, K. and Bellomaria, B. (1987) Ricerche sull'olio essenziale di alcune specie del genere *Thymus* nell'Italia Centrale. *Informatore Botanico Italiano*, 19, 270–279.

Van den Broucke, C.O. (1983) The therapeutic value of *Thymus* species. *Fitoterapia*, 54, 171–174.

Van den Dries, J.M.A. and Baerheim Svendsen, A. (1989) A simple method for detection of glycosidic bound monoterpenes and other volatile compounds occurring in fresh plant material. *Flavour Fragr. J.*, 4, 59–61.

Velasco Negueruela, A. and Pérez Alonso, J. (1984) Aceites esenciales de tomillos ibéricos. III. Contribución al estudio de quimiotipos en el grupo *Thymus zygis* L. *Anal. Bromatol.*, 36, 301–308.

Velasco Negueruela, A. and Pérez Alonso, M.J. (1985a) Essential oils of Iberian species of Thyme II. Contribution to the knowledge of the essential oil of *Thymus lacaitae* Pau. *Anales Jardin Bot. Madrid*, 42, 159–164.

Velasco Negueruela, A. and Pérez Alonso, M.J. (1985b) Aceites esenciales de tomillos ibéricos. I. Contribución al conocimiento del aceite esencial de *Thymus orospedanus* H. del Villar. *Anales Jard. Bot. Madrid*, 41, 337–340.

Velasco Negueruela, A. and Pérez Alonso, M.J. (1986) Aceites esenciales de tomillos ibéricos. VI. Contribución al estudio quimiotaxonómico (Terpenoides) del género *Thymus* L. *Trab. Dep. Botánica*, 13, 115–133.

Velasco Negueruela, A. and Pérez Alonso, J. (1987) Aceites esenciales de Tomillos ibéricos. V. Contribución al conocimiento del aceite esencial de *Thymus camphoratus* Hoffmanns. & Link. *Anales Jard. Bot. Madrid*, 43, 383–386.

Velasco Negueruela, A., Pérez Alonso, M.J. and Burzaco, A. (1991a) Natural aroma compounds from vegetal origin essential oils from *Thymus riatarum* and *Origanum elongatum*. *Anal. Bromatol.*, 43, 395–400.

Velasco Negueruela, A., Pérez Alonso, J. and Burzaco (1991b) Aceites esenciales de tomillos ibéricos. VI. Contribución al conocimiento del aceite esencial de *Thymus capitellatus* Hoffmanns. & Link. *Anales Jard. Bot. Madrid*, 49, 77–81.

Velasco Negueruela, A., Pérez Alonso, M.J., Carraquilla C.B. and Samaniego, N.M. (1992) Datos sobre la composicion química (terpenoides) de plantas aromáticas de la provincia de Toledo. In Ministerio de Agricultura, Pesca y Alimentación (eds), *I Jornadas Ibéricas de Plantas Medicinales, Aromáticas y de Aceites Esenciales*. Instituto Nacional de Investigación y Tecnología Agraria y Alimentaria, Madrid, pp. 291–301.

Vernin, G., Ghiglione, C. and Parkanyi, C. (1994) GC-MS-SPECMA bank analysis of *Thymus serpyllum praecox* (Opiz) Wollm (wild thyme) from Hautes Alpes (France). In G. Charalambous (ed.), *Developments in Food Science – Spices, Herbs and Edible Fungi*. Elsevier, Amsterdam, London, New York, Tokyo, pp. 501–515.

Vila, R., Adzet, T. and Ibáñez, C. (1987) Analysis por GC-MS del aceite esencial de *Thymus moroderi, Thymus membranaceus* y su híbrido. *Actas III Congreso Internat. Ciencias Farmaceuticas, Barcelona*.

Vila, R., Vicario, G., Cañigueral, S. and Adzet, T. (1991a) Constituents of the essential oil of *Thymus antoninae*. *Planta Med. Suppl. 2*, 57, A 90.

Vila, R., Freixa, B., Cañigueral, S., Adzet, T., Tomàs, X. and Molins, J.J. (1995) Composition and variability of the essential oil of *Thymus funkii*. *Flavour Fragr. J.*, 10, 379–383.

Von Schantz, M. and Ivars, L. (1964) Über die Zusammensetzung des ätherischen Öles von *Thymus serpyllum* ssp. *tanaensis* (Hyl.) Jalas. *Ann. Univ. Turku. A. II.*, 32, 301–307.

Weiss, B. and Flück, H. (1970) Untersuchungen über die Variabilität von Gehalt und Zusammensetzung des ätherischen Öles in Blatt- und Krautdrogen von *Thymus vulgaris* L. *Pharm. Acta Helv.*, 45, 169–183.

Yamaura, T., Tanaka, S. and Tabata, M. (1992) Localization of the biosynthesis and accumulation of monoterpenoids in glandular trichomes of Thyme. *Planta Med.*, 58, 153–158.

Yamaura, T., Tanaka, S. and Tabata, M. (1989) Light-dependent formation of glandular trichomes and monoterpenes in thyme seedlings. *Phytochemistry*, 28, 741–744.

Yamaura, T., Tanaka, S. and Tabata, M. (1991) Participation of phytochrome in the photo-regulation of terpenoid synthesis in Thyme seedlings. *Plant Cell Physiol.*, 32, 603–607.

Zafra-Polo, M.C., Blázquez, M.A. and Villar, A. (1988a) Variations in the composition of the essential oils from *Thymus leptophyllus* Lange and *Thymus webbianus* Rouy. *Plant. Méd. Phytothér.*, 22, 189–194.

Zafra-Polo, M.C., Blázquez, M.A. and Villar, A. (1988b) Volatile constituents of *Thymus webbianus*. *Plant. Méd. Phytothér.*, 22, 184–188.

Zarzuelo, A., Navarro, C., Crespo, M.E., Ocete, M.A., Jiménez, J. and Cabo, J. (1987) Spasmolytic activity of *Thymus membranaceus* essential oil. *Phytother. Res.*, 1, 114–116.

Zhang Hongli, Wang Youmin and ZhangZhenjie (1992) Study on chemical constituents of essential oil from *Thymus mongolicus* Ronn. *Acta Bot. Boreal.-Occident. Sin.*, 12, 245–248.

4 Essential oil polymorphism in the genus *Thymus*

Francisco Sáez and Elisabeth Stahl-Biskup

INTRODUCTION

The complexity and diversity of living or even of extinct organisms have always attracted man, since basic differences are recognized as essential for understanding the evolutionary development of life. Scientists studying this variability have used the knowledge and techniques available at their time to enhance their knowledge of the living creatures. Early taxonomists interested in structuring the increasing complexity of the plant kingdom initially discussed whether a certain form was or was not a true species. From this debate the question developed if this form was sufficiently constant and distinct from other forms, and whether the differences were sufficiently important to deserve a specific name. Within the classical taxonomy based on plant morphology this happened many times, as long as a genus was imperfectly known, until the limits among various species were considered to be clearly established. The enhancement of this knowledge resulted in the fact that more individuals were put into intermediate positions, thus demanding a revision of previous concepts.

In the plant kingdom chemical polymorphism is well known and is seen in an infraspecific variability of the chemical patterns of individuals or even of populations. Tétényi (1970) coined the term 'polychemism' for this phenomenon, and it was he, in the early 1970s, who showed 750 plant species of 106 families which were known to be chemically polymorphous (Tétényi, 1970). In species containing essential oils the phenomenon of polychemism seems to be widespread; in 1975 Tétényi had already estimated 360 species of 36 families, and it was Lawrence (1980) who gave a first report on the infraspecific differences in several genera of the Labiatae. In his publication, he mentioned four genera which were found to be polymorphous, namely *Mentha*, *Monarda*, *Pycnanthemum*, and *Thymus*. Nowadays, as a result of the understanding that it is a fundamental requirement to analyse infraspecific differences, we encounter a flood of reports on polychemism, the genus *Thymus* being one of the most frequently investigated and the most detailed research regarding this phenomenon.

The studies on the polymorphism of the genus *Thymus* can be said to start with the publications by Granger and Passet (1971, 1973), who reported 6 chemotypes for *T. vulgaris* after studying several populations and many individuals in the south of France. In contrast, the majority of other initial efforts in this field must be compared with the first steps of morphological taxonomy, mentioned above. As scientists tried to characterise the essential oil of a species after analysing only a few samples, it was impossible to know where in the 'cloud' of chemical variability these samples should be placed.

During the next decade some variations in essential oil composition were published by authors such as Adzet *etal.* (1976), Bellomaria *etal.* (1981), Benjilali *etal.* (1987), and it became evident that the real limits for this variability were still unknown. At the same time new questions and doubts were arising, namely, whether a correlation between classical taxonomy and chemical taxonomy could be found at the genus level; or, to which extent one chemotype described for one species was exclusive or widely spread over a group of them, or even over the whole genus.

This way the phenomenon became more and more interesting, and in the 1990s, taking advantage of the technical improvements made during the previous decades, numerous studies discussed the problem of polychemism in *Thymus*. It was accepted that a greater number of samples taken under homogeneous ecological conditions had to be analysed and that the flood of data obtained needed to be examined with the help of specific statistical techniques. Thus, Salgueiro, dedicated to the study of *Thymus* in western Iberia, García-Vallejo, Arrebola, Cañigueral, Blanquer, Sáez processed samples from central, eastern and southern Iberia, Stahl-Biskup focused on the variability in northern European latitudes.

When one looks at the taxonomical diversity described in Chapter 1 and sees how few of these taxa have been intensively studied from the chemical point of view, the feeling arises that still much research is necessary before we can develop a realistic impression about chemical polymorphism in *Thymus*. Thus, the present chapter will focus on the highlights found by the researchers interested in the phenomenon of polychemism so far, with the conviction that the years to come will significantly improve our understanding of the matter. The methods, techniques, and procedures used by the research teams during the last 30 years or so have produced a bewildering variety of data that cannot easily be brought together with at least a minimum of scientific confidence in the accuracy of the conclusions. It does not make any sense to mix in the same pot data obtained from a few samples and a few compounds identified in the essential oil with studies whose conclusions are based on statistical analysis of a representative number of individuals and whose essential oils components have been adequately established.

GENETIC ORIGIN OF CHEMICAL POLYMORPHISM

Based on the distribution of different *Thymus* species in northern Africa, the Canary Islands, the Iberian Peninsula and the Balearic Islands, and using today's knowledge of the geological evolution of these areas during the last 5 million years from late Miocene onwards, Morales (1986) explains the possible early evolution of the genus saying that the diversification and expansion occurred mainly after the separation of the Iberian peninsula from Africa. The origin would be the Tertiary xerophytic flora, with a great evolutionary success achieved as new arid periods were encountered, especially during the Pliocene and onwards up to now. The section *Serpyllum* would play an important role during the periods that showed a withdrawal of the ice cover with great diversification during the cold phases of the Quaternary.

Genetic differentiation within a population may basically have developed in three ways. (a) It may be a result of the isolation of individuals from a genetically variable parental pool at the periphery of a population, with any new contact producing a clinal intergradation. Individuals of different genotypes may have selective advantage in different places within the total area of a population, forming a pattern of genetic polymorphism in a patchy but stable environment. (b) Where intrinsic barriers to gene exchange arise in an environment that is changing in a particular direction (for instance, warming-up), individuals whose genotype provides a better adaptation to the new circumstances are selected, producing a sorting of variability. (c) Stabilizing or normalizing selection produce uniformity in an already genetically variable population which is well adapted to its environment. When this is not changing directionally and fundamentally, new individuals that deviate significantly from the mean have less chance to survive than the ones better adapted.

Apart from these gradual processes of separation from ancestral species, there is another possibility for developing new taxa, namely abrupt speciation, by which new species suddenly arise. Within the plant kingdom this is mostly due to polyploidy, and *Thymus* owes a good part of its variability to this phenomenon, since several species have been found to be polyploid.

All these events mentioned above may have occurred profusely along the Mediterranean and adjacent areas, affected by a quite variable environment, resulting in today's diversity of morphological and chemical taxa in *Thymus*. This speciation process cannot be regarded as concluded, with precise and well-developed barriers among the different species. Several reports from Morales and Sáez on Spanish *Thymus* and from Stahl-Biskup for northern species show that intense interspecific relationships can be noticed in regions where distribution areas for different species overlap and climatic conditions allow simultaneous blooming and interchange of genetic material between them. This is especially achievable for species included in the section *Thymus*.

In this chapter we provide a review of studies published on chemical polymorphism of *Thymus*. They have been grouped regionally, reflecting both species distribution and the different approaches that several authors have made to the problem.

THE SITUATION OF *THYMUS* IN SOUTHEASTERN SPAIN

The chemical polymorphism of the genus *Thymus* in southeastern Spain was studied in detail by Sáez (1996), who sampled 13 species living there, showing high taxonomical diversity, perhaps influenced by the high variability in ecological conditions for such a small area. These species are classified in the following four sections: (a) Section *Thymus*: *T. hyemalis, T. zygis, T. vulgaris, T. baeticus, T. orospedanus, T. serpylloides* ssp. *gadorensis*. (b) Section *Pseudothymbra*: *T. membranaceus, T. longiflorus, T. funkii, T. moroderi, T. antoninae*. (c) Section *Piperella*: *T. piperella* (monospecific), (d) Section *Mastichina*: *T. mastichina* (monospecific).

The essential oils of a total number of 327 individual plants of southeastern *Thymus* were analysed using gas chromatography (GC), and the results were studied with different statistical methods to detect similarities/dissimilarities among them. In order to quantify the chemical polymorphism realised at the species, section and genus level, three different sets of extensive statistics were put together: (a) Analysis of many individual plants of each species to find out the infraspecific variability; the number of samples investigated per species is directly related to its abundance and size of distribution area. (b) Cluster analysis including all individual plants belonging to the same section disregarding their affiliation to a distinct species. The results are presented in the form of tables, one for the section *Thymus* (Table 4.1) and another for the section *Pseudothymbra* (Table 4.2). (c) Principal component analysis of the complete data set of the genus.

Table 4.1 Section *Thymus* – the result of a cluster analysis of 211 individual samples

Chemotypes	Number of individuals	Species					
		T. baeticus	T. serpylloides ssp. gadorensis	T. hyemalis	T. orospedanus	T. vulgaris	T. zygis
Thymol/p-cymene	74	–	8	35	–	–	31
1,8-Cineole	30	1	–	–	–	27	2
Myrcene or terpinen-4-ol	24	11	5	–	7	–	1
Carvacrol/p-cymene	20	1	5	12	–	–	2
Camphor	20	–	–	–	–	20	–
Linalool	16	–	7	–	2	1	6
Linalool/linalyl acetate	7	1	1	4	1	–	–
Linalool/p-cymene	4	–	1	2	–	–	1
Geraniol	3	–	3	–	–	–	–
Linalool/camphor	2	–	–	–	–	2	–
Several*	11	2	4	–	2	1	2
Total	211	16	34	53	12	51	45

Note
* This group contains samples chemically quite different with unusual combinations of compounds.

Table 4.2 Section *Pseudothymbra* – the result of a cluster analysis of 85 individual samples

Chemotypes	Number of individuals	Species				
		T. antoninae	T. funkii	T. longiflorus	T. membranaceus	T. moroderi
1,8-cineole (50–72%)	24	–	6	5	10	3
1,8-cineole (33–50%)/ camphor (10–27%)	18	1	3	7	3	4
1,8-cineole (22–51%)	15	3	4	1	5	2
camphor/borneol	12	2	2	4	3	1
1,8-cineole (73–82%)	5	–	3	1	1	–
several*	11	–	4	1	1	5
total	85	6	22	19	23	15

Note
* This group contains samples chemically quite different, with unusual combinations of compounds.

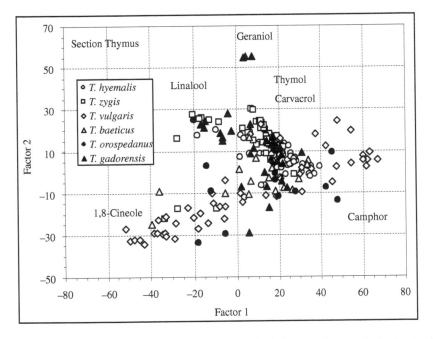

Figure 4.1 Principal component analysis of essential oils from southeastern Spain. Section *Thymus*.

Figures 4.1–4.4 reflect the results with respect to each section, section *Thymus* (Figure 4.1), section *Pseudothymbra* (Figure 4.2), section *Mastichina* (Figure 4.3), and section *Piperella* (Figure 4.4).

Some aspects related to the whole genus are highlighted here before focusing on the peculiarities of the species: (1) The homogeneity of the monospecific sections *Mastichina* and *Piperella* contrasts with the more polymorphic sections *Thymus* and *Pseudothymbra*. From the Figures 4.1–4.4, it can be derived that there is not only more variability in the latter two at the section level but also at the species level since *T. mastichina* or *T. piperella* are not so widely dispersed. (2) The location of the samples containing

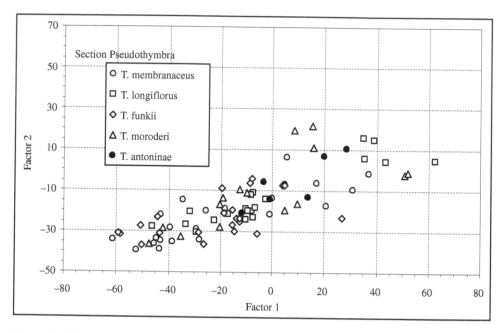

Figure 4.2 Principal component analysis of essential oils from southeastern Spain. Section *Pseudothymbra*.

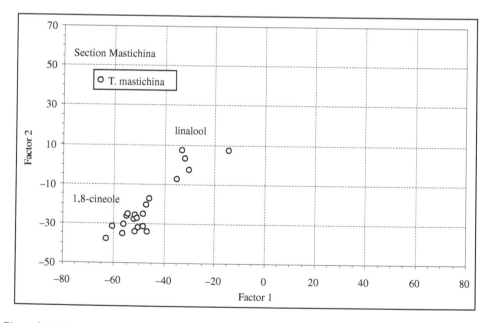

Figure 4.3 Principal component analysis of essential oils from southeastern Spain. Section *Mastichina*.

phenolic chemotypes can be clearly determined by comparing the sections *Thymus* (Figure 4.1) and *Pseudothymbra* (Figure 4.2). Indeed, most samples from *T. hyemalis* and *T. zygis* plus some from *T. serpylloides* ssp. *gadorensis*, are concentrated in a relatively small area and they mainly differ only in the concentration that phenols reach in the

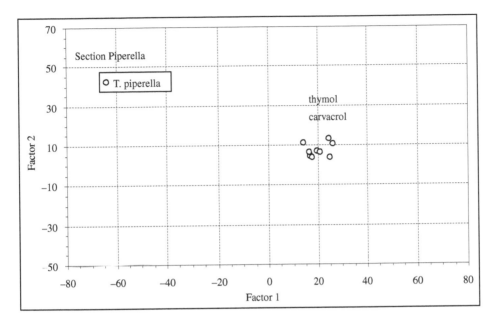

Figure 4.4 Principal component analysis of essential oils from southeastern Spain. Section *Piperella*.

essential oils. (3) In the section *Thymus* a small group of samples with multiple taxonomical adscription is characterized by the presence of linalool. (4) The geraniol chemotype realized in three samples from *T. serpylloides* ssp. *gadorensis* is represented in the most positive values for Factor 2, thus reflecting the low frequency for this compound in the essential oils studied. (5) The deviations from the main trends in the essential oils of a distinct species can easily be detected. This is the case of the two samples of *T. zygis* that were placed in the area for high 1,8-cineole, reflecting the influence of *T. vulgaris*. (6) *T. orospedanus* shows an interesting pattern of chemotype distribution, with highly dispersed samples through the variability defined by the genus. Something similar happens to all the species in the section *Pseudothymbra*: there is no area in the 'cloud' of samples which is exclusive to one of the species, but all contribute in a similar way to develop its shape.

Description of chemical polymorphism on the species level

T. antoninae is a tetraploid species that is found in a very small area. Sáez (1996) described all the essential oils as being characterized by the presence of 1,8-cineole (max. 44.3 per cent) and camphor (max. 35.8 per cent) with similar proportions. Camphene and borneol show lower levels, from 10 to 20 per cent, and myrcene reached up to 8.2 per cent. Cañigueral *et al.* (1994) only found a little chemical polymorphism with respect to β-elemol, β-eudesmol and an unidentified sesquiterpene. Figure 4.2 shows this species in about the center of the diagram for the section *Pseudothymbra*, meaning comparatively that not very high percentages for 1,8-cineole or camphor were found.

T. baeticus is spread along the central and eastern parts of southern Spain. It is an erect shrub characterised by the greyish colour of the leaves and a globular inflorescence and is of some economical importance. Sáez (1999) studied the eastern-most communities and found the remarkable presence of terpinen-4-ol (max. 28.8 per cent) appearing together with α-terpineol, linalool or geranyl acetate. Borneol (max. 53.6 per cent) and 1,8-cineole (max. 56.2 per cent) characterise other groups of samples, and it is worth mentioning a content of 14.9 per cent geranial in one sample, and 20.1 per cent *tr*-sabinene hydrate in another. The presence of the precursors of thymol or carvacrol (*p*-cymene, γ-terpinene) was recorded in quantities up to 37 per cent, but the phenols themselves did not appear in high quantities. This high variability can be observed in Figure 4.1, although in the cluster analysis performed for the section *Thymus* (Table 4.1), most of the samples for this species are included in one main group characterised by either myrcene or terpinen-4-ol.

T. *funkii* is characterised by the small size of the bracts at the inflorescence and a high morphological variability compared with other species from the section *Pseudothymbra*. Sáez (1996) studied 22 individuals whose essential oils were mostly characterised by 1,8-cineole (max. 78.8 per cent) present in all the samples. Camphor was also found frequently, but in lower percentages, up to 19.1 per cent. Myrcene (max. 20.4 per cent) appeared sporadically, and camphene and borneol showed even lower levels. The percentage of 21.4 per cent thymol in one sample is remarkable and explained as an introgression with T. *zygis*. Figure 4.2 shows the samples for T. *funkii* mostly situated at the 1,8-cineole end. Vila *et al.* (1995) studied two populations using both collective and individual samples. They noticed chemical differences between the two communities, but only with respect to some compounds that presented low percentages in the essential oils, such as α-eudesmol or caryophyllene oxide. No significant chemical polymorphism was found in this study.

T. *hyemalis* grows in areas with moderate winter temperatures, near sea shores. It is of very high economical importance due to the absolute predominance of phenols in the essential oil. Thus, thymol (max. 36.7 per cent) and carvacrol (max. 41.3 per cent) chemotypes are the most important ones, but the high ability to interbreed with other species from the section *Thymus* leads to the appearance of compounds such as 1,8-cineole (max. 25.7 per cent) and borneol (20.4 per cent) in restricted areas (Sáez 1995a). The case of one population with some individuals presenting a linalool chemotype and some others a thymol chemotype is interesting, given the tendency of the linalool chemotype to appear in colder areas. Other compounds with less significance with regard to the percentage in the essential oil (about 10 per cent) are α-pinene, camphene and α-terpineol. Figure 4.1 and Table 4.1 clearly reflect this separation between phenolic and linalool chemotypes in T. *hyemalis*.

The essential oil of T. *longiflorus* was found to be characterised mainly by either 1,8-cineole (max. 74.7 per cent) or camphor (max. 46.6 per cent). Camphene and borneol reached maximum percentages of about 20 per cent. These four compounds dominated, in different combinations, in all the essential oils studied. Phenols or their precursors were practically absent, with a maximum of 3.5 per cent thymol in one sample. Morales (1986) described one sample containing 26.2 per cent of terpinen-4-ol. Figure 4.2 shows the samples clearly split into two groups, and the one characterised by camphor/borneol in Table 4.2 is the more common.

T. *mastichina* (Spanish marjoram) has attracted economical interests and has a wide distribution area across Spain. In southeastern Spain Sáez (1996) found its essential oil characterised by 1,8-cineole (max. 79.9 per cent) and linalool (max. 56.6 per cent). The chemotypes detected were either characterised by 1,8-cineole/linalool, or – less frequently – by 1,8-cineole alone, but no individual containing linalool without 1,8-cineole was found. Sabinene, *tr*-sabinene hydrate, camphor and borneol reached percentages up to 5 per cent in different samples. Linalyl acetate scored 35.1 per cent in one sample due to the early stage of development achieved at the time of collection. The situation in *T. mastichina* is reflected in Figure 4.3 with the 1,8-cineole chemotype represented more frequently. García-Vallejo *et al.* (1984) studied a wider area, with 228 samples and described three chemotypes: a 1,8-cineole-type, a linalool-type, and a mixed 1,8-cineole/linalool-type.

T. membranaceus is recognised by an inflorescence with broad, pale-yellow bracts. Sáez (1996) described the essential oils as mostly characterised by 1,8-cineole (max. 82.2 per cent; min.13 per cent), with camphene (max. 15.7 per cent), camphor (max. 35.6 per cent) and borneol (max. 23.6 per cent) usually presenting lower levels. The oil of one sample was described to contain 39.9 per cent linalool. It was collected in an area where also *T. zygis* grows showing the same chemotype, thus suggesting a relation with this species. Phenols were found to be almost absent (max. 1.5 per cent for both thymol and carvacrol), but Zarzuelo *et al.* (1987) reported quantities of about 3.5 per cent for these compounds. Studying *T. membranaceus* on the section and genus level (Table 4.2 and Figure 4.1), these observations are supported.

T. moroderi is characterised by its deep purple bracts, and is of some economic interest as a raw material for producing a liqueur. The essential oil is characterised by 1,8-cineole (max. 68.1 per cent), camphor (max. 34.6 per cent) and borneol (max. 23.3 per cent). Camphene reached percentages of about 20 per cent in two individuals from the same population. Thymol (14.4 per cent) and carvacrol (19.0 per cent) appeared in two different populations where this species may have introgreded with *T. hyemalis*. Previous studies by Cañigueral *et al.* (1994) revealed the existence of chemical polymorphism based on the presence of either an unidentified oxygenated sesquiterpene or on the simultaneous occurrence of β-eudesmol, ledol, α-elemol and β-elemol. But it is not clear if this species presents a well-defined set of chemotypes. Figure 4.2 and Table 4.2 show this species, placing its samples all over the range of diversity for the section *Pseudothymbra*.

T. orospedanus presents an erect habit, with dense and short branching to protect itself from low winter temperatures in the medium–high mountainous areas where it grows, in a relatively small area. Twelve individuals have been chemically studied, and they show the important presence of myrcene (max. 27.4 per cent) and 1,8-cineole (max. 34.3 per cent). The highest score obtained for linalool was 87.6 per cent. Caryophyllene oxide, β-caryophyllene, borneol, camphene and *tr*-sabinene hydrate presented percentages from 10–20 per cent in different samples. No significant quantities of phenols or their precursors were found by Sáez (1996). Although Table 4.1 collects most of the samples studied into one chemical group, their situation within Figure 4.1 supports the idea of a chemically variable species.

T. piperella presents broad-round to heart-shaped leaves and other morphological features that makes this species quite distinct from the rest. The essential oil is typically phenolic, and Sáez (1996) described, for southern populations of the species, maximum values of 14.4 per cent thymol and 10.2 per cent carvacrol, with higher percentages of *p*-cymene (65.1 per cent). The presence of α-pinene, limonene, terpinen-4-ol and 1,8-cineole in quantities from 5–10 per cent did not represent important alternatives to the phenolic character of the essential oil. A similar situation was found by Blanquer *et al.* (1998), who described three chemotypes based on 31 individuals investigated:

One *p*-cymene/thymol-type, one *p*-cymene/carvacrol-type, and one *p*-cymene/carvacrol/ γ-terpinene-type. They also found geographical separation among the chemotypes. Figure 4.4 shows this species as the least variable one when studied within the whole genus context.

T. serpylloides ssp. *gadorensis* can be found in the upper-most mountain ranges, exposed to very low winter temperatures, thus showing a procumbent habit. It is of little economic interest. The studies from Sáez (2001) show that it is a mainly phenolic species, with both thymol (max. 56.9 per cent) and carvacrol (max. 27.7 per cent) chemotypes, usually excluding each other but sometimes present at the same time in one plant. A linalool chemotype is frequently found, with percentages up to 79.7 per cent, and linalyl acetate reached 39.4 per cent. With concentrations up to 30.4 per cent and 79.9 per cent respectively myrcene and geraniol chemotypes were determined in restricted areas. *tr*-Sabinene hydrate (28.9 per cent) and caryophyllene oxide (14.1 per cent) characterised some samples but cannot be regarded as common within the species. Arrebola *et al.* (1995) found thymol levels up to 15.8 per cent and carvacrol up to 73.5 per cent. The idea of a quite variable species is supported by Figure 4.1 and Table 4.1 (*T. gadorensis*), both also showing the clear separation of a geraniol chemotype.

T. vulgaris is well known and was the first *Thymus* species to receive attention on its chemical variability. Six chemotypes have been described in the south of France: thymol, carvacrol, *tr*-sabinene hydrate/terpinen-4-ol, linalool, α-terpineol and geraniol. Another chemotype has been found vastly dispersed in Spain, 1,8-cineole, reported by several authors (Adzet *et al.*, 1977). The highest levels in the essential oils of plants from southeastern Spain were about 70 per cent, usually accompanied by camphene (max. 24.2 per cent), camphor (max. 38.6 per cent) and borneol (max. 34.2 per cent) as secondarily important compounds after 1,8-cineole. But mixed chemotypes camphor/ camphene with 1,8-cineole almost absent were also found. A linalool chemotype was recorded only at a few points in the area studied, with up to 41.5 per cent. The absence of phenols is noticeable (max. 1.1 per cent thymol), although a 32.2 per cent *p*-cymene was recorded in one sample, perhaps reflecting some introgression with *T. hyemalis* growing nearby. Comparing Figures 4.1 and 4.2 it can be concluded that the distribution of the samples from *T. vulgaris* is similar to that of the whole section *Pseudothymbra*, thus demonstrating how both are highly characterised by 1,8-cineole. It is remarkable that no well-defined groups can be seen for *T. vulgaris* (Figure 4.1), but rather a progressive substitution of 1,8-cineole by other compounds such as camphor or borneol.

T. zygis is represented by two subspecies in southeastern Spain. The ssp. *gracilis* is widespread and presents a more erect habit than ssp. *sylvestris* which grows far away from the coast in wetter and colder environments (Sáez, 1995b). Thymol (max. 72.9 per cent) and carvacrol (max. 22.8 per cent) chemotypes are the ones most frequently found in the area. A linalool chemotype (max. 91.4 per cent) is predominant in places of high altitudes and low temperatures, where individuals show smaller and denser habit than those living in the foothills, which present phenolic chemotypes. Geranyl acetate was found to reach 24.3 per cent in the oil of one sample, and limonene, camphene, camphor and terpinen-4-ol registered maximum percentages between 10–20 per cent. The remarkable presence of 1,8-cineole in the oils of some samples seems to be connected with the presence of *T. vulgaris* within the population. The distribution pattern of *T. zygis* in Figure 4.1 reflects the important chemical partitions commented on here.

THE SITUATION OF *THYMUS* IN WESTERN IBERIA

The essential oil of *Thymus* species from Portugal and the western part of Spain has been studied by Salgueiro and her co-workers during the last two decades. Special attention has been given to the chemical polymorphism, having even been established in experimental cultures for some species. The area shows a high taxonomical variability, with eight species from four different sections studied: (a) Section *Thymus*: *T. zygis*, *T. carnosus*, *T. camphoratus*; (b) Section *Pseudothymbra*: *T. lotocephalus*, *T. villosus*; (c) Section *Mastichina*: *T. mastichina*, *T. albicans*; (d) Section *Micantes*: *T. caespititius*. These species were studied from collective samples that provided the general characteristics of the populations, and from individual samples, to determine the chemical variability.

Section *Thymus* presents two subsections in the area, mainly differing in the presence/absence of floral bracts. *T. camphoratus* is spread along the southern-most part of Portugal and belongs to the subsection *Thymastra*, thus presenting floral bracts. The analysis of 72 individual plants (Salgueiro *et al.*, 1997a) showed four main groups characterised by (a) linalool (max. 21.0 per cent in the oil), (b) borneol (max. 24.0 per cent), (c) 1,8-cineole (max. 20.0 per cent), d) 1,8-cineole/borneol. Terpinen-4-ol and *tr*-sabinene hydrate were found in lower concentrations with maximum values of 10.2 per cent and 10.8 per cent respectively. All these percentages were achieved from collective samples obtained at the same localities as the individual ones. The absence of phenols or their precursors in significant quantities should be noticed.

T. capitellatus is also endemic in southern Portugal. The presence of ovate floral bracts place this species within the subsection *Thymastra*. The chemical composition of the essential oils resembles that of *T. camphoratus*. Three chemotypes could be described, a 1,8-cineole-type, a camphene/1,8-cineole/borneol-type and a linalool/linalyl acetate-type (Salgueiro, 1992).

T. carnosus and *T. zygis* do not present floral bracts, which puts them in the subsection *Thymus*. The essential oil of *T. carnosus* was studied; on analysis of 83 samples (Salgueiro *et al.*, 1995) showed a division into three main groups: (a) a large group of individuals was characterised by borneol/*cis*-sabinene hydrate/terpinen-4-ol; (b) a group with linalool/*tr*-sabinene hydrate, (c) a small group of samples was characterised by borneol/camphene. The highest percentages obtained from collective samples were 32.0 per cent for borneol, 25.5 per cent for linalool, 17.0 per cent for *tr*-sabinene hydrate, 13.0 per cent for camphene, 11.2 per cent for *cis*-sabinene hydrate and 11.1 per cent for terpinen-4-ol. Phenolic chemotypes were absent.

T. zygis ssp. *zygis* and ssp. *sylvestris* differ mainly in their indumentum and the distribution area, ssp. *sylvestris* being hairier and found farther south than ssp. *zygis*. The latter proved to be polymorphous showing five chemotypes (Salgueiro and Proença da Cunha, 1989), a linalool-type (max. 87.0 per cent linalool), a thymol-type (49.2 per cent thymol), a geraniol/geranyl acetate-type (52.5 per cent and 38.0 per cent respectively), a 1,8-cineole/linalool-type (32.5 per cent and 42.3 per cent respectively), and a 1,8-cineole/thymol-type (29.2 per cent and 25.6 per cent respectively).

In western Iberia there are two species in the section *Pseudothymbra*, *T. lotocephalus* and *T. villosus*, the latter with the ssp. *villosus* and ssp. *lusitanicus*. Salgueiro *et al.* (1997b) studied the essential oil of *T. villosus* ssp. *villosus*, finding four groups of individuals characterised by (a) *p*-cymene/camphor/linalool, (b) *p*-cymene/borneol, (c) linalool/geraniol/geranyl acetate, (d) α-terpineol/camphor/myrcene. They reported maximum percentages in collective samples as follows: 40.0 per cent for *p*-cymene, 19.0 per cent for camphor, 23.7 per cent for linalool, 30.2 per cent for α-terpineol, 18.5 per cent for myrcene, 15.5 per cent for geraniol and 12.9 per cent for geranyl acetate. Maximum percentages for thymol and carvacrol were only 11.7 per cent and 5.5 per cent, despite high levels of *p*-cymene.

The essential oils of *T. villosus* ssp. *lusitanicus* were studied by Salgueiro *et al.* (2000a), who found that the samples could be classified into five main groups, with either (a) linalool/terpinen-4-ol/*tr*-sabinene hydrate, (b) linalool/1,8-cineole, (c) linalool, (d) geranyl acetate/geraniol, e) geranyl acetate/geraniol/1,8-cineole. *T. lotocephalus* presented 1,8-cineole, camphor, linalool, linalyl acetate and α-pinene as the main constituent(s) in one sample (Salgueiro, 1992). The analyses of four populations of this species demonstrated the existence of groups characterised by either linalool, 1,8-cineole, linalool/1,8-cineole, linalyl acetate/linalool, or geranyl acetate (Salgueiro *et al.*, 2000b).

In Western Iberia the section *Mastichina* shows two species: *T. albicans* and *T. mastichina*, the latter with two subspecies, ssp. *mastichina* and ssp. *donyanae*. The chemical variability of these three taxa is described in Salgueiro *et al.* (1997c). They identified 77 compounds in 304 individuals of *T. mastichina* ssp. *mastichina*, in 15 individuals of *T. mastichina* ssp. *donyanae* and in 43 plants of *T. albicans*. The ssp. *mastichina* presents individuals arranged in three clusters, characterised by either 1,8-cineole (max. 64.2 per cent), linalool (max. 45.0 per cent), or plants with similar quantities of both 1,8-cineole and linalool. With respect to ssp. *donyanae*, the main characteristic is a higher level of borneol (max. 15.3 per cent) than in ssp. *mastichina*, although 1,8-cineole predominates (max. 38.4 per cent), and no clear chemical polymorphism was detected among individuals. Finally, for *T. albicans* they found a chemical pattern similar to *T. mastichina* ssp. *mastichina*, with three groups of individuals differing in the relative proportions of 1,8-cineole (max. 42.9 per cent) and linalool (max. 22.0 per cent). All these percentages were obtained in population samples, although the chemical groupings were detected in individual samples. Trace quantities of phenols or their precursors were reported.

Section *Micantes* presents one species in the Iberian Peninsula, *T. caespititius*, studied by Salgueiro *et al.* (1997d) by means of 91 plants and collective samples, provenant from northwest Portugal and the Azores. These two locations showed important chemical differences since the samples collected on the mainland were characterized by α-terpineol (max. 40.5 per cent), *p*-cymene (max 9.1 per cent) and T-cadinol (max 8.7 per cent), while the Azores sample registered carvacrol (36.3 per cent), thymol (16.1 per cent), carvacrol acetate (8.3 per cent) and *p*-cymene (6.8 per cent). It is worth mentioning that the presence of *tr*-dihydroagarofuran (max. 3.0 per cent), an oxygenated sesquiterpene, was recorded in all samples, this compound not having been previously described within the essential oils of the genus *Thymus*.

THE SITUATION OF *THYMUS* IN NORTHERN EUROPE

The essential oil chemistry of *Thymus* species of northern Europe and Greenland has been studied intensively by Stahl-Biskup and her co-workers as revealed in a series of publications which appeared from 1984 to 1998. Their papers also contain detailed studies on the chemical polymorphism of the species concerned. The experimental concepts fulfilled all requirements for studies of the chemical variation of a taxon, the

Table 4.3 *Thymus* species of northern Europe and Greenland investigated concerning their chemical polymorphism

Thymus species	Country	Number of individuals analyzed	References
T. praecox ssp. arcticus	Norway	52	Stahl-Biskup, 1986a
	Iceland	108	Stahl, 1998
	Greenland	17	Stahl, 1984
	Scotland	377	Bischof-Deichnik, 1997
	South of England	303	Schmidt, 1998
	Ireland	52	Schmidt, 1998
T. serpyllum ssp. serpyllum	Finland	52	Stahl-Biskup and Laakso, 1990
T. serpyllum ssp. tanaensis	Finland	133	Stahl-Biskup and Laakso, 1990
T. pulegioides	Norway	79	Stahl-Biskup, 1986b
	South of England	85	Schmidt, 1998

analysis of individual plants being the most important. Table 4.3 presents a list of the investigated species, the regions, where the plant material was collected, and the number of individuals analysed.

The *Thymus* species studied belong to the section *Serypyllum* which contains the largest number of species (see Chapter 1) and is the most extensive section covering Central and North Europe, the eastern Mediterranean region and the Middle East extending over eastern Asia. Whereas the southern species of the section *Serpyllum* grow as small subshrubs and are woody at the base, the northern species are hardly lignified, procumbent and herbaceous. In general all species of the section *Serpyllum* are characterised by a high morphological variety. Intrasectional as well as intersectional hybridisation is common, which makes the taxonomy of this phylogenetically young section more difficult, this being one of the reasons why this section has often been revised.

The chemical characteristics of the essential oils of the northern species can be defined as exclusively terpenoid including mono- and sesquiterpenes. The quantitatively most important compounds of *T. praecox* ssp. *arcticus* are the monoterpenes linalyl acetate and linalool, accompanied by some sesquiterpene alcohols more exceptional in *Thymus* species namely hedycaryol, nerolidol, T-cadinol, α-cadinol, germacra-1(10),4-dien-6-ol, and germacra-1(10),5-dien-4-ol (Stahl, 1984b). The essential oil of *T. serpyllum* ssp. *tanaensis* resembles that of the former with high contents of linalool and/or linalyl acetate and the two germacradienols as minor compounds (Stahl-Biskup and Laakso, 1990). In *T. serpyllum* ssp. *serpyllum* the monoterpenes 1,8-cineole and myrcene are dominant and are accompanied by germacrene D and the more exclusive sesquiterpene alcohols hedycaryol, germacra-1(10),4-dien-6-ol, and germacra-1(10),5-dien-4-ol (Stahl-Biskup and Laakso, 1990). The oil of *T. pulegioides* (Stahl-Biskup, 1986b; Schmidt, 1998) differs from the other oils, containing the terpene phenols, thymol and carvacrol, which occur only sporadically in *T. praecox* ssp. *arcticus* and which are lacking in *T. serpyllum* ssp. *serpyllum* and ssp. *tanaensis*. Further components of importance are linalool and geraniol. The variation of the essential oil composition within the species will be described in the following paragraphs.

Thymus praecox ssp. *arcticus*

T. praecox Opiz ssp. *arcticus* (E. Durand) Jalas is a tetraploid species with chromosome numbers 2n=50–56 (Jalas and Kaleva, 1967), 2n=50–58 (Pigott, 1955), 2n=56, 58 (Schmidt, 1968), and 2n=50, 51, 54 (Jalas, 1972). In literature various synonyms exist; the one most commonly used is *T. drucei* which was established by Ronniger for the population of the British Isles. It is mainly indigenous to the European North Atlantic region reaching from Iceland, the Faeroes via the British Isles to the west coast of Norway and Greenland. It has been supposed that this subspecies survived the Ice Age on ice-free areas. Only in the south of England it is associated with *T. pulegioides*, whereas in the other regions it grows apart from other *Thymus* species. *T. praecox* ssp. *arcticus* is a plant with long, somewhat woody, creeping branches, non-flowering or with a terminal inflorescence, flowering stems are born in rows (Jalas, 1972).

A compilation on the early findings concerning the chemical polymorphism of *Thymus praecox* ssp. *arcticus* was published in 1984b when Stahl presented 8 chemotypes of this species evaluating essential oil data from 177 individual plants. Seven chemotypes contained essential oils with high percentages of linalyl acetate (about 70 per cent). The chemotype characterising compounds were the sesquiterpene alcohols nerolidol, hedycaryol, and T-cadinol which occurred in different combinations within the oils (2–11 per cent). One type was totally lacking these sesquiterpenoids. One chemotype did not contain linalyl acetate but hedycaryol in high percentages (42 per cent). Two further facts were remarkable: (a) in Iceland all the 8 chemotypes were present, in Norway 6 with the hedycaryol- and nerolidol-type lacking, in Greenland only 4 chemotypes could be found; (b) within all the populations the different chemotypes were growing side by side.

In the 1990s, *T. praecox* ssp. *arcticus* was studied again. There were 377 individuals analysed from Scotland (Bischof-Deichnik, 1997), 303 individuals from South England, and 52 individuals from Ireland (Schmidt, 1998). At that time multivariate statistical analysis was accessible and applied to evaluate the flood of quantitative GC data obtained. The oils of the Scottish population proved to be chemically similar to the oils of Iceland, Norway and Greenland with some further compounds, e.g. neral, geranial, citronellol, thymol, carvacrol, γ-terpinene, and *tr*, *tr*-farnesol. As a result of a multivariate statistical evaluation of the individual oils, 22 chemotypes could be established, with a linalyl acetate/linalool-type as the most frequent one (105 individuals), followed by a hedycaryol type (35), a germacra-1(10),4-dien-6-ol/linalool/linalyl acetate type (34), a *tr*-nerolidol-type (39), a germacra-1(10),4-dien-6-ol-type (21), a germacra-1(10), 5-dien-4-ol/germacra-1(10),4-dien-6-ol-type (20), a monoterpene hydrocarbon-type (17), a *tr*-ocimene-type (17), and a borneol-type (11). The further chemotypes occurred only in 10 individuals or fewer.

New aspects concerning the number of chemotypes arose when Schmidt (1998) evaluated the essential oil data from 52 individuals from Ireland and 303 individuals from the south of England. Her aim was to present a compilation of all the oil data available for *T. praecox* ssp. *arcticus* in the North Atlantic region, applying neuronal nets for the formation of groups. In comparison to the multivariate statistical analysis this method produced more plausible results because a greater value was placed on the quantitative presence of substances, not only on the fact that it was present. The results are more compatible with our subjective perception by including all 909 individuals in the calculation with the neuronal nets; 17 chemotypes of *T. praecox* ssp. *arcticus* were found. Again the linalyl acetate/linalool-type (36 per cent of the individuals), the hedycaryol-type (20 per cent), and a germacra-1(10),4-dien-6-ol-type (14 per cent) were the most abundant types, followed by the *tr*-nerolidol-type (5.3 per cent), a T-cadinol/hedycaryol-type (5 per cent), a β-caryophyllene-type (4.7 per cent), and a linalool-type (3 per cent). All the other chemotypes were represented by fewer than 2 per cent of the individuals, of which a thymol-type is worth mentioning.

As a result one can say that chemical polymorphism in the northern *T. praecox* ssp. *arcticus* with 17 chemotypes is more highly developed than in the southern species. The revised definition of the oil types revealed the following frequency of oil types within the countries: Greenland 2, Iceland 5, Norway 1, Scotland 13, Ireland 11, south of England 17. A north-south gradient of the linalyl acetate/linalool-type with higher frequency in the north and a contrary pattern of the thymol-type with higher frequency in the south is striking. The existence of a thymol-type in *Thymus praecox* ssp. *arcticus* gives reason to discuss a relation with the also polymorphous *T. praecox* ssp. *polytrichus* of the Tyrolean Alps which was proved to show 12 chemotypes, one of them a thymol-type which 33 per cent of the investigated plants of this region belonged to (Bischof-Deichnik *et al.*, 2000).

T. SERPYLLUM SSP. SERPYLLUM AND T. SERPYLLUM SSP. TANAENSIS

T. serpyllum L. ssp. *serpyllum* (syn. ssp. *angustifolius* (Pers.) Arcangel) is widespread in the sandy soils in southern and central Finland but rare north of the 62nd parallel, whereas *T. serpyllum* ssp. *tanaensis* grows north of the polar circle in two areas of Lapland: in the northeastern part of Finland and the area around Kuusamo as well as in the north on either bank of the Tana river, which forms the border to Norway. The chromosome number for both subspecies is $2n=24$. The polymorphism of both taxa was investigated by Stahl-Biskup and Laakso (1990) evaluating the oil composition of 52 and 133 individual plants respectively. At that time neither multivariate statistical analysis nor neuronal nets were available; therefore the group-forming method was based on the subjective evaluation of the peak patterns.

The pattern of the above-mentioned sesquiterpene alcohols again gave reason to distinguish four different chemotypes of *T. serpyllum* L. ssp. *serpyllum* in Finland, namely a hedycaryol-type, a germacra-1(10),4-dien-6-ol-type, a germacra-1(10),5-dien-4-ol-type and one type lacking those alcohols. It must be stressed that these chemotype characterising compounds were not the main constituents of the oils. 1,8-cineole and myrcene were the main compounds. *T. serpyllum* ssp. *tanaensis* also showed 4 chemotypes, two of them characterised again by the germacradienols, the two further by high percentages of linalool and linalyl acetate, respectively.

A chemical overlap of the chemotypes of these two Finnish subspecies is remarkable and one can speak of about 6 chemotypes of *Thymus serpyllum* s.l. in Finland if the subspecies level is not considered. Once more a certain north-south gradient is noticeable, again with the linalyl acetate type only present in the north (northern Lapland) and the linalool-type only present in southern Lapland. The germacradienol-types as well as the hedycaryol-type become more abundant from the north to the south, the latter totally lacking in Lapland. The assumption discussed in the past that *T. serpyllum* ssp. *tanaensis* belongs to a group of plants which immigrated from the Northeast, the

Eurasian Taiga, whereas the other northern *Thymus* species originated from the Mediterranean center of the genus (Meusel *et al.*, 1978), cannot be derived from the chemical patterns which are similar in both subspecies.

Thymus pulegioides

T. pulegioides L. (syn. *T. chamaedrys* Fries) is widely distributed on the European continent south to the Mediterranean isles. Chromosome numbers are 2n=28 (Pigott, 1955; Schmidt, 1968), 2n=28, 30 (Jalas, 1972). Northern occurrences are in the south of Sweden, in the south of Norway near the Oslo Fjord, and in the south of England. It is a more upright growing *Thymus* species, 25–40 cm high, somewhat woody at the base, branched flowering stems, long creeping branches absent (Jalas, 1972). The populations of Norway (Stahl-Biskup, 1986b) and the south of England were studied (Schmidt, 1998).

Contrary to the other northern *Thymus* species the essential oil of *T. pulegioides* contains monoterpene phenols, thymol and carvacrol, as well as their precursors *p*-cymene and γ-terpinene. These chemical characteristics make it resemble the southern *Thymus* where the phenols are abundant and characteristic compounds of many species. With regard to polymorphism, when analysing 79 individual plants, *T. pulegioides* turned out to be less polymorphous than *T. arcticus*. In Norway only two chemotypes could be found, a carvacrol-type with average percentages of about 37 per cent caravcrol in the oils, and a thymol-type with 35 per cent thymol. Three quarters of the plants belong to the carvacrol type, one quarter to the thymol type (Stahl-Biskup, 1986b).

On evaluating the oil data of 85 individual plants collected in southern England (Schmidt, 1998), *T. pulegioides* was found to comprise 4 chemotypes, a thymol-type, a linalool-type, a geraniol-type and a carvacrol/γ-terpinene-type. The thymol-type was the most abundant including 65 per cent of the individuals followed by the linalool-type with 29 per cent of the plants. The geraniol- and the carvacrol/γ-terpinene-type occurred only sporadically with 4 per cent and 2 per cent of the plants respectively. Within the populations the different chemotypes grew side by side as was found within the other northern *Thymus* species. The chemotypes found agree with those detected in *T. pulegioides* ssp. *chamaedrys* in Slovakia (Mártonfi, 1992). He analysed 181 samples using a principal component analysis which resulted in 5 chemotypes: a thymol chemotype (20.8 per cent thymol in average), a carvacrol chemotype (32.9 per cent carvacrol in average), a citral/geraniol chemotype (29.1 per cent citral and 22.4 per cent geraniol), a linalool chemotype (54.8 per cent linalool), and a fenchone chemotype (33.9 per cent fenchone). The latter had never been found before in a *Thymus* species. In a further paper the chemotype pattern differentiation on different substrates is studied (Mártonfi *et al.*, 1994).

CHEMICAL POLYMORPHISM IN THE GENUS *THYMUS*: A RÉSUMÉ

Despite some reservations concerning the comparability of the experimental concepts of the polymorphism studies it can be postulated that in *Thymus* two forms of polymorphism are manifested: some species occur with only few chemotypes, such as *T. pulegioides*, *T. vulgaris*, *T. mastichina*, other species show more than seven or even an uncertain number of chemotypes. The latter may include *T. praecox* ssp. *arcticus*, *T. baeticus*, *T. camporatus*,

T. herba-barona, T. tosevii ssp. *tosevii* and *T. zygis* ssp. *sylvestris*. Some species have been found to be tetraploid, such as *T. herba-barona, T. zygis* ssp. *sylvestris, T. praecox* ssp. *arcticus* or *T. mastichina* ssp. *mastichina* among others (Jalas and Kaleva, 1967; Morales, 1986), but it is still unclear whether in some species chemical polymorphism and high ploidy levels are related. The highest chemical variability seems to concentrate in species from section *Serpyllum* and section *Thymus*.

REFERENCES

Adzet, T., Granger, R., San Martin, R. and Simeon de Bouchberg, M. (1976) Les huiles essentielles de *Thymus vulgaris* spontané de France et d'Espagne. *Pharmacia Mediterranea*, 9, 1–6.

Adzet, T., Granger, R., Passet, J. and San Martin, R. (1977) Le polymorphisme chimique dans le genre *Thymus*: sa signification taxonomique. *Biochem. Syst. Ecol.*, 5, 269–272.

Arrebola M.L., Navarro, M.C. and Jiménez, J. (1995) Variations in yield and composition of the essential oil of *Thymus serpylloides* Bory subsp. *gadorensis* (Pau) Jalas. *J. Essent. Oil Res.* 7, 369–374.

Bellomaria, B., Hruska, K. and Valentini, G. (1981) Composizione degli olii essenziali di *Thymus longicaulis* C. Presl in varie località dell'Italia Centrale. *Giorn. Bot. Ital.*, 115, 17–27.

Benjilali, B., Hammoumi, M. and Richard, H. (1987) Polymorphisme chimique des huiles essentielles de thym du Maroc. 1. Charactérisation des composants. *Sci. Aliment.*, 7, 77–91.

Bischof-Deichnik, C. (1997) *Das ätherische Öl der schottischen Population von Thymus praecox ssp. arcticus (E. Durand) Jalas (Lamiaceae)*. Doctoral thesis, University of Hamburg.

Bischof-Deichnik, C., Stahl-Biskup, E. and Holthuijzen, J. (2000) Multivariate statistical analysis of the essential oil data from *T. praecox* ssp. *polytrichus* of the Tyrolean Alps. *Flavour Fragr. J.*, 15, 1–6.

Blanquer, A., Boira, H., Soler, V. and Pérez, I. (1998) Variability of the essential oil of *Thymus piperella*. *Phytochemistry*, 47, 1271–1276.

Cañigueral, S., Vila, R., Vicario, G., Tomàs, X. and Adzet, T. (1994) Chemometrics and essential oil analysis: chemical polymorphism in two *Thymus* species. *Biochem. Syst. Ecol.*, 22, 307–315.

García-Vallejo, M.C., García, D. and Muñoz, F. (1984) Avance de un estudio sobre las esencias de *Thymus mastichina* español (mejorana de España). *Anales INIA / Serie Forestal*, 8, 201–218.

Granger, R. and Passet, J. (1971) Types chimiques (chémotypes) de l'espèce *Thymus vulgaris* L. *C. R. Acad. Sci. Paris*, 273, 2350–2353.

Granger, R. and Passet, J. (1973) *Thymus vulgaris* spontane de France: Races chimiques et chemotaxonomie. *Phytochemistry*, 12, 1683–1691.

Jalas J. and Kaleva, K. (1967) Chromosome studies in *Thymus* L. (Labiatae). V. *Ann. Bot. Fenn.*, 4, 74–80.

Jalas, J. (1972) In T.G. Tutin, V.H. Heywood, N.A. Burges, D.M. Moore and D.H. Valentine (eds), *Flora Europaea*, Vol. 3, Cambridge University Press, Cambridge, pp. 172–183.

Lawrence, B.M. (1980) The existence of infraspecific differences in specific genera in the Labiatae family. In Annales techniques, VIIe Congrès International des Huiles Essentielles, Octobre, Cannes-Grasse, pp.118–131.

Mártonfi, P. (1992) Polymorphism of essential oil in *Thymus pulegioides* subsp. *chamaedrys* in Slovakia. *J. Essent. Oil Res.*, 4, 173–179.

Mártonfi, P., Grejtoský, A. and Repcák, M. (1994) Chemotype pattern differentiation of *Thymus pulegioides* on different substrates. *Biochem. Syst. Ecol.*, 22, 819–825.

Meusel, H., Jäger, E., Rauschert, S. and Weinert, E. (1978) *Vergleichende Chorologie der Zentraleuropäischen Flora*, Vol. II, VEB Gustav Fischer, Jena, pp. 111–112, p. 385.

Morales, R. (1986) Taxonomía de los géneros *Thymus* (excluida la sección Serpyllum) y *Thymbra* en la Península Ibérica. *Ruizia*, 3, 1–324.

Pigott, C.D. (1955) *Thymus* L. *J. Ecol.*, 43, 365–387.

Sáez, F. (1995a) Essential oil variability of *Thymus hyemalis* growing wild in Southeastern Spain. *Biochem. Syst. Ecol.*, 23, 431–438.

Sáez, F. (1995b) Essential oil variability of *Thymus zygis* growing wild in southeastern Spain. *Phytochemistry*, 40, 819–825.

Sáez, F. (1996) *El género Thymus en el Sureste Ibérico: estudios biológicos y taxonómicos.* Ph.D. Thesis. University of Murcia. Spain.

Sáez, F. (1999). Essential oil variability of *Thymus baeticus* growing wild in southeastern Spain. *Biochem. Syst. Ecol.*, 27, 269–276.

Sáez, F. (2001). Volatile oil variability in *Thymus serpylloides* ssp. *gadorensis* growing wild in southeastern Spain. *Biochem. Syst. Ecol.*, 29, 198.

Salgueiro, L. and Proença da Cunha, A. (1989) Determinaçâo de quimiotipos no *Thymus zygis* L. subsp. *sylvestris* (Hoffmanns & Link) Brot. ex Coutinho da regiâo de Eiras-Coimbra. *Rev. Port. Farm.*, 39, 19–27.

Salgueiro, L. (1992) Essential oils of endemic *Thymus* species from Portugal. *Flavour Fragr. J.*, 7, 159–162.

Salgueiro, L., Vila, R., Tomàs, X., Tomi, F., Cañigueral, S., Casanova, J., Proença da Cunha, A. and Adzet, T. (1995) Chemical polymorphism of the essential oil of *Thymus carnosus* from Portugal. *Phytochemistry*, 38, 391–396.

Salgueiro, L., Vila, R. Tomi, F., Figueiredo, A.C., Barroso, J.C., Cañigueral, S., Casanova, J., Proença da Cunha, A. and Adzet, T. (1997d) Variability of essential oils of *Thymus caespititius* from Portugal. *Phytochemistry*, 45, 307–311.

Salgueiro, L., Proença da Cunha, A., Tomàs, X., Cañigueral, S., Adzet, T. and Vila, R. (1997b) The essential oil of *Thymus villosus* L. ssp. *villosus* and its chemical polymorphism. *Flavour Fragr. J.*, 12, 117–122.

Salgueiro, L., Vila, R., Tomi, F., Tomàs, X., Cañigueral, S., Casanova, J., Proença da Cunha, A. and Adzet, T. (1997a) Composition and infraspecific variability of essential oil from *Thymus camphoratus*. *Phytochemistry*, 45, 1177–1183.

Salgueiro, L., Vila, R., Tomàs, X., Cañigueral, S., Proença da Cunha, A. and Adzet, T. (1997c). Composition and variability of the essential oils of *Thymus* species from Section Mastichina from Portugal. *Biochem. Syst. Ecol.*, 25, 659–672.

Salgueiro, L., Vila, R., Tomàs, X., Cañigueral, S., Paiva, J., Proença da Cunha, A. and Adzet, A. (2000a) Chemotaxonomic study on *Thymus villosus* from Portugal. *Biochem. Syst. Ecol.*, 28, 471–482.

Salgueiro, L., Vila, R., Tomàs, X., Cañigueral, S., Paiva, J., Proença da Cunha, A and Adzet, T. (2000b) Essential oil composition and variability of two endemic taxa from Portugal, *Thymus lotocephalus* and *Thymus* x *mourae*. *Biochem. Syst. Ecol.*, 28, 457–470.

Schmidt, P. (1968) Beitrag zur Kenntnis der Gattung *Thymus* L. in Mitteldeutschland. *Hercynia*, 5, 385–419.

Schmidt, A. (1998) *Polychemismus bei den ätherisches Öl führenden Arten* Thymus pulegioides *L. und* Thymus praecox *Opiz ssp.* arcticus *(E. Durand) Jalas (Lamiaceae) im nordatlantischen Europa*. Doctoral thesis, University of Hamburg.

Stahl, E. (1984a) Chemical polymorphism of essential oil in *Thymus praecox* ssp. *arcticus* (Lamiaceae) from Greenland. *Nord. J. Bot.*, 4, 597–600.

Stahl, E. (1984b) Das ätherische Öl aus *Thymus praecox* ssp. *arcticus* isländischer Herkunft. *Planta Med.*, 50, 157–160.

Stahl-Biskup, E. (1986a) Das ätherische Öl norwegischer Thymusarten. I. *Thymus praecox* ssp. *arcticus*. *Planta Med.*, 52, 36–38.

Stahl-Biskup, E. (1986b) Das ätherische Öl norwegischer Thymianarten II. *Thymus pulegioides*. *Planta Med.*, 52, 223–235.

Stahl-Biskup, E. and Laakso, I. (1990) Essential oil polymorphism in Finnish *Thymus* species. *Planta Med.*, 56, 464–468.

Tétényi, P. (1970) *Infraspecific Chemical Taxa of Medicinal Plants.* Akadémiai Kiadó, Budapest (1970).

Tétényi, P. (1975) Polychemismus bei ätherischölhaltigen Pflanzenarten. *Planta Med.*, **28**, 244–256.

Vila, R., Freixa, B., Cañigueral, S. and Adzet, T. (1995) Composition of the variability of the essential oil of *Thymus funkii* Cosson. *Flavour Fragr. J.*, **10**, 379–383.

Zarzuelo, A., Navarro, C., Crespo, M.E., Ocete, M.A., Jiménez, J. and Cabo, J. (1987). Spasmolytic activity of *Thymus membranaceus* essential oil. *Phytother. Res.*, **1**, 114–116.

5 Flavonoids and further polyphenols in the genus *Thymus*

Roser Vila

INTRODUCTION

Flavonoids constitute one of the main groups of natural phenolic compounds, being widely extended among green plants, where they can be found in different organs: leaves, flowers, barks, fruits, etc. Flavonoid aglycones may have several types of structures, all of them with a 15 carbon nucleus arranged in a C_6-C_3-C_6 disposition, that is: two aromatic rings linked by a three carbon chain, that may or may not form a third ring. The main flavonoid aglycone structures are related by a common biosynthetic pathway that involves precursors from both the shikimic and polyketide route (Ebel and Hahlbrock, 1982; Grisebach, 1985; Hahlbrock and Grisebach, 1975). The first flavonoid is formed immediately after the confluence of the two ways, and it seems to be the chalcone, from which all the other structures derive. During the biosynthetic process, several reactions of either addition or loss of hydroxyl groups, methylation or isoprenylation, dimerization, bisulphate formation, and, what is more important, glycosylation of either the hydroxyls or the flavonic nucleus, may occur at different levels. All this, will lead to a great diversity of structures (about 2000 flavonoids are known at present) compiled in several revisions (Harborne and Mabry, 1982; Harborne *et al.*, 1975; Wollenweber and Dietz, 1981).

Flavonoids can be found as aglycones or, more frequently, as glycosides either O- or C-glycosides. Although any flavonoid position may be glycosylated, some have more probabilities, such as 7-OH of flavones, flavanones and isoflavones, 3-OH and 7-OH of flavonols and dihydroflavonols, and 6- and/or 8-C in C-glycosides, being glucose the most usual sugar found in them. Glycosylation, as well as methylation, occurs in the latest stages of biosynthesis and is catalyzed by high specific enzymes. Sometimes glycosides may have acyl-substituents linked to one or more hydroxyl groups of the sugars by an ester bond, or more rarely directly linked to the flavonoid molecule (Wollenweber, 1985b). Among these acyl-substituents there are aliphatic acids such as acetic, malonic, or succinic acid, and aromatic acids like benzoic, gallic, *p*-coumaric, and caffeic acid (Aguinagalde and Pero Martínez, 1982; Harborne, 1986; Wollenweber *et al.*, 1978).

The role of flavonoids and, in general, polyphenolic compounds in plants is not completely established. Their pigmentary function, responsible for the attraction of zoopollinators and zoodispersors, is well known. Other functions as antioxidants, antimutagenics, on plant growth regulation and on resistance to plant diseases have also been attributed to this group of natural polyphenols (Harborne, 1985; McClure, 1975).

Flavonoids, for their structural variability, their physiological and chemical stability, their wide distribution among plants and their relatively easy detection, constitute one of the main chemotaxonomic markers. Their importance in this respect has been the object of a large number of publications (Bate-Smith, 1962, 1963; Harborne, 1966, 1967). The more or less restricted distribution of specific types of flavonoids or substitution patterns in certain systematic groups is what determines their chemotaxonomic and possibly phylogenetic application (Harborne, 1975; Harborne and Turner, 1984; Swain, 1975). The latter is based on accepting the fact that plants which are able to synthesize structures placed in advanced stages of the biogenetic pathways will have a superior and more complex enzymatic supply. In general, evolution involves an increase of the number of flavonoid types present in each systematic group, and concomitantly their structural complexity increases.

Among Lamiaceae the presence of flavonoids is well known (Adzet and Martínez, 1981a; Barberán, 1986; Harborne, 1967; Hegnauer, 1966, 1989; Semrau, 1958; Zinchenko and Bandyukova, 1969). Especially during the last two decades several authors have revealed a wide range of substitution patterns with chemotaxonomic significance from both supra- and infrageneric point of view (Tomás-Barberán and Wollenweber, 1990; Tomás-Barberán *et al.*, 1988a,b). Particularly, the genus *Thymus* has been shown to have a noteworthy flavonoid composition of valuable taxonomic importance (Adzet and Martínez, 1981b; Adzet *et al.*, 1988; Hernández *et al.*, 1987; Litvinenko and Zoz, 1969; Martínez, 1980; Simonyan, 1972; Simonyan and Litvinenko, 1971; Vila, 1987). Thus, flavonoid aglycones play an important role to separate *T. capitatus* (now *Thymbra capitata*) from other species of this genus (Adzet and Martínez, 1981a,b; Barberán *et al.*, 1986); luteolin and 6-hydroxyluteolin, two important taxonomic markers (Semrau, 1958; Hegnauer, 1966; Harborne and Williams, 1971), have been found in several *Thymus* taxa (Adzet and Martínez, 1980a), while Section *Pseudothymbra* is characterized by a high content of methoxylated flavones (Adzet *et al.*, 1988).

FLAVONOID DISTRIBUTION AND OTHER POLYPHENOLS IN THE GENUS *THYMUS*

Within the genus *Thymus*, many more flavonoids, especially aglycones, have been described, than other polyphenols, which only include phenolic acids. In Tables 5.1, 5.2 and 5.3 the substitution patterns and frequency of flavonoid aglycones, flavonoid glycosides and phenolic acids found in the genus *Thymus* are shown, respectively. As it can be seen, the former have been largely investigated in thyme plants, while glycosides and phenolic acids have been studied to a lesser extent. A brief discussion of each table follows below.

Flavonoid aglycones

Among the aglycones (Table 5.1), 32 flavones, 4 flavanones, 2 flavonols and 2 dihydro-flavonols have been described, luteolin and apigenin being, by far, the more frequently found (in 100 and 99 taxa, respectively), followed by scutellarein (in 55 species). Up to now, no isoflavonoids have been reported in *Thymus* taxa.

Flavones

Concerning the substitution pattern of these compounds (Table 5.1), flavone aglycones found in the genus *Thymus* may have a 5,7-, 5,6,7-, 5,7,8- or 5,6,7,8- substituted A-ring, the C-5 position being always hydroxylated and the C-8 methoxylated, while the C-6 and C-7 positions may have both types of substitution. Most of the flavones found in *Thymus* species have methoxyl groups in the A- and/or B-rings. Some of them are highly methoxylated flavonoids, as for example: 5-desmethylnobiletin with five methoxy groups in 6, 7, 8, 3' and 4' (reported in 28 taxa), 8-OMe-cirsilineol with a 6,7,8,3'-(OMe)$_4$-substitution (reported in 31 taxa), 5-desmethylsinensetin with a 6,7,3',4'-(OMe)$_4$-substitution (found in 28 taxa), gardenin B with a 6,7,8,4'-(OMe)$_4$-substituted structure (reported in 5 taxa), and 5,6-(OH)$_2$-7,8,3',4'-(OMe)$_4$-flavone (described in only one taxon, *T. piperella*). The 5,7,8-trisubstituted A-ring has only been found once in 5,4'-(OH)$_2$-7,8-(OMe)$_2$-flavone (8-OMe-genkwanin), in *T. moroderi* (Vila, 1987).

Table 5.1 Substitution pattern and frequency of flavonoid aglycones found in *Thymus* sp.

-OH	-OMe	Name	N° of taxa
(A) Flavones and flavonols			

	-OH	-OMe	Name	N° of taxa
1	5,7	4'	Acacetin	9
2	5,7,4'	–	Apigenin	99
3	5,7,4'	3'	Chrysoeriol	1
4	5,4'	6,7,3'	Cirsilineol	32
5	5,3',4'	6,7	Cirsiliol	1
6	5,4'	6,7,8,3'-	8-OMe-Cirsilineol	31
7	5,4'	6,7	Cirsimaritin	33
8	5	6,7,8,3',4'	5-Desmethylnobiletin	28
9	5	6,7,3',4'	5-Desmethylsinensetin	28
10	5,7,3'	4'	Diosmetin	10
11	5	6,7,8,4'	Gardenin B	5
12	5,4'	7	Genkwanin	13
13	3,5,7,4'	–	Kaempferol	1
14	5,6	7,4'	Ladanein	1
15	5,7,3',4'	–	Luteolin	100
16	5,6,7,3',4'	–	6-OH-Luteolin	16
17	5,3',4'	7	7-OMe-Luteolin	2
18	5,3'	7,4'	Pilloin	1
19	5,6	7,8,4'	Pebrellin	1
20	3,5,7,3',4'	–	Quercetin	1
21	5	6,7,4'	Salvigenin	6

22	5,6,7,4′	–	Scutellarein	55
23	5,3′,4′	6,7,8	Sideritoflavone	30
24	5,6,4′	7	Sorbifolin	2
25	5,6,4′	7,8,3′	Thymonin	24
26	5,6,4′	7,8	Thymusin	29
27	5,4′	6,7,8	Xanthomicrol	34
28	5	7,4′	4′-OMe-Genkwanin	4
29	5,4′	7,8	8-OMe-Genkwanin	1
30	5,6	7,3′,4′	–	3
31	5,6	7,8,3′,4′	–	1
32	5,6,4′	7,3′	–	4

(B) Flavanones and dihydroflavonols

1	3,5,7,4′	–	Dihydrokaempferol	7
2	5,7,3′,4′	–	Eriodictyol	25
3	5,7,4′	–	Naringenin	23
4	5,4′	7	Sakuranetin	11
5	3,5,7,3′,4′	–	Taxifolin	13
6	5,4′	6,7,8	Dihydroxanthomicrol	8

(C) Anthocyanidins

| 1 | 3,5,7,3′,4′ | – | Cyanidin | 2 |

With respect to the B ring of thyme flavones, it can be 4′-monosubstituted or 3′,4′-disubstituted, either by hydroxy and/or methoxy groups.

Other flavones that are widely reported within the *Thymus* taxa investigated are (Table 5.1): xanthomicrol (in 34 taxa), cirsimaritin (in 33 taxa), cirsilineol (in 32 taxa), sideritoflavone (in 30 taxa), thymusin (in 29 taxa) and thymonin (in 24 taxa), whereas chrysoeriol, cirsiliol, ladanein, pilloin and pebrellin have only been described in one taxon.

Flavonols

Kaempferol and quercetin are the only flavonols described for *Thymus* species They have only been found once in *T. moroderi* (Vila, 1987) and *T. vulgaris* (Morimitsu *et al.*, 1995), respectively.

Flavanones and dihydroflavonols

Concerning to flavanones and dihydroflavonols, they are less widespread within the genus *Thymus* than flavones (Table 5.1). Among the former, eriodictyol, naringenin, sakuranetin and dihydroxanthomicrol have been reported in 25, 23, 11 and 8 taxa, respectively, while only dihydrokaempferol and taxifolin have been described among the latter, in 8 and 7 taxa, respectively. The A- and B-rings of these flavonoids are mainly characterized by having a 5,7,3'- and/or 4'-substitution pattern, these substituents usually being hydroxy groups, except in sakuranetin, which has a methoxy group at C-7 position, and in dihydroxanthomicrol, which represents a $5,4'-(OH)_2-6,7,8-(OMe)_3$-flavanone.

Acylated flavonoid aglycones

Although acylated flavonoid aglycones have been rarely described in the Labiatae family, unusual 8-C-*p*-hydroxybenzyl-derivatives of several flavone (apigenin, luteolin, diosmetin) and flavonol (quercetin, kaempferol) aglycones have been identified in *T. hirtus*, an Algerian taxon (Merghem *et al.*, 1995). This is the only report of acyl substituents directly linked to the flavonoid skeleton in the genus *Thymus*.

Anthocyanidins

Anthocyanidins and other flavonoid-related structures have scarcely been found in the genus *Thymus*, the anthocyanidin cyanidin (Table 5.1) being the only one reported in two species: *T. pulegioides* and *T. vulgaris* (Stoess, 1972).

Flavonoid glycosides

Flavonoid glycosides of *Thymus* have been less intensively investigated than aglycones. Only 16 different structures of this group of flavonoids, particularly hydroxylated flavone-glycosides, have been reported in the reviewed literature (Table 5.2), all of them being *O*-glycosides with one exception. Those derived from luteolin and apigenin are the most widespread, especially luteolin-7-*O*-glucoside and apigenin-7-*O*-glucoside, which have been found in ten and eight species, respectively. All the other flavonoid-*O*-glycosides included in Table 5.2 have been reported in only one or two species of *Thymus*, mainly *T. membranaceus* (Tomás *et al.*, 1985), *T. moroderi* (Vila, 1987) and *T. serpyllum* (Olechnowicz-Stepien and Lamer-Zarawska, 1975). Flavonol and isoflavonoid glycosides have not been previously reported in this genus.

The flavonoid aglycone more frequently found in its glycosidic form is luteolin, from which nine different 7-*O*-glycosides have been described in *Thymus*. More rarely,

Table 5.2 Substitution pattern and frequency of flavonoid glycosides in *Thymus* sp.

	Structures	Aglycone	N° taxa
1	7-O-glucoside	Apigenin	8
2	4'-O-p-cumaroyl-glucoside	Apigenin	1
3	7-O-glucuronide	Diosmetin	1
4	3'-O-alloside	Luteolin	1
5	galactoarabinoside	Luteolin	1
6	7-O-glucoside	Luteolin	10
7	7-O-diglucoside	Luteolin	2
8	7-O-glucuronide	Luteolin	2
9	7-O-neohesperidoside	Luteolin	1
10	7-O-rutinoside	Luteolin	2
11	7-O-sambubioside	Luteolin	1
12	7-O-xyloside	Luteolin	2
13	7-O-glucoside	6-OH-Luteolin	2
14	glucosylglucuronide	Scutellarein	1
15	7-O-glucosyl(1– 4)rhamnoside	Scutellarein	1
16	6,8-di-C-glucoside (Vicenin-2)	Apigenin	20

6-OH-luteolin-, scutellarein- and diosmetin-derived glycosides have been also identified. Glycosides from A-ring methoxylated flavones have not been reported in thyme plants.

Sugars found in *Thymus* flavonoid glycosides are usually linked to the aglycone through the hydroxy group in the C-7 position of the flavonoid skeleton, being either a monosaccharide (glucoside, xyloside, alloside, glucuronide) or a disaccharide (diglucoside, galactoarabinoside, neohesperidoside, rutinoside, sambubioside, glucosylglucuronide, glucosylrhamnoside). Sometimes the linkage data of the sugar chain are not completely described by the authors. The presence of acyl-substituents in the sugar moiety has been reported only once, particularly apigenin-4'-O-p-cumaroyl-glucoside, in *T. serpyllum* (Washington and Saxena, 1983).

Concerning C-glycosides, vicenin-2 (apigenin-6,8-di-C-glucoside) is the only one found in *Thymus* sp. It is the flavonoid glycoside most frequently reported in this genus, particularly in 20 taxa (Table 5.2). It has been observed that while flavone O-glycosides occur ubiquitously among several Labiatae genera, vicenin-2 occurred only in certain taxonomic groups, for instance, in the sections *Pseudothymbra* and *Thymus* of the genus *Thymus* (Husain and Markham, 1981).

Extraction procedures used as well as difficulties in structure elucidation, particularly sugar linkages, may have been the cause of the too few results reported on flavonoid glycosides in *Thymus* compared with those on aglycones.

Phenolic acids

Nine different phenolic acids have been reported in the genus *Thymus* (Table 5.3), caffeic and rosmarinic acids being those more frequently found (in 29 and 20 species, respectively). The others have only been detected in one or two taxa. Particularly, chlorogenic, p-coumaric, 3,5-dicaffeoylquinic, protocatechuic and syringic acid have been identified in *T. webbianus* (Blázquez *et al.*, 1994), while caffeic, p-coumaric, p-hydroxybenzoic, syringic and vanillic acid have been found in *T. carnosus* (Marhuenda *et al.*, 1987b).

Table 5.3 Frequency of phenolic acids in *Thymus* sp.

	Name	N° taxa
1	Caffeic acid	29
2	Chlorogenic acid	1
3	*p*-Coumaric acid	2
4	3,5-Dicaffeoylquinic acid	1
5	*p*-Hydroxybenzoic acid	1
6	Protocatechuic acid	1
7	Rosmarinic acid	20
8	Syringic acid	2
9	Vanillic acid	1

Compilation of flavonoids and phenolic acids in *Thymus*

Table 5.4 includes a review of the literature published on the polyphenols found in *Thymus* species between 1969 and 1999, with some significative preliminary references from 1958 and 1959. This wide group of secondary metabolites has been investigated in 120 *Thymus* taxa. Usually, leaves are the part of the plant investigated, although in some cases the plant material is not well indicated by the authors.

It is important to consider that the information given in this table comes from very heterogeneous sources. First, it has to be taken into account that the authors have used different extraction methods which determine the compounds to be found in the final extracts. Thus, in some cases extraction was performed with solvents of increasing polarity or more frequently, after removing the more lipophilic compounds with petroleum ether or hexane, a methanolic or hydroalcoholic extract was obtained and successively partitioned by solvents of increasing polarity. In addition, the more polar extracts were sometimes submitted to acidic hydrolysis, consequently the free flavonoid aglycones found probably occur in their glycosidic form in the plant. Sometimes only the flavonoid aglycones from leaf surfaces (exudate flavonoids) were investigated, in these cases the plant material was rinsed with lipophilic solvents such as chloroform. As a matter of fact, the extraction procedure obviously restricts the compounds to be found, meaning that those flavonoids which are not described in a species are not necessarily absent.

Second, while some authors isolated the flavonoids or other polyphenolic constituents and determined their structures from their spectroscopic data (UV/Vis, MS and/or nuclear magnetic resonance (NMR) data), others analysed the extracts by high-performance liquid chromatography (HPLC) and/or several thin-layer chromatography (TLC) systems, comparing retention indices with those of reference substances previously isolated from other extracts.

EXUDATE FLAVONOID AGLYCONES: INFLUENCE OF THE HABITAT AND GENETIC FACTORS ON THEIR PRODUCTION

The exudate flavonoid aglycones from leaves and stems surfaces have been shown to play an important ecological role. Particularly, a correlation between the preferred habitat of the plant and the production of excreted flavonoids has been reported, the species from (semi-)arid habitats being those which generally accumulate external flavonoids (Tomás-Barberán and Wollenweber, 1990; Wollenweber, 1985a). These are usually

Table 5.4 Polyphenols in *Thymus* sp.

Species	Polyphenols	References
T. aestivus Reut. *ex* Willk.	Apigenin	Adzet *et al.*, 1988
	Cirsilineol	Adzet *et al.*, 1988
	8-OMe-Cirsilineol	Adzet *et al.*, 1988
	Cirsimaritin*	Adzet *et al.*, 1988
	5-Desmethylnobiletin	Adzet *et al.*, 1988
	2,3-Dihydrokaempferol	Adzet *et al.*, 1988
	2,3-Dihydroxanthomicrol	Adzet *et al.*, 1988
	Eriodictyol	Adzet *et al.*, 1988
	Genkwanin	Adzet *et al.*, 1988
	Luteolin	Adzet *et al.*, 1988
	Naringenin	Adzet *et al.*, 1988
	Sakuranetin	Adzet *et al.*, 1988
	Salvigenin	Adzet *et al.*, 1988
	Sideritoflavone	Adzet *et al.*, 1988
	Taxifolin	Adzet *et al.*, 1988
	Xanthomicrol	Adzet *et al.*, 1988
	Caffeic acid	Adzet *et al.*, 1988
	Rosmarinic acid	Adzet *et al.*, 1988
T. albanus H. Braun	Apigenin	Kulevanova *et al.*, 1997
	Luteolin	Kulevanova *et al.*, 1997
	Caffeic acid	Kulevanova *et al.*, 1997
T. algeriensis Boiss.	Eriodictyol	El-Domiaty *et al.*, 1997
	Taxifolin	El-Domiaty *et al.*, 1997
	5,6-$(OH)_2$-7,3',4'-$(OMe)_3$-flavone	El-Domiaty *et al.*, 1997
	5,6,4'-$(OH)_3$-7,3'-$(OMe)_2$-flavone	El-Domiaty *et al.*, 1997
T. alsarensis Ronn.	Apigenin	Kulevanova *et al.*, 1997
	Eriodictyol	Kulevanova *et al.*, 1997
	Luteolin	Kulevanova *et al.*, 1997
	Caffeic acid	Kulevanova *et al.*, 1997
	Rosmarinic acid	Kulevanova *et al.*, 1997
T. alternus	Apigenin	Litvinenko and Zoz, 1969
	Luteolin	Litvinenko and Zoz, 1969
	Scutellarein	Litvinenko and Zoz, 1969
T. amictus Klok.	Apigenin	Litvinenko and Zoz, 1969
	Luteolin	Litvinenko and Zoz, 1969
	Scutellarein	Litvinenko and Zoz, 1969
T. antoninae Rouy et Coincy	Apigenin	Adzet *et al.*, 1988
	Cirsilineol†	Adzet *et al.*, 1988; Hernández *et al.*, 1987
	8-OMe-Cirsilineol‡	Adzet *et al.*, 1988; Hernández *et al.*, 1987
	Cirsimaritin	Adzet *et al.*, 1988; Hernández *et al.*, 1987
	5-Desmethylnobiletin	Adzet *et al.*, 1988; Hernández *et al.*, 1987
	5-Desmethylsinensetin	Hernández *et al.*, 1987
	2,3-Dihydroxanthomicrol	Adzet *et al.*, 1988
	Eriodictyol	Adzet *et al.*, 1988
	Genkwanin	Adzet *et al.*, 1988
	Luteolin	Adzet *et al.*, 1988
	Naringenin	Adzet *et al.*, 1988
	Sakuranetin	Adzet *et al.*, 1988
	Sideritoflavone	Adzet *et al.*, 1988; Hernández *et al.*, 1987
	Taxifolin	Adzet *et al.*, 1988

Table 5.4 (Continued)

Species	Polyphenols	References
T. antoninae Rouy et Coincy (Continued)	Thymusin	Hernández *et al.*, 1987
	Xanthomicrol	Adzet *et al.*, 1988; Hernández *et al.*, 1987
	Vicenin-2	Husain and Markham, 1981
	Caffeic acid	Adzet *et al.*, 1988
	Rosmarinic acid	Adzet *et al.*, 1988
T. aranjuezii Jalas (*T. lacaitae* Pau)	Cirsilineol	Hernández *et al.*, 1987
	8-OMe-Cirsilineol	Hernández *et al.*, 1987
	Cirsimaritin	Adzet and Martínez, 1981b; Hernández *et al.*, 1987
	5-Desmethylnobiletin	Hernández *et al.*, 1987
	5-Desmethylsinensetin	Hernández *et al.*, 1987
	Luteolin	Adzet and Martínez, 1981b
	6-OH-Luteolin	Adzet and Martínez, 1981b
	Sideritoflavone	Hernández *et al.*, 1987
	Thymonin	Hernández *et al.*, 1987
	Thymusin	Hernández *et al.*, 1987
	Xanthomicrol	Adzet and Martínez 1981b; Hernández *et al.*, 1987
T. ararati-minoris Klok. et Schost.	Apigenin	Simonyan and Litvinenko, 1971
	Luteolin	Simonyan and Litvinenko, 1971
	Scutellarein	Simonyan and Litvinenko, 1971
T. attenuatus	Apigenin	Litvinenko and Zoz, 1969
	Luteolin	Litvinenko and Zoz, 1969
	Scutellarein	Litvinenko and Zoz, 1969
T. baeticus Boiss. *ex* Lacaitae	Apigenin	Adzet and Martínez, 1981b; Adzet *et al.*, 1988
	Cirsilineol	Adzet *et al.*, 1988; Hernández *et al.*, 1987
	8-OMe-Cirsilineol	Adzet *et al.*, 1988; Hernández *et al.*, 1987
	Cirsimaritin	Adzet and Martínez 1980b, 1981b; Adzet *et al.*, 1988; Hernández *et al.*, 1987
	5-Desmethylnobiletin	Adzet *et al.*, 1988; Hernández *et al.*, 1987
	5-Desmethylsinensetin	Hernández *et al.*, 1987
	2,3-Dihydrokaempferol	Adzet *et al.*, 1988
	Eriodictyol	Adzet *et al.*, 1988
	Genkwanin	Adzet and Martínez, 1980b, 1981b; Adzet *et al.*, 1988
	Luteolin	Adzet and Martínez, 1980a, 1981b; Adzet *et al.*, 1988
	6-OH-Luteolin	Adzet and Martínez, 1980a, 1981b
	Naringenin	Adzet *et al.*, 1988
	Sakuranetin	Adzet *et al.*, 1988
	Sideritoflavone	Adzet *et al.*, 1988; Hernández *et al.*, 1987
	Taxifolin	Adzet *et al.*, 1988
	Thymonin	Hernández *et al.*, 1987
	Thymusin	Hernández *et al.*, 1987
	Xanthomicrol	Adzet and Martínez, 1980b, 1981b; Adzet *et al.*, 1988; Hernández *et al.*, 1987
	Vicenin-2	Husain and Markham, 1981
	Caffeic acid	Adzet *et al.*, 1988
	Rosmarinic acid	Adzet *et al.*, 1988
T. balcanus Borb.	Apigenin	Kulevanova *et al.*, 1997
	Luteolin	Kulevanova *et al.*, 1997
	Caffeic acid	Kulevanova *et al.*, 1997

T. bashkiriensis Klok. et Schost.	Apigenin	Kurkin *et al.*, 1988
	Luteolin	Kurkin *et al.*, 1988
	Caffeic acid	Kurkin *et al.*, 1988
	Rosmarinic acid	Kurkin *et al.*, 1988
T. borysthenicus Klok. et Schost.	Apigenin	Litvinenko and Zoz, 1969
	Luteolin	Litvinenko and Zoz, 1969
	Scutellarein	Litvinenko and Zoz, 1969
T. bracteatus Lange *ex* Cutanda	Cirsilineol	Hernández *et al.*, 1987
	8-OMe-Cirsilineol	Hernández *et al.*, 1987
	Cirsimaritin	Hernández *et al.*, 1987
	5-Desmethylnobiletin	Hernández *et al.*, 1987
	5-Desmethylsinensetin	Hernández *et al.*, 1987
	Sideritoflavone	Hernández *et al.*, 1987
	Thymonin	Hernández *et al.*, 1987
	Thymusin	Hernández *et al.*, 1987
	Xanthomicrol	Hernández *et al.*, 1987
T. caespititius Brot.	Apigenin	Adzet and Martínez, 1981b
	Cirsilineol	Hernández *et al.*, 1987
	8-OMe-Cirsilineol	Hernández *et al.*, 1987
	Luteolin	Adzet and Martínez, 1980a, 1981b
	6-OH-Luteolin	Adzet and Martínez, 1980a, 1981b
	Thymonin	Hernández *et al.*, 1987
	Thymusin	Hernández *et al.*, 1987
	Xanthomicrol	Hernández *et al.*, 1987
T. calcareus Klok. et Schost.	Luteolin	Litvinenko and Zoz, 1969
	Scutellarein	Litvinenko and Zoz, 1969
T. callieri Borbás *ex* Velen.	Acacetin	Litvinenko and Zoz, 1969
	Apigenin	Litvinenko and Zoz, 1969
	Luteolin	Litvinenko and Zoz, 1969
	Scutellarein	Litvinenko and Zoz, 1969
T. camphoratus Hoffmanns. et Link	Apigenin	Adzet and Martínez, 1981b; Adzet *et al.*, 1988
	Cirsilineol	Hernández *et al.*, 1987
	8-OMe-Cirsilineol	Hernández *et al.*, 1987
	Cirsimaritin	Hernández *et al.*, 1987
	5-Desmethylnobiletin	Hernández *et al.*, 1987
	5-Desmethylsinensetin	Hernández *et al.*, 1987
	Luteolin	Adzet and Martínez, 1981b; Adzet *et al.*, 1988
	6-OH-Luteolin	Adzet and Martínez, 1981b
	Naringenin	Adzet *et al.*, 1988
	Sideritoflavone	Hernández *et al.*, 1987
	Taxifolin	Adzet *et al.*, 1988
	Thymonin	Hernández *et al.*, 1987
	Thymusin	Hernández *et al.*, 1987
	Xanthomicrol	Adzet *et al.*, 1988; Hernández *et al.*, 1987
	Vicenin-2	Husain and Markham, 1981
	Caffeic acid	Adzet *et al.*, 1988
	Rosmarinic acid	Adzet *et al.*, 1988
T. capitellatus Hoffmanns. et Link	Apigenin	Adzet *et al.*, 1988
	Cirsilineol	Hernández *et al.*, 1987
	8-OMe-Cirsilineol	Hernández *et al.*, 1987
	Cirsimaritin	Hernández *et al.*, 1987
	5-Desmethylnobiletin	Hernández *et al.*, 1987

Table 5.4 (Continued)

Species	Polyphenols	References
T. capitellatus Hoffmanns. et Link (Continued)	5-Desmethylsinensetin	Hernández *et al.*, 1987
	Eriodictyol	Adzet *et al.*, 1988
	Luteolin	Adzet *et al.*, 1988
	Naringenin	Adzet *et al.*, 1988
	Sideritoflavone	Hernández *et al.*, 1987
	Taxifolin	Adzet *et al.*, 1988
	Thymonin	Hernández *et al.*, 1987
	Thymusin	Hernández *et al.*, 1987
	Xanthomicrol	Adzet *et al.*, 1988; Hernández *et al.*, 1987
	Vicenin-2	Husain and Markham, 1981
	Caffeic acid	Adzet *et al.*, 1988
	Rosmarinic acid	Adzet *et al.*, 1988
T. carnosus Boiss.	Apigenin	Marhuenda *et al.*, 1987a
	Cirsilineol	Hernández *et al.*, 1987; Marhuenda *et al.*, 1987a
	8-OMe-Cirsilineol	Hernández *et al.*, 1987
	Cirsimaritin	Hernández *et al.*, 1987
	5-Desmethylnobiletin	Hernández *et al.*, 1987
	5-Desmethylsinensetin	Hernández *et al.*, 1987
	Luteolin	Marhuenda *et al.*, 1987a
	6-OH-Luteolin	Marhuenda *et al.*, 1987a
	Sideritoflavone	Hernández *et al.*, 1987
	Thymonin	Hernández *et al.*, 1987
	Thymusin	Hernández *et al.*, 1987
	Xanthomicrol	Hernández *et al.*, 1987
	Caffeic acid	Marhuenda *et al.*, 1987b
	p-Coumaric acid	Marhuenda *et al.*, 1987b
	p-Hydroxybenzoic acid	Marhuenda *et al.*, 1987b
	Syringic acid	Marhuenda *et al.*, 1987b
	Vanillic acid	Marhuenda *et al.*, 1987b
T. caucasicus Willd.	Apigenin	Simonyan and Litvinenko, 1971
	Luteolin	Simonyan and Litvinenko, 1971
	Scutellarein	Simonyan and Litvinenko, 1971
T. cephalotos L.	Vicenin-2	Husain and Markham, 1981
T. cherlerioides Vis.	Vicenin-2	Husain and Markham, 1981
T. ciliatissimus Klok.	Apigenin	Litvinenko and Zoz, 1969
	Luteolin	Litvinenko and Zoz, 1969
	Scutellarein	Litvinenko and Zoz, 1969
T. cinerascens	Apigenin	Semrau, 1958
T. circumcinctus Klok.	Apigenin	Litvinenko and Zoz, 1969
	Luteolin	Litvinenko and Zoz, 1969
	Scutellarein	Litvinenko and Zoz, 1969
T. collinus Bieb.	Apigenin	Simonyan and Litvinenko, 1971
	Luteolin	Simonyan and Litvinenko, 1971
	Scutellarein	Simonyan and Litvinenko, 1971
	Apigenin-7-*O*-β-D-glucoside	Simonyan, 1972
	Luteolin-7-*O*-β-D-glucoside	Simonyan, 1972
T. cretaceus Klok. et Schost.	Apigenin	Litvinenko and Zoz, 1969
	Luteolin	Litvinenko and Zoz, 1969

	Scutellarein	Litvinenko and Zoz, 1969
T. czernajevii Klok. et Schost.	Apigenin	Litvinenko and Zoz, 1969
	Luteolin	Litvinenko and Zoz, 1969
	Scutellarein	Litvinenko and Zoz, 1969
T. dagestanicus Klok. et Schost.	Apigenin	Litvinenko and Zoz, 1969
	Luteolin	Litvinenko and Zoz, 1969
	Scutellarein	Litvinenko and Zoz, 1969
T. decussatus L.	Apigenin	Khodair *et al.*, 1993
	Thymonin	Khodair *et al.*, 1993
	5,6,4'-(OH)$_3$-7,3'-(OMe)$_2$-flavone	Khodair *et al.*, 1993
T. desjatovae Ronn.	Apigenin	Litvinenko and Zoz, 1969
	Luteolin	Litvinenko and Zoz, 1969
	Scutellarein	Litvinenko and Zoz, 1969
T. dolopicus Form.	Vicenin-2	Husain and Markham, 1981
T. dimorphus Klok. et Schost.	Acacetin	Litvinenko and Zoz, 1969
	Apigenin	Litvinenko and Zoz, 1969; Simonyan and Litvinenko, 1971
	Luteolin	Litvinenko and Zoz, 1969; Simonyan and Litvinenko, 1971
	Scutellarein	Litvinenko and Zoz, 1969; Simonyan and Litvinenko, 1971
T. dzevanovskyi Klok. et Schost.	Apigenin	Litvinenko and Zoz, 1969
	Luteolin	Litvinenko and Zoz, 1969
	Scutellarein	Litvinenko and Zoz, 1969
T. elisabethae Klok. et Schost.	Apigenin	Simonyan and Litvinenko, 1971
	Scutellarein	Simonyan and Litvinenko, 1971
T. eupatoriensis Klok. et Schost.	Scutellarein	Litvinenko and Zoz, 1969
T. fontqueri (Jalas) Molero et Rovira	Cirsilineol	Hernández *et al.*, 1987
	8-OMe-Cirsilineol	Hernández *et al.*, 1987
	Cirsimaritin	Hernández *et al.*, 1987
	5-Desmethylnobiletin	Hernández *et al.*, 1987
	5-Desmethylsinensetin	Hernández *et al.*, 1987
	Diosmetin	Hernández *et al.*, 1987
	Gardenin-B	Hernández *et al.*, 1987
	Salvigenin	Hernández *et al.*, 1987
	Sideritoflavone	Hernández *et al.*, 1987
	Thymonin	Hernández *et al.*, 1987
	Thymusin	Hernández *et al.*, 1987
	Xanthomicrol	Hernández *et al.*, 1987
T. fominii Klok. et Schost.	Apigenin	Simonyan and Litvinenko, 1971
	Luteolin	Simonyan and Litvinenko, 1971
	Scutellarein	Simonyan and Litvinenko, 1971
T. funkii Coss.	Apigenin	Adzet *et al.*, 1988
	Cirsilineol	Adzet *et al.*, 1988; Hernández *et al.*, 1987
	8-OMe-Cirsilineol	Adzet *et al.*, 1988; Hernández *et al.*, 1987
	Cirsimaritin	Adzet *et al.*, 1988; Hernández *et al.*, 1987
	5-Desmethylnobiletin	Adzet *et al.*, 1988; Hernández *et al.*, 1987
	5-Desmethylsinensetin	Hernández *et al.*, 1987
	2,3-Dihydroxanthomicrol	Adzet *et al.*, 1988
	Eriodictyol	Adzet *et al.*, 1988

Table 5.2 (Continued)

Species	Polyphenols	References
T. funkii Coss. (Continued)	Genkwanin	Adzet *et al.*, 1988
	Luteolin	Adzet *et al.*, 1988
	Naringenin	Adzet *et al.*, 1988
	Sakuranetin	Adzet *et al.*, 1988
	Salvigenin	Adzet *et al.*, 1988
	Sideritoflavone	Adzet *et al.*, 1988; Hernández *et al.*, 1987
	Thymusin	Hernández *et al.*, 1987
	Xanthomicrol	Adzet *et al.*, 1988; Hernández *et al.*, 1987
	Caffeic acid	Adzet *et al.*, 1988
T. glandulosus Lag. *ex* H. Del Villar	Apigenin	Adzet *et al.*, 1988
	Cirsilineol	Adzet *et al.*, 1988
	8-OMe-Cirsilineol	Adzet *et al.*, 1988
	Cirsimaritin	Adzet *et al.*, 1988
	2,3-Dihydroxanthomicrol	Adzet *et al.*, 1988
	Eriodictyol	Adzet *et al.*, 1988
	Luteolin	Adzet *et al.*, 1988
	Naringenin	Adzet *et al.*, 1988
	Sakuranetin	Adzet *et al.*, 1988
	Sideritoflavone	Adzet *et al.*, 1988
	Taxifolin	Adzet *et al.*, 1988
	Xanthomicrol	Adzet *et al.*, 1988
	Caffeic acid	Adzet *et al.*, 1988
	Rosmarinic acid	Adzet *et al.*, 1988
T. granatensis Boiss.	Apigenin	Adzet and Martínez, 1981b
	Luteolin	Adzet and Martínez, 1981b
	6-OH-Luteolin	Adzet and Martínez, 1981b
T. graniticus Klok. et Schost.	Acacetin	Litvinenko and Zoz, 1969
	Apigenin	Litvinenko and Zoz, 1969
	Luteolin	Litvinenko and Zoz, 1969
	Scutellarein	Litvinenko and Zoz, 1969
T. herba-barona Loisel.	Apigenin	Corticchiato *et al.*, 1995
	Cirsiliol	Corticchiato *et al.*, 1995
	Cirsilineol	Corticchiato *et al.*, 1995
	8-OMe-Cirsilineol	Corticchiato *et al.*, 1995
	Cirsimaritin	Corticchiato *et al.*, 1995
	Eriodictyol	Corticchiato *et al.*, 1995
	Genkwanin	Corticchiato *et al.*, 1995
	Luteolin	Corticchiato *et al.*, 1995
	Naringenin	Corticchiato *et al.*, 1995
	Sideritoflavone	Corticchiato *et al.*, 1995
	Sorbifolin	Corticchiato *et al.*, 1995
	Thymusin	Corticchiato *et al.*, 1995
	Xanthomicrol	Corticchiato *et al.*, 1995
T. hirsutus Bleb.	Acacetin	Litvinenko and Zoz, 1969
	Apigenin	Litvinenko and Zoz, 1969; Semrau, 1958
	Luteolin	Litvinenko and Zoz, 1969
	Scutellarein	Litvinenko and Zoz, 1969
T. hirtellus	Acacetin	Litvinenko and Zoz, 1969
	Apigenin	Litvinenko and Zoz, 1969

	Luteolin	Litvinenko and Zoz, 1969
	Scutellarein	Litvinenko and Zoz, 1969
T. hirtus Willd.	Apigenin	Merghem *et al.*, 1995
	Diosmetin	Merghem *et al.*, 1995
	Luteolin	Merghem *et al.*, 1995
	8-C-*p*-Hydroxybenzyl-apigenin	Merghem *et al.*, 1995
	8-C-*p*-Hydroxybenzyl-diosmetin	Merghem *et al.*, 1995
	8-C-*p*-Hydroxybenzyl-kaempferol	Merghem *et al.*, 1995
	8-C-*p*-Hydroxybenzyl-luteolin	Merghem *et al.*, 1995
	8-C-*p*-Hydroxybenzyl-quercetin	Merghem *et al.*, 1995
	Vicenin-2	Husain and Markham, 1981
T. hyemalis Lange	Apigenin	Adzet and Martínez, 1981b
	Cirsilineol	Hernández *et al.*, 1987
	8-OMe-Cirsilineol	Hernández *et al.*, 1987
	Cirsimaritin	Adzet and Martínez, 1981b; Hernández *et al.*, 1987
	5-Desmethylnobiletin	Hernández *et al.*, 1987
	5-Desmethylsinensetin	Hernández *et al.*, 1987
	Genkwanin	Adzet and Martínez, 1981b
	Luteolin	Adzet and Martínez, 1980a, 1981b
	6-OH-Luteolin	Adzet and Martínez, 1980a, 1981b
	Sideritoflavone	Hernández *et al.*, 1987
	Thymonin	Hernández *et al.*, 1987
	Thymusin	Hernández *et al.*, 1987
	Xanthomicrol	Adzet and Martínez, 1981b; Hernández *et al.*, 1987
T. jajlae (Klok. et Schost.) Starkov	Acacetin	Litvinenko and Zoz, 1969
	Luteolin	Litvinenko and Zoz, 1969
	Scutellarein	Litvinenko and Zoz, 1969
	Vicenin-2	Husain and Markham, 1981
T. jankae var. *jankae* Celak.	Apigenin	Kulevanova *et al.*, 1997, 1998
	Diosmetin	Kulevanova *et al.*, 1997, 1998
	Luteolin	Kulevanova *et al.*, 1998
	Naringenin	Kulevanova *et al.*, 1998
	Caffeic acid	Kulevanova *et al.*, 1997, 1998
T. jankae var. *pantotrichus* Ronn.	Apigenin	Kulevanova *et al.*, 1997, 1998
	Diosmetin	Kulevanova *et al.*, 1997, 1998
	Luteolin	Kulevanova *et al.*, 1998
	Naringenin	Kulevanova *et al.*, 1998
	Caffeic acid	Kulevanova *et al.*, 1997, 1998
T. jankae var. *patentipilus* Lyka	Apigenin	Kulevanova *et al.*, 1998
	Diosmetin	Kulevanova *et al.*, 1998
	Eriodyctiol	Kulevanova *et al.*, 1998
	Luteolin	Kulevanova *et al.*, 1998
	Naringenin	Kulevanova *et al.*, 1998
	Caffeic acid	Kulevanova *et al.*, 1998
T. kalmiussicus Klok. et Schost.	Acacetin	Litvinenko and Zoz, 1969
	Apigenin	Litvinenko and Zoz, 1969
	Luteolin	Litvinenko and Zoz, 1969
	Scutellarein	Litvinenko and Zoz, 1969

Table 5.4 (Continued)

Species	Polyphenols	References
T. karamarjanicus Klok. et Schost.	Apigenin	Simonyan and Litvinenko, 1971
	Luteolin	Simonyan and Litvinenko, 1971
T. kostchyanus Boiss. et Hohen.	Apigenin	Simonyan and Litvinenko, 1971
	Luteolin	Simonyan and Litvinenko, 1971
	Scutellarein	Simonyan and Litvinenko, 1971
T. latifolius (Besser) Andrz.	Apigenin	Litvinenko and Zoz, 1969
	Luteolin	Litvinenko and Zoz, 1969
	Scutellarein	Litvinenko and Zoz, 1969
T. leptophyllus Lange	Cirsilineol	Hernández *et al.*, 1987
	8-OMe-Cirsilineol	Hernández *et al.*, 1987
	Cirsimaritin	Hernández *et al.*, 1987
	5-Desmethylnobiletin	Hernández *et al.*, 1987
	5-Desmethylsinensetin	Hernández *et al.*, 1987
	Sideritoflavone	Hernández *et al.*, 1987
	Thymonin	Hernández *et al.*, 1987
	Thymusin	Hernández *et al.*, 1987
	Xanthomicrol	Hernández *et al.*, 1987
T. leucotrichus Halácsy	Vicenin-2	Husain and Markham, 1981
T. littoralis Klok. et Schost.	Apigenin	Litvinenko and Zoz, 1969
	Luteolin	Litvinenko and Zoz, 1969
	Scutellarein	Litvinenko and Zoz, 1969
T. loevianus Opiz	Apigenin	Litvinenko and Zoz, 1969
	Luteolin	Litvinenko and Zoz, 1969
	Scutellarein	Litvinenko and Zoz, 1969
T. longidens var. *dassareticus* Ronn.	Apigenin	Kulevanova *et al.*, 1997
	Diosmetin	Kulevanova *et al.*, 1997
	Eriodictyol	Kulevanova *et al.*, 1997
	Naringenin	Kulevanova *et al.*, 1997
	Caffeic acid	Kulevanova *et al.*, 1997
T. longidens var. *lanicaulis* Ronn.	Apigenin	Kulevanova *et al.*, 1997, 1998
	Eriodyctiol	Kulevanova *et al.*, 1997, 1998
	Luteolin	Kulevanova *et al.*, 1997, 1998
	Naringenin	Kulevanova *et al.*, 1997, 1998
	Caffeic acid	Kulevanova *et al.*, 1997, 1998
	Rosmarinic acid	Kulevanova *et al.*, 1997, 1998
T. longiflorus Boiss.	Apigenin	Adzet *et al.*, 1988
	Cirsilineol	Adzet *et al.*, 1988; Hernández *et al.*, 1987
	8-OMe-Cirsilineol	Adzet et al., 1988; Hernández *et al.*, 1987
	Cirsimaritin	Adzet *et al.*, 1988; Hernández *et al.*, 1987
	5-Desmethylnobiletin	Adzet *et al.*, 1988; Hernández *et al.*, 1987
	5-Desmethylsinensetin	Hernández *et al.*, 1987
	2,3-Dihydrokaempferol	Adzet *et al.*, 1988
	2,3-Dihydroxanthomicrol	Adzet *et al.*, 1988
	Eriodictyol	Adzet *et al.*, 1988
	Genkwanin	Adzet *et al.*, 1988
	4'-O-Me-Genkw anin	Adzet *et al.*, 1988
	Luteolin	Adzet *et al.*, 1988
	Naringenin	Adzet *et al.*, 1988
	Sakuranetin	Adzet *et al.*, 1988

	Sideritoflavone	Adzet *et al.*, 1988; Hernández *et al.*, 1987
	Taxifolin	Adzet *et al.*, 1988
	Thymusin	Hernández *et al.*, 1987
	Xanthomicrol	Adzet *et al.*, 1988; Hernández *et al.*, 1987
	Vicenin-2	Husain and Markham, 1981
	Caffeic acid	Adzet *et al.*, 1988
	Rosmarinic acid	Adzet *et al.*, 1988
T. loscosii Willk.	Luteolin	Adzet and Martínez, 1980a, 1981b
	6-OH-Luteolin	Adzet and Martínez, 1980a, 1981b
	Luteolin-7-*O*-β-D-glucoside	Adzet *et al.*, 1982
	6-OH-Luteolin-7-*O*-β-D-glucoside	Adzet *et al.*, 1982
T. macedonicus (Deg. et Ur.) Ronn.	Apigenin	Kulevanova *et al.*, 1997
	Eriodictyol	Kulevanova *et al.*, 1997
	Luteolin	Kulevanova *et al.*, 1997
	Caffeic acid	Kulevanova *et al.*, 1997
	Rosmarinic acid	Kulevanova *et al.*, 1997
T. marschallianus Willd.	Apigenin	Litvinenko and Zoz, 1969; Simonyan and Litvinenko, 1971
	Luteolin	Litvinenko and Zoz, 1969; Simonyan and Litvinenko, 1971
	Scutellarein	Litvinenko and Zoz, 1969
	Apigenin-7-*O*-β-D-glucoside	Simonyan, 1972
	Luteolin-7-*O*-β-D-glucoside	Simonyan, 1972
T. mastichina L.	Apigenin	Adzet and Martínez, 1981b
	Cirsimaritin	Hernández *et al.*, 1987
	5-Desmethylnobiletin	Hernández *et al.*, 1987
	5-Desmethylsinensetin	Hernández *et al.*, 1987
	Genkwanin	Adzet and Martínez, 1981b
	Luteolin	Adzet and Martínez, 1980a, 1981b
	6-OH-Luteolin	Adzet and Martínez, 1980a, 1981b
	Thymusin	Hernández *et al.*, 1987
	Xanthomicrol	Hernández *et al.*, 1987
T. mastigophorus Lacaita	Cirsilineol	Hernández *et al.*, 1987
	8-OMe-Cirsilineol	Hernández *et al.*, 1987
	Cirsimaritin	Hernández *et al.*, 1987
	5-Desmethylnobiletin	Hernández *et al.*, 1987
	5-Desmethylsinensetin	Hernández *et al.*, 1987
	Gardenin-B	Hernández *et al.*, 1987
	Sideritoflavone	Hernández *et al.*, 1987
	Thymonin	Hernández *et al.*, 1987
	Thymusin	Hernández *et al.*, 1987
	Xanthomicrol	Hernández *et al.*, 1987
	Vicenin-2	Husain and Markham, 1981
T. membranaceus Boiss.	Apigenin	Adzet and Martínez, 1981b; Adzet *et al.*, 1988; Ferreres *et al.*, 1985a
	Cirsilineol	Adzet *et al.*, 1988; Hernández *et al.*, 1987
	8-OMe-Cirsilineol	Adzet *et al.*, 1988; Ferreres *et al.*, 1985a; Hernández *et al.*, 1987
	Cirsimaritin	Adzet and Martínez 1981b; Adzet *et al.*, 1988; Ferreres *et al.*, 1985a; Hernández *et al.*, 1987
	Chrysoeriol	Ferreres *et al.*, 1985a
	5-Desmethylnobiletin	Adzet *et al.*, 1988; Ferreres *et al.*, 1985a; Hernández *et al.*, 1987

Table 5.4 (Continued)

Species	Polyphenols	References
	5-Desmethylsinensetin	Ferreres *et al.*, 1985a; Hernández *et al.*, 1987
	2,3-Dihydrokaempferol	Adzet *et al.*, 1988
	2,3-Dihydroxanthomicrol	Adzet *et al.*, 1988
	Eriodictyol	Adzet *et al.*, 1988
	Eupatorin	Ferreres *et al.*, 1985a
	Genkwanin	Adzet and Martínez, 1981b; Adzet *et al.*, 1988; Ferreres *et al.*, 1985a
	Luteolin	Adzet and Martínez 1980a, 1981b; Adzet *et al.*, 1988; Ferreres *et al.*, 1985a
	6-OH-Luteolin	Adzet and Martínez 1980a, 1981b
	7-O-Me-Luteolin	Ferreres *et al.*, 1985a
	Naringenin	Adzet *et al.*, 1988; Ferreres *et al.*, 1985a
	Sakuranetin	Adzet *et al.*, 1988
	Sideritoflavone	Adzet *et al.*, 1988; Ferreres *et al.*, 1985a; Hernández *et al.*, 1987
	Taxifolin	Adzet *et al.*, 1988
	Thymusin	Ferreres *et al.*, 1985b; Hernández *et al.*, 1987
	Xanthomicrol	Adzet and Martínez 1981b; Adzet *et al.*, 1988; Ferreres *et al.*, 1985a; Hernández *et al.*, 1987
	Apigenin-7-*O*-β-D-glucoside	Tomás *et al.*, 1985
	Luteolin-7-*O*-β-D-glucoside	Tomás *et al.*, 1985
	Luteolin-7-*O*-β-D-glucuronide	Tomás *et al.*, 1985
	Luteolin-7-*O*-β-D-neohesperidoside	Tomás *et al.*, 1985
	Luteolin-7-*O*-β-D-rutinoside	Tomás *et al.*, 1985
	Luteolin-7-*O*-β-D-sambubioside	Tomás *et al.*, 1985
	Luteolin-7-*O*-xyloside	Tomás *et al.*, 1985
	6-OH-Luteolin-7-*O*-β-D-glucoside	Tomás *et al.*, 1985
	Vicenin-2	Husain and Markham, 1981; Tomás *et al.*, 1985
	Caffeic acid	Adzet *et al.*, 1988
	Rosmarinic acid	Adzet *et al.*, 1988
T. migricus Klok. et Schost.	Apigenin	Simonyan and Litvinenko, 1971
	Luteolin	Simonyan and Litvinenko, 1971
	Scutellarein	Simonyan and Litvinenko, 1971
T. moesiacus Velen	Apigenin	Kulevanova *et al.*, 1997
	Luteolin	Kulevanova *et al.*, 1997
	Caffeic acid	Kulevanova *et al.*, 1997
T. moldavicus Klok. et Schost.	Apigenin	Litvinenko and Zoz, 1969
	Luteolin	Litvinenko and Zoz, 1969
	Scutellarein	Litvinenko and Zoz, 1969
T. moroderi Pau *ex* Martínez	Apigenin	Adzet *et al.*, 1988
	Cirsilineol	Adzet *et al.*, 1988; Hernández *et al.*, 1987
	8-OMe-Cirsilineol	Adzet *et al.*, 1988; Hernández *et al.*, 1987
	Cirsimaritin	Adzet *et al.*, 1988; Hernández *et al.*, 1987
	5-Desmethylnobiletin	Adzet *et al.*, 1988; Hernández *et al.*, 1987
	5-Desmethylsinensetin	Hernández *et al.*, 1987

	2,3-Dihydrokaempferol	Adzet *et al*., 1988
	2,3-Dihydroxanthomicrol	Adzet *et al*., 1988
	Eriodictyol	Adzet *et al*., 1988
	Genkwanin	Adzet *et al*., 1988
	8-OMe-Genkwanin	Vila, 1987
	Kaempferol	Vila, 1987
	Luteolin	Adzet *et al*., 1988
	6-OH-Luteolin	Vila, 1987
	7-*O*-Me-Luteolin	Vila, 1987
	Naringenin	Adzet *et al*., 1988
	Pilloin	Vila, 1987
	Sakuranetin	Adzet *et al*., 1988
	Sideritoflavone	Adzet *et al*., 1988; Hernández *et al*., 1987
	Sorbifolin	Vila, 1987
	Taxifolin	Adzet *et al*., 1988
	Thymusin	Hernández *et al*., 1987; Vila, 1987
	Xanthomicrol	Adzet *et al*., 1988; Hernández *et al*., 1987
	Luteolin-3′-*O*-alloside	Vila, 1987
	Luteolin-7-*O*-glucoside	Vila, 1987
	Luteolin-7-*O*-glucuronide	Vila, 1987
	Luteolin-7-*O*-xyloside	Vila, 1987
	Vicenin-2	Vila, 1987
	Caffeic acid	Adzet *et al*., 1988
	Rosmarinic acid	Adzet *et al*., 1988
T. nervosus Gay *ex* Willk.	Apigenin	Adzet and Martínez, 1981b
	Cirsilineol	Hernández *et al*., 1987
	8-OMe-Cirsilineol	Hernández *et al*., 1987
	Cirsimaritin	Hernández *et al*., 1987
	5-Desmethylnobiletin	Hernández *et al*., 1987
	5-Desmethylsinensetin	Hernández *et al*., 1987
	Diosmetin	Hernández *et al*., 1987
	Gardenin-B	Hernández, *et al*., 1987
	Luteolin	Adzet and Martínez, 1981b
	Salvigenin	Hernández *et al*., 1987
	Sideritoflavone	Hernández *et al*., 1987
	Thymonin	Hernández *et al*., 1987
	Thymusin	Hernández *et al*., 1987
	Xanthomicrol	Hernández *et al*., 1987
T. nummularius Bieb.	Apigenin	Simonyan and Litvinenko, 1971
	Luteolin	Simonyan and Litvinenko, 1971
	Scutellarein	Simonyan and Litvinenko, 1971
	Apigenin-7-*O*-β-D-glucoside	Simonyan, 1972
	Luteolin-7-*O*-β-D-glucoside	Simonyan, 1972
T. orospedanus H. Del Villar	Apigenin	Adzet *et al*., 1988
	Cirsilineol	Adzet *et al*., 1988; Hernández *et al*., 1987
	8-OMe-Cirsilineol	Adzet *et al*., 1988; Hernández *et al*., 1987
	Cirsimaritin	Adzet *et al*., 1988; Hernández *et al*., 1987
	5-Desmethylnobiletin	Adzet *et al*., 1988; Hernández *et al*., 1987
	5-Desmethylsinensetin	Hernández *et al*., 1987
	Diosmetin	Hernández *et al*., 1987
	Eriodictyol	Adzet *et al*., 1988
	Genkwanin	Adzet *et al*., 1988
	Luteolin	Adzet *et al*., 1988
	Naringenin	Adzet *et al*., 1988
	Sakuranetin	Adzet *et al*., 1988

Table 5.4 (Continued)

Species	Polyphenols	References
T. orospedanus H. Del Villar (Continued)	Sideritoflavone	Adzet *et al.*, 1988; Hernández *et al.*, 1987
	Taxifolin	Adzet *et al.*, 1988
	Thymonin	Hernández *et al.*, 1987
	Thymusin	Hernández *et al.*, 1987
	Xanthomicrol	Adzet *et al.*, 1988; Hernández *et al.*, 1987
	Caffeic acid	Adzet *et al.*, 1988
	Rosmarinic acid	Adzet *et al.*, 1988
T. pallasianus H. Braun	Luteolin	Litvinenko and Zoz, 1969
	Scutellarein	Litvinenko and Zoz, 1969
T. pannonicus All.	Apigenin	Litvinenko and Zoz, 1969
	Luteolin	Litvinenko and Zoz, 1969
	Scutellarein	Litvinenko and Zoz, 1969
T. pastoralis Turrill, *non* Iljn	Apigenin	Simonyan and Litvinenko, 1971
	Luteolin	Simonyan and Litvinenko, 1971
	Scutellarein	Simonyan and Litvinenko, 1971
T. piperella L.	Apigenin	Adzet and Martínez, 1981b; Barberán *et al.*, 1985
	Eriodictyol	Barberán *et al.*, 1985
	Ladanein	Barberán *et al.*, 1985; Hernández *et al.*, 1987
	Luteolin	Adzet and Martínez, 1980a, 1981b
	6-OH-Luteolin	Adzet and Martínez, 1980a, 1981b
	Naringenin	Barberán *et al.*, 1985
	Pebrellin	Barberán *et al.*, 1985; Hernández *et al.*, 1987
	5-OH-7,4'-(OMe)$_2$-flavone	Barberán *et al.*, 1985
	5,6-(OH)$_2$-7,3',4'-(OMe)$_3$-flavone	Barberán *et al.*, 1985; Hernández *et al.*, 1987
	5,6-(OH)$_2$-7,8,3',4'-(OMe)$_4$-flavone	Barberán *et al.*, 1985; Hernández *et al.*, 1987
	Apigenin-7-*O*-β-D-glucoside	Barberán *et al.*, 1985
	Luteolin-7-*O*-β-D-glucoside	Barberán *et al.*, 1985
	Vicenin-2	Barberán *et al.*, 1985
T. platyphyllus	Apigenin	Litvinenko and Zoz, 1969
	Luteolin	Litvinenko and Zoz, 1969
	Scutellarein	Litvinenko and Zoz, 1969
T. podolicus Klok. et Schost.	Apigenin	Litvinenko and Zoz, 1969
	Luteolin	Litvinenko and Zoz, 1969
	Scutellarein	Litvinenko and Zoz, 1969
T. polessicus	Apigenin	Litvinenko and Zoz, 1969
	Luteolin	Litvinenko and Zoz, 1969
	Scutellarein	Litvinenko and Zoz, 1969
T. praecox Opiz	Apigenin	Adzet and Martínez, 1981b
	Cirsilineol	Hernández *et al.*, 1987
	8-OMe-Cirsilineol	Hernández *et al.*, 1987
	Cirsimaritin	Hernández *et al.*, 1987
	5-Desmethylnobiletin	Hernández *et al.*, 1987
	5-Desmethylsinensetin	Hernández *et al.*, 1987
	Luteolin	Adzet and Martínez, 1981b

T. praecox Opiz (Continued)	6-OH-Luteolin	Adzet and Martínez, 1981b
	Sideritoflavone	Hernández *et al.*, 1987
	Thymonin	Hernández *et al.*, 1987
	Thymusin	Hernández *et al.*, 1987
	Xanthomicrol	Hernández *et al.*, 1987
T. pseudograniticus Klok. et Schost.	Acacetin	Litvinenko and Zoz, 1969
	Apigenin	Litvinenko and Zoz, 1969
	Luteolin	Litvinenko and Zoz, 1969
	Scutellarein	Litvinenko and Zoz, 1969
T. pseudohumillimus Klok. et Schost.	Apigenin	Litvinenko and Zoz, 1969
	Luteolin	Litvinenko and Zoz, 1969
	Scutellarein	Litvinenko and Zoz, 1969
T. pseudonummularius Klok. et Schost.	Apigenin	Simonyan and Litvinenko, 1971
	Luteolin	Simonyan and Litvinenko, 1971
	Scutellarein	Simonyan and Litvinenko, 1971
T. pulegioides L.	Apigenin	Adzet and Martínez, 1981b; Stoess, 1972; Van den Broucke *et al.*, 1982a
	Cyanidin	Stoess, 1972
	Cirsilineol	Hernández *et al.*, 1987
	8-OMe-Cirsilineol	Hernández *et al.*, 1987
	Cirsimaritin	Hernández *et al.*, 1987
	5-Desmethylnobiletin	Hernández *et al.*, 1987
	5-Desmethylsinensetin	Hernández *et al.*, 1987
	Luteolin	Adzet and Martínez, 1980a, 1981b; Stoess, 1972; Van den Broucke *et al.*, 1982a
	6-OH-Luteolin	Adzet and Martínez, 1980a, 1981b
	Sideritoflavone	Hernández *et al.*, 1987
	Thymonin	Hernández *et al.*, 1987
	Thymusin	Hernández *et al.*, 1987
	Xanthomicrol	Hernández *et al.*, 1987
T. rariflorus C. Koch	Apigenin	Simonyan and Litvinenko, 1971
	Luteolin	Simonyan and Litvinenko, 1971
	Scutellarein	Simonyan and Litvinenko, 1971
T. richardii Pers. ssp. *ebusitanus* (Font Quer) Jalas	Apigenin	Adzet and Martínez, 1981b
	Luteolin	Adzet and Martínez, 1980a, 1981b
	6-OH-Luteolin	Adzet and Martínez, 1980a, 1981b
T. richardii Pers. ssp. *richardii*	Apigenin	Adzet and Martínez, 1981b
	Luteolin	Adzet and Martínez, 1980a, 1981b
	6-OH-Luteolin	Adzet and Martínez, 1980a, 1981b
T. satureioides Coss.	Apigenin	Van den Broucke *et al.*, 1982a
	Cirsilineol	Van den Broucke *et al.*, 1982a; Voirin *et al.*, 1985
	8-OMe-Cirsilineol	Van den Broucke *et al.*, 1982a; Voirin *et al.*, 1985
	Cirsimaritin	Voirin *et al.*, 1985
	5-Desmethylsinensetin	Voirin *et al.*, 1985
	Luteolin	Van den Broucke *et al.*, 1982a
	Thymonin	Van den Broucke *et al.*, 1982a; Voirin *et al.*, 1985

Table 5.4 (Continued)

Species	Polyphenols	References
T. satureioides Coss. (Continued)	Xanthomicrol	Voirin *et al.*, 1985
	5,6,4′-(OH)$_3$-7,3′-(OMe)$_2$-flavone	Voirin *et al.*, 1985
T. serpylloides Bory	Apigenin	Adzet and Martínez, 1981b
	Luteolin	Adzet and Martínez, 1980a, 1981b
	6-OH-Luteolin	Adzet and Martínez, 1980a, 1981b
T. serpylloides Bory ssp. *serpylloides*	Cirsilineol	Hernández *et al.*, 1987
	8-OMe-Cirsilineol	Hernández *et al.*, 1987
	5-Desmethylnobiletin	Hernández *et al.*, 1987
	Sideritoflavone	Hernández *et al.*, 1987
	Thymonin	Hernández *et al.*, 1987
	Thymusin	Hernández *et al.*, 1987
	Xanthomicrol	Hernández *et al.*, 1987
T. serpylloides Bory ssp. *gadorensis* (Pau) Jalas	Cirsilineol	Hernández *et al.*, 1987
	8-OMe-Cirsilineol	Hernández *et al.*, 1987
	Cirsimaritin	Hernández *et al.*, 1987
	5-Desmethylnobiletin	Hernández *et al.*, 1987
	5-Desmethylsinensetin	Hernández *et al.*, 1987
	Sideritoflavone	Hernández *et al.*, 1987
	Thymonin	Hernández *et al.*, 1987
	Thymusin	Hernández *et al.*, 1987
	Xanthomicrol	Hernández *et al.*, 1987
T. serpyllum L.	Apigenin	Litvinenko and Zoz, 1969; Olechnowicz-Stepien and Lamer-Zarawska, 1975; Semrau, 1958; Van den Broucke *et al.*, 1982a
	Diosmetin	Olechnowicz-Stepien and Lamer-Zarawska, 1975
	Luteolin	Litvinenko and Zoz, 1969; Olechnowicz-Stepien and Lamer-Zarawska, 1975; Van den Broucke *et al.*, 1982a
	Scutellarein	Litvinenko and Zoz, 1969; Olechnowicz-Stepien and Lamer-Zarawska, 1975
	Apigenin-7-O-β-D-glucoside	Olechnowicz-Stepien and Lamer-Zarawska, 1975
	Apigenin-4′-O-β-D-p-cumaroyl-glucoside	Washington and Saxena, 1983
	Diosmetin-7-O-β-D-glucuronide	Olechnowicz-Stepien and Lamer-Zarawska, 1975
	Luteolin-galactoarabinoside	Olechnowicz-Stepien and Lamer-Zarawska, 1975
	Luteolin-7-O-β-D-glucoside	Olechnowicz-Stepien and Lamer-Zarawska, 1975
	Luteolin-7-O-β-D-diglucoside	Olechnowicz-Stepien and Lamer-Zarawska, 1975
	Scutellarein-glucosyl glucuronide	Olechnowicz-Stepien and Lamer-Zarawska, 1975
	Scutellarein-7-O-β-D-glucosyl (1–4)α-L-rhamnoside	Washington and Saxena, 1986

T. sosnowskyi Grossh.	Apigenin	Simonyan and Litvinenko, 1971
	Luteolin	Simonyan and Litvinenko, 1971
	Scutellarein	Simonyan and Litvinenko, 1971
T. striatus Vahl.	Apigenin	Van den Broucke *et al.*, 1982a
T. subalpestris Klok.	Apigenin	Litvinenko and Zoz, 1969
	Luteolin	Litvinenko and Zoz, 1969
	Scutellarein	Litvinenko and Zoz, 1969
T. tauricus Klok. et Schost.	Apigenin	Litvinenko and Zoz, 1969
	Luteolin	Litvinenko and Zoz, 1969
	Scutellarein	Litvinenko and Zoz, 1969
T. tiflisiensis Klok. et Schost.	Apigenin	Simonyan and Litvinenko, 1971
	Luteolin	Simonyan and Litvinenko, 1971
	Scutellarein	Simonyan and Litvinenko, 1971
T. tosevii ssp. *substriatus* (Borb.) Matevski	Apigenin	Kulevanova *et al.*, 1997
	Eriodictyol	Kulevanova *et al.*, 1997
	Luteolin	Kulevanova *et al.*, 1997
	Caffeic acid	Kulevanova *et al.*, 1997
	Rosmarinic acid	Kulevanova *et al.*, 1997
T. tosevii ssp. *tosevii* var. *degenii* Ronn.	Apigenin	Kulevanova *et al.*, 1997
	Eriodictyol	Kulevanova *et al.*, 1997
	Luteolin	Kulevanova *et al.*, 1997
	Caffeic acid	Kulevanova *et al.*, 1997
	Rosmarinic acid	Kulevanova *et al.*, 1997
T. tosevii ssp. *tosevii* var. *longifrons* Ronn.	Apigenin	Kulevanova *et al.*, 1997
	Eriodictyol	Kulevanova *et al.*, 1997
	Naringenin	Kulevanova *et al.*, 1997
	Luteolin	Kulevanova *et al.*, 1997
	Caffeic acid	Kulevanova *et al.*, 1997
	Rosmarinic acid	Kulevanova *et al.*, 1997
T. tosevii ssp. *tosevii* var. *tosevii* Velen.	Apigenin	Kulevanova *et al.*, 1997, 1998
	Eriodictyol	Kulevanova *et al.*, 1997, 1998
	Naringenin	Kulevanova *et al.*, 1997, 1998
	Luteolin	Kulevanova *et al.*, 1997, 1998
	Caffeic acid	Kulevanova *et al.*, 1997, 1988
	Rosmarinic acid	Kulevanova *et al.*, 1997, 1998
T. transcaucasicus Ronn.	Apigenin	Simonyan and Litvinenko, 1971
	Luteolin	Simonyan and Litvinenko, 1971
	Scutellarein	Simonyan and Litvinenko, 1971
T. trautvetteri Klok. et Schost.	Apigenin	Simonyan and Litvinenko, 1971
	Luteolin	Simonyan and Litvinenko, 1971
	Scutellarein	Simonyan and Litvinenko, 1971
T. ucrainicus Klok. et Schost.	Apigenin	Litvinenko and Zoz, 1969
	Luteolin	Litvinenko and Zoz, 1969
	Scutellarein	Litvinenko and Zoz, 1969
T. villosus L.	Apigenin	Adzet and Martínez, 1981b
	Cirsilineol	Hernández *et al.*, 1987
	8-OMe-Cirsilineol	Hernández *et al.*, 1987
	Cirsimaritin	Hernández *et al.*, 1987
	5-Desmethylnobiletin	Hernández *et al.*, 1987
	5-Desmethylsinensetin	Hernández *et al.*, 1987

Table 5.4 (Continued)

Species	Polyphenols	References
T. villosus L. (Continued)	Gardenin-B	Hernández *et al.*, 1987
	Luteolin	Adzet and Martínez, 1981b
	6-OH-Luteolin	Adzet and Martínez, 1981b
	Salvigenin	Hernández *et al.*, 1987
	Sideritoflavone	Hernández *et al.*, 1987
	Thymusin	Hernández *et al.*, 1987
	Xanthomicrol	Hernández *et al.*, 1987
	Vicenin-2	Husain and Markham, 1981
T. vulgaris L.	Apigenin	Adzet and Martínez, 1981b; Adzet *et al.*, 1988; Awe *et al.*, 1959; Kümmell, 1959; Olechnowicz-Stepien and Lamer-Zarawska, 1975; Semrau, 1958; Stoess, 1972; Van den Broucke *et al.*, 1982a
	Cyanidin	Stoess, 1972
	Cirsilineol	Adzet *et al.*, 1988; Hernández *et al.*, 1987; Miura and Nakatani, 1989; Morimitsu *et al.*, 1995; Van den Broucke *et al.*, 1982b
	8-OMe-Cirsilineol	Adzet *et al.*, 1988; Hernández *et al.*, 1987; Miura and Nakatani, 1989; Van den Broucke *et al.*, 1982b
	Cirsimaritin	Adzet and Martínez, 1981b; Adzet *et al.*, 1988; Hernández *et al.*, 1987; Miura and Nakatani, 1989
	5-Desmethylnobiletin	Adzet *et al.*, 1988; Hernández *et al.*, 1987
	5-Desmethylsinensetin	Hernández *et al.*, 1987
	2,3-Dihydrokaempferol	Adzet *et al.*, 1988
	2,3-Dihydroxanthomicrol	Adzet *et al.*, 1988
	Eriodictyol	Adzet *et al.*, 1988; Morimitsu *et al.*, 1995
	Gardenin-B	Hernández *et al.*, 1987
	Genkwanin	Adzet *et al.*, 1988; Miura and Nakatani, 1989
	Luteolin	Adzet and Martínez, 1980a, 1981b; Adzet *et al.*, 1988; Awe *et al.*, 1959; Kümmell, 1959; Olechnowicz-Stepien and Lamer-Zarawska, 1975; Semrau, 1958; Stoess, 1972; Van den Broucke *et al.*, 1982a
	6-OH-Luteolin	Adzet and Martínez, 1980a, 1981b
	Naringenin	Adzet *et al.*, 1988; Stoess, 1972
	Quercetin	Morimitsu *et al.*, 1995
	Sakuranetin	Adzet *et al.*, 1988
	Salvigenin	Hernández *et al.*, 1987
	Sideritoflavone	Adzet *et al.*, 1988; Hernández *et al.*, 1987
	Taxifolin	Adzet *et al.*, 1988
	Thymonin	Hernández *et al.*, 1987; Morimitsu *et al.*, 1995; Van den Broucke *et al.*, 1982b
	Thymusin	Hernández *et al.*, 1987
	Xanthomicrol	Adzet and Martínez, 1981b; Adzet *et al.*, 1988; Hernández *et al.*, 1987; Miura and Nakatani, 1989

	5-OH-7,4'-(OMe)₂-flavone	Miura and Nakatani, 1989
	Apigenin-7-O-β-D-glucoside	Olechnowicz-Stepien and Lamer-Zarawska, 1975
	Luteolin-7-O-β-D-glucoside	Olechnowicz-Stepien and Lamer-Zarawska, 1975
	Luteolin-7-O-β-D-diglucoside	Olechnowicz-Stepien and Lamer-Zarawska, 1975
	Vicenin-2	Husain and Markham, 1981
	Caffeic acid	Adzet *et al.*, 1988
	Rosmarinic acid	Adzet *et al.*, 1988
T. vulgaris L. ssp. *erycoides*	Cirsilineol	Hernández *et al.*, 1987
	8-OMe-Cirsilineol	Hernández *et al.*, 1987
	Cirsimaritin	Hernández *et al.*, 1987
	5-Desmethylnobiletin	Hernández *et al.*, 1987
	5-Desmethylsinensetin	Hernández *et al.*, 1987
	Sideritoflavone	Hernández *et al.*, 1987
	Thymonin	Hernández *et al.*, 1987
	Thymusin	Hernández *et al.*, 1987
	Xanthomicrol	Hernández *et al.*, 1987
T. webbianus Rouy	Apigenin	Blázquez *et al.*, 1990, 1994
	Cirsimaritin	Blázquez *et al.*, 1990
	Eriodictyol	Blázquez *et al.*, 1994
	Genkwanin	Blázquez *et al.*, 1990
	Luteolin	Blázquez *et al.*, 1990, 1994
	Naringenin	Blázquez *et al.*, 1994
	Thymonin	Blázquez *et al.*, 1990
	5-OH-7,4'-(OMe)₂-flavone	Blázquez *et al.*, 1990
	Apigenin-7-O-glucoside	Blázquez *et al.*, 1994
	Luteolin-7-O-glucoside	Blázquez *et al.*, 1994
	Vicenin-2	Blázquez *et al.*, 1994
	Chlorogenic acid	Blázquez *et al.*, 1994
	p-Coumaric acid	Blázquez *et al.*, 1994
	3,5-Dicaffeoylquinic acid	Blázquez *et al.*, 1994
	Protocatechuic acid	Blázquez *et al.*, 1994
	Syringic acid	Blázquez *et al.*, 1994
T. willkomii Ronn.	Apigenin	Adzet *et al.*, 1988
	Cirsilineol	Adzet *et al.*, 1988
	8-OMe-Cirsilineol	Adzet *et al.*, 1988
	Cirsimaritin	Adzet *et al.*, 1988
	2,3-Dihydrokaempferol	Adzet *et al.*, 1988
	Eriodictyol	Adzet *et al.*, 1988
	Luteolin	Adzet *et al.*, 1988
	Naringenin	Adzet *et al.*, 1988
	Sakuranetin	Adzet *et al.*, 1988
	Sideritoflavone	Adzet *et al.*, 1988
	Taxifolin	Adzet *et al.*, 1988
	Xanthomicrol	Adzet *et al.*, 1988
	Caffeic acid	Adzet *et al.*, 1988
	Rosmarinic acid	Adzet *et al.*, 1988
T. zeleneitzkyi Klok. et Schost.	Luteolin	Litvinenko and Zoz, 1969
	Scutellarein	Litvinenko and Zoz, 1969
T. ziaratinus Klok. et Schost.	Apigenin	Simonyan and Litvinenko, 1971
	Luteolin	Simonyan and Litvinenko, 1971
	Scutellarein	Simonyan and Litvinenko, 1971

Table 5.4 (Continued)

Species	Polyphenols	References
T. zygis L.	Apigenin	Adzet and Martínez, 1981b
	Cirsimaritin	Adzet and Martínez, 1981b
	Luteolin	Adzet and Martínez, 1980a, 1981b
	6-OH-Luteolin	Adzet and Martínez, 1980a, 1981b
	Xanthomicrol	Adzet and Martínez, 1981b
	Vicenin-2	Husain and Markham, 1981
T. zygis L. ssp. *sylvestris* (Hoffmanns. et Link.) Brot. ex Coutinho	Cirsilineol	Hernández *et al.*, 1987
	8-OMe-Cirsilineol	Hernández *et al.*, 1987
	Cirsimaritin	Hernández *et al.*, 1987
	5-Desmethylnobiletin	Hernández *et al.*, 1987
	5-Desmethylsinensetin	Hernández *et al.*, 1987
	Sideritoflavone	Hernández *et al.*, 1987
	Thymonin	Hernández *et al.*, 1987
	Thymusin	Hernández *et al.*, 1987
	Xanthomicrol	Hernández *et al.*, 1987
T. zygis Loefl. ex L. ssp. *zygis*	Cirsilineol	Hernández *et al.*, 1987
	8-OMe-Cirsilineol	Hernández *et al.*, 1987
	Cirsimaritin	Hernández *et al.*, 1987
	5-Desmethylnobiletin	Hernández *et al.*, 1987
	5-Desmethylsinensetin	Hernández *et al.*, 1987
	Sideritoflavone	Hernández *et al.*, 1987
	Thymonin	Hernández *et al.*, 1987
	Thymusin	Hernández *et al.*, 1987
	Xanthomicrol	Hernández *et al.*, 1987

Notes
* Cirsimaritin coeluted together with 8-OMe-Genkwanin in the TLC and HPLC systems used by Adzet *et al.* (1988).
† Cirsilineol coeluted together with Eupatorin in the TLC and HPLC systems used by Hernández *et al.* (1987).
‡ 8-OMe-Cirsilineol coeluted together with Gardenin D in the TLC and HPLC systems used by Hernández *et al.* (1987).

constituted by apolar methoxylated aglycones, with different A-ring substitution patterns of taxonomic importance.

In particular, the genus *Thymus* is rich in exudate flavonoids, especially the sections *Pseudothymbra*, *Thymus*, *Piperella* and *Mastichina*, which grow in semi-arid habitats. In this way, it has been found that species of the section *Pseudothymbra* (*T. membranaceus*, *T. moroderi*, *T. funkii* and *T. longiflorus*) from southeastern Spain, or *T. herba-barona* (section *Serpyllum*) from Corsica, produce higher levels of excreted flavonoids than species growing in alpine habitats, as it has been reported for some other taxa of the section *Serpyllum* (*T. praecox*, *T. pulegioides* and *T. nervosus*) growing in the Pyrenees. This fact supports the influence of ecological factors on the excretion of flavonoids. Although *T. herba-barona* and the species of the section *Pseudothymbra* have been found to produce external flavonoids with different A-ring susbstitution patterns, none of them being thymonin, the latter can be characterized by the presence of 4'-,6- and 6,8-di-substituted flavones which are absent from *T. herba-barona* (Corticchiato *et al.*, 1995; Hernández *et al.*, 1987).

Furthermore, Corticchiato *et al.* (1995) also studied the distribution of the exudate flavonoids in seven essential oil chemotypes of *T. herba-barona*. The presence of three different flavonoid patterns among them confirms the inherent chemical variability in this species.

Not only the habitat but also the genetic factors may influence the production of external flavonoids, as it has been reported by Hernández *et al.* (1987). The authors found that thyme species growing under the same ecological conditions show very different amounts of exudate flavonoids, which must be explained on the basis of the metabolic capability for their production. That was the case of *T. camphoratus*, *T. capitellatus* and *T. carnosus*, originally included in the section *Thymus* subsection *Thymastra*, which grow in sandy areas near the sea in Portugal. While the former two species contained only trace amounts of these compounds, the latter showed a very high concentration.

In general, although the exudate flavonoids in thyme taxa are quite different in structure, the most remarkable feature is the presence of 5,6-dihydroxy-7,8-dimethoxyflavones (thymonin, thymusin or pebrellin) and the nearly absence of 5,6-dihydroxy-7-methoxy-flavones, which are usually co-occurring with the former in related genera (Tomás-Barberán and Wollenweber, 1990). These two types of flavonoids have been only found together in four *Thymus* species: in *T. piperella* (Barberán *et al.*, 1985) and *T. herba-barona* (Corticchiato *et al.*, 1995) as exudate flavonoids, and in *T. satureioides* (Voirin *et al.*, 1985) and *T. moroderi* (Vila, 1987) as constituents of lipophilic extracts of ground plant material (Table 5.5).

FLAVONOID PATTERN: INFLUENCE OF THE ENVIRONMENTAL CONDITIONS

Within the genus *Thymus* the influence of the environmental conditions on the pattern of flavonoids has been investigated. Analysis of the flavonoid composition of individuals belonging to the same species coming from different localities, as well as studies of seasonal variations, have been performed in order to establish whether a genetic control is directly related to it or not (Gil, 1993).

The fact that the flavonoid composition of different thyme species is clearly different, although they grow under the same environmental conditions, supports a direct relation of the flavonoid pattern with the genetic features of these plants.

Furthermore, thyme plants of the same species coming from different geographical localities showed no variations of their flavonoid composition but of their total amount. It has been found that even though differences in altitude may influence the total flavonoid production as well as the relative amount of each one, they did not cause any

Table 5.5 Co-occurrence of 5,6-(OH)$_2$-7,8-(OMe)$_2$- and 5,6-(OH)$_2$-7-OMe-flavones in some *Thymus* taxa

Thymus taxa	*Flavones according substitution pattern*		*References*
	5,6-(OH)$_2$-7,8-(OMe)$_2$	*5,6-(OH)$_2$-7-OMe*	
T. piperella	Pebrellin	5,6-(OH)$_2$-7,3′,4′-(OMe)$_3$-flavone	Barberán *et al.*, 1985
	5,6-(OH)$_2$-7,8,3′4′-(OMe)$_4$-flavone	Ladanein	
T. satureioides	Thymusin	5,6,4′-(OH)$_3$-7,3′-(OMe)$_2$-flavone	Voirin *et al.*, 1985
T. moroderi	Thymusin	Sorbifolin	Vila, 1987
T. herba-barona	Thymusin	Sorbifolin	Corticchiato *et al.*, 1995

Table 5.6 Flavonoid Composition of *Thymbra capitata* (syn. *Thymus capitatus*)

Flavonoids	References
Acacetin	Adzet and Martínez, 1981b
Apigenin	Adzet and Martínez, 1981b
Diosmetin	Barberán *et al.*, 1986
Luteolin	Adzet and Martínez, 1981b; Barberán *et al.*, 1986
5,6-(OH)$_2$-7,3',4'-(OMe)$_3$-flavone	Barberán *et al.*, 1986
5,6,4'-(OH)$_3$-7,3'-(OMe)$_2$-flavone	Barberán *et al.*, 1986
Luteolin-7-O-rutinoside	Barberán *et al.*, 1986
Vicenin-2	Barberán *et al.*, 1986; Husain and Markham, 1981

changes in the pattern of flavonoids, as it occurs in *T. serpylloides*. In this species, the total flavonoid content is much lower in plants growing at a high altitude than in those growing in a low altitude, the same flavonoids being found in all of them (Hernández, 1985).

The monthly evolution of the flavone pattern from several thyme species growing under different climatic conditions (alpine and xeric habitats) has also been evaluated in order to establish the influence of the seasonal variations on the flavonoid composition. The results showed that the pattern of flavonoids did not undergo any changes during the year in any of them, as for example in *T. serpylloides*, which showed the same pattern in winter when it was covered by snow as in summer during flowering time (Gil, 1993; Hernández, *et al.*, 1987).

CHEMOTAXONOMIC VALUE OF THYME FLAVONOIDS

Five papers (Adzet and Martínez, 1981b; Adzet *et al.*, 1988; Hernández *et al.*, 1987; Kulevanova *et al.*, 1997, 1998) are devoted to a comparative TLC and/or HPLC analysis of flavonoids of several *Thymus* species from different sections: *Pseudothymbra*, *Thymus*, *Mastichina*, *Micantes*, *Piperella*, *Hyphodromi* and *Serpyllum*, mostly Iberian or Balearic endemisms, and *Marginati*, from Macedonia. The results obtained give valuable information from the chemotaxonomic point of view, which allow us to arrive at interesting conclusions that support recent re-classifications of this genus.

In this sense, *T. capitatus* (*Thymbra capitata*), a Mediterranean plant which constitutes a taxonomic problem extensively discussed (Adzet and Martínez, 1981b; Elena-Rosselló, 1976; Morales, 1985), was found to lack 6-OH-luteolin, present in almost all the species of the subgenus *Thymus*, but to contain acacetin, a flavonoid of restricted distribution within *Thymus* species (Adzet and Martínez, 1981b). In addition, Barberán *et al.* (1986) found that this taxon externally accumulated the same unusual 5,6-dihydroxy-7-methoxy-flavonoids (Table 5.6) previously isolated from *Thymbra spicata* (Miski *et al.*, 1983) and occasionally reported in the genus *Thymus*. This fact supports the separation of this taxon from the genus *Thymus* and its inclusion in the genus *Thymbra* on the basis of morphological and caryological data (Barberán *et al.*, 1986; Morales, 1985). The absence of 6-hydroxyflavone glycosides, universally present in *Thymus* species, from *T. capitatus* (Tomás-Barberán *et al.*, 1988b), also supports its inclusion in the genus *Thymbra* as has been proposed also on the basis of its flavonoid aglycones.

Excluding *T. capitatus*, Hernández *et al.* (1987) distinguished two well-defined groups among the sections of the genus *Thymus* according to their flavonoid pattern, particularly to the presence or absence of the highly methoxylated 5,6-dihydroxyflavones thymusin and/or thymonin, which are the most characteristic flavones of the genus. One group is constituted by the section *Piperella* (*T. piperella*), characterized by the presence of pebrellin, ladanein, 5,6-$(OH)_2$-7,8,3′,4′-$(OMe)_4$-flavone and 5,6-$(OH)_2$-7,3′,4′-$(OMe)_3$-flavone and by the lack of thymusin and thymonin. The second group includes all the other sections of the genus, in which thymusin and/or thymonin are always present.

Although some sections (i.e. *Piperella*, *Mastichina*, *Micantes*) do not pose taxonomic problems, their caryological and morphological data being in accordance with their flavonoid pattern, in several species of other sections, like *Pseudothymbra* and *Thymus*, the flavonoid composition may help to resolve some taxonomic uncertainties.

Section *Pseudothymbra* is divided into two subsections: *Anomalae*, which only includes *T. antoninae*, and *Pseudothymbra* (Morales, 1985). The most remarkable feature of the flavonoid pattern of this section is the high content of apolar methoxylated flavones, especially xanthomicrol, and the presence of flavanones, mainly naringenin, and dihydroflavonols (Adzet *et al.*, 1988). Thymusin was detected in *T. membranaceus*, *T. longiflorus*, *T. moroderi*, *T. funkii*, *T. antoninae* and *T. villosus*, which however lacked thymonin. Significant amounts of the latter were found in *T. mastigophorus*, its flavonoid composition being very similar to that of *T. lacaitae*, supporting the inclusion of *T. mastigophorus*, originally included in section *Pseudothymbra* (Jalas, 1972), in section *Hyphodromi* as suggested by Morales (1985). Furthermore, similarities between the flavonoid pattern of *T. villosus* and *T. antoninae* might allow to join them in the subsection *Anomalae* of this section, being clearly differentiated from the rest of the species of the subsection *Pseudothymbra* (Hernández *et al.*, 1987).

Section *Thymus*, which is morphologically very heterogeneous, comprises two subsections: *Thymastra*, with *T. carnosus*, *T. camphoratus* and *T. capitellatus*, and *Thymus*, which includes the rest of the species of the section (Jalas and Kaleva, 1970). The analysis of their flavonoid composition allows to differentiate two groups of taxa. One of them with practically no methoxylated flavonoids (neither flavones nor flavanones), only traces of xanthomicrol, that includes *T. camphoratus* and *T. capitellatus* belonging to the subsection *Thymastra*. The other one, which comprises taxa of the subsection *Thymus* (*T. vulgaris*, *T. vulgaris* ssp. *aestivus*, *T. glandulosus*, *T. baeticus*, *T. orospedanus*, *T. hyemalis*, *T. zygis* and *T. serpylloides*) as well as *T. carnosus*, turned out to contain a great variety of non-polar flavones, flavanones and dihydroflavonols (Adzet *et al.*, 1988; Hernández *et al.*, 1987). These findings are in accordance with morphological results reported by Morales (1985), which separated *T. carnosus* from the subsection *Thymastra*.

Within section *Serpyllum*, the largest one of the genus *Thymus* (Jalas, 1971), only the flavonoid pattern of few taxa has been investigated, particularly of *T. nervosus* and *T. praecox* (subsection *Pseudomarginati*), *T. pulegioides* (subsection *Alternantes*), *T. willkomii* (subsection *Insulares*) and *T. herba-barona* (subsection *Pseudopiperellae*). In general, those of alpine occurrence, the first four species, show low levels of high methoxylated flavones, thymusin and thymonin being reported in all of them except in *T. willkomii* (Hernández *et al.*, 1987). The latter, an Iberian northeastern endemism which grows in restricted areas, shows a flavonoid pattern mainly characterized by flavanones and dihydroflavonols, the most important being naringenin and sakuranetin (Adzet *et al.*, 1988). In contrast, *T. herba-barona* growing in xeric habitats produces several methoxylated flavones among which thymusin but not thymonin was found (Corticchiato *et al.*, 1995).

The main features of the flavonoid pattern of the sections *Pseudothymbra, Thymus,* and *Serpyllum,* according to the results provided by Adzet *et al.* (1988), Corticchiato *et al.* (1995) and Hernández *et al.* (1987), are summarized in Table 5.7.

The flavonoid composition of 14 taxa belonging to the subsections *Verticillati* and *Marginati* of the section *Marginati* (A. Kerner) A. Kerner (nowadays included in the section *Serpyllum* (Jalas, 1971)) from the Macedonian flora has also been investigated (Kulevanova *et al.,* 1997, 1998). Apigenin and luteolin are present in each taxa, the latter being the major one in all of them, except in *T. jankae* var. *jankae* and *T. jankae* var. *pantotrichus.* They both contain diosmetin instead of luteolin as the principal flavone. No methoxylated flavones other than diosmetin were detected in the taxa investigated. Furthermore, the distribution of phenolic acids shows that caffeic acid is present in the taxa of the two subsections, whereas rosmarinic acid is only detected in those of the subsection *Verticillati.* In addition, the authors found great resemblance between the flavonoid pattern of *T. moesiacus, T. albanus* and *T. balcanus* from Macedonia, and *T. zheguliensis* and *T. bashkiriensis* from Russia, all of them closely related from a botanical point of view.

Finally, although in general flavonoid glycosides are not considered to have as much chemotaxonomic significance as aglycones, some interesting results on the distribution of 6-hydroxy-, 6-methoxy- and 8-hydroxyflavone glycosides in the Labiatae, Scrophulariaceae and related families have been reported (Tomás-Barberán *et al.,* 1988a). Particularly, in the genus *Thymus,* only 6-hydroxyflavone glycosides (mainly 6-hydroxyluteolin glycosides) were found in twenty-two taxa belonging to several sections (*Mastichina, Micantes, Piperella, Pseudothymbra, Thymus, Hyphodromi, Serpyllum*).

All these findings support the fact that flavonoids are valuable taxonomic markers that particularly in the genus *Thymus* have provided worthy information that together with those obtained from caryological and morphological studies, allow one to resolve taxonomic doubts and improve the classification of the genus. Despite the great work carried out, much more research should be done in order to complete the characterization of the genus.

Table 5.7 Main features of the flavonoid pattern of the sections *Pseudothymbra, Thymus,* and *Serpyllum* (Adzet *et al.,* 1988; Corticchiato *et al.,* 1995; Hernández *et al.,* 1987)

Section	Subsection	Highly methoxylated flavones	Thymusin	Thymonin	Exudate flavonoids level
Pseudothymbra	Anomalae	↑	Yes	No	↓
	Pseudothymbra	↑	Yes	No	↑
Thymus	Thymastra	↓	Yes	Yes	↓
	Thymus	↑	Yes	Yes	↑
Serpyllum	Alternantes	↓	Yes	Yes	↓
	Pseudomarginati	↓	Yes	Yes	↓
	Pseudopiperellae	↑	Yes	No	↑
	Insulares	↓	–*	–	–

* Not reported.

REFERENCES

Adzet, T. and Martínez, F. (1980a) Luteolin and 6-hydroxyluteolin: taxonomically important flavones in the genus *Thymus. Planta Med. (Suppl.)*, 52–55.

Adzet, T. and Martínez, F. (1980b) Sur les flavones méthylées du *Thymus baeticus* Boiss. ex Lacaita (Labiatae). *Plant. Méd. Phytothér.*, 14, 8–19.

Adzet, T. and Martínez, F. (1981a) Aglicons flavònics de les Labiades. *Butll. Inst. Cat. Hist. Nat.*, 46 (Sec. Bot., 4), 25–49.

Adzet, T. and Martínez, F. (1981b) Flavonoids in the leaves of *Thymus*: a chemotaxonomic survey. *Biochem. Syst. Ecol.*, 9, 293–295.

Adzet, T., Martínez, F. and Zamora, I. (1982) Hétérosides flavoniques du *Thymus loscosii* Willk. *Plant. Méd. Phytothér.*, 16, 116–119.

Adzet, T., Vila, R. and Cañigueral, S. (1988) Chromatographic analysis of polyphenols of some Iberian *Thymus. J. Ethnopharmacol.*, 24, 147–154.

Aguinagalde, I. and Pero Martínez, M.A. (1982) The occurrence of the acylated flavonol glycosides in the Cruciferae. *Phytochemistry*, 21, 2875–2878.

Awe, W., Schaller, J.F. and Kümmell, H.J. (1959) The flavones from *Thymus vulgaris. Naturwissensch.*, 46, 558.

Barberán, F.A.T. (1986) The flavonoid compounds from the Labiatae. *Fitoterapia*, 57, 67–95.

Barberán, F.A.T., Hernández, L., Ferreres, F. and Tomás, F. (1985) Highly methylated 6-hydroxyflavones and other flavonoids from *Thymus piperella. Planta Med.*, 51, 452–454.

Barberán, F.A.T., Hernández, L. and Tomás, F. (1986) A chemotaxonomic study of flavonoids in *Thymbra capitata. Phytochemistry*, 25, 561–562.

Bate-Smith, E.C. (1962) The phenolic constituents of plants and their significance. 1. Dicotyledons. *J. Linn. Soc. (Bot.)*, 58, 95–173.

Bate-Smith, E.C. (1963) Usefulness of chemistry in plant taxonomy as illustrated by the flavonoid constituents. In T. Swain (ed.), *Chemical plant taxonomy*, Academic Press, London, pp. 127–139.

Blázquez, M.A., Zafra-Polo, M.C. and Máñez, S. (1990) Flavones from *Thymus webbianus* and their chemotaxonomic significance. *Planta Med.*, 56, 581–582.

Blázquez, M.A., Máñez, S. and Zafra-Polo, M.C. (1994) Further flavonoids and other phenolics of *Thymus webbianus. Z. Naturforsch., C: Biosci.*, 49, 687–688.

Corticchiato, M., Bernardini, A., Costa, J., Bayet, C., Saunois, A. and Voirin, B. (1995) Free flavonoid aglycones from *Thymus herba-barona* and its monoterpenoid chemotypes. *Phytochemistry*, 40, 115–120.

Ebel, J. and Hahlbrock, K. (1982) Biosynthesis. In J.B. Harborne and T.J. Mabry (eds), *The Flavonoids: Advances in Research*, Chapman and Hall, London, pp. 641–679.

El-Domiaty, M.N., El-Shafae, A.M. and Abdel-Aal, M.M. (1997) A flavanol, flavanone, and highly-oxygenated flavones from *Thymus algeriensis* Boiss. *J. Pharm. Sci.*, 11, 13–17.

Elena-Rosselló, J.A. (1976) *Projet d'une étude de Taxonomie Expérimentale du Genre Thymus*. Thèse de Doctorat. Université des Sciences et Techniques du Languedoc, Montpellier.

Ferreres, F., Tomás, F., Barberán, F.A.T. and Hernández, L. (1985a) Free flavone aglycones from *Thymus membranaceus* Boiss. subsp. *membranaceus. Plant. Méd. Phytothér.*, 19, 89–97.

Ferreres, F., Barberán, F.A.T. and Tomás, F. (1985b) 5,6,4'-Trihydroxy-7,8-dimethoxyflavone from *Thymus membranaceus. Phytochemistry*, 24, 1869–1871.

Gil, M.I. (1993) *Contribución al Estudio Fitoquímico y Quimiosistemático de Flavonoides en la Familia Labiatae*. Tesis Doctoral. Universidad de Murcia. Murcia.

Grisebach, H. (1985) Topics in flavonoid biosynthesis. In C.F. Van Summere and P.J. Lea (eds), *Annual Proceedings of the Phytochemical Society of Europe. Vol. 25. The Biochemistry of Plant Phenolics*, Clarendon Press, Oxford, pp. 183–198.

Hahlbrock, K. and Grisebach, H. (1975) Biosynthesis of flavonoids. In J.B. Harborne, T.J. Mabry and H. Mabry (eds), *The Flavonoids*, Chapman and Hall, London, pp. 866–915.

Harborne, J.B. (1966) The evolution of flavonoid pigments in plants. In T. Swain (ed.), *Comparative Phytochemistry*, Academic Press, London, pp. 271–295.

Harborne, J.B. (1967) *Comparative Biochemistry of the Flavonoids*, Academic Press, London.

Harborne, J.B. (1975) The biochemical systematics of flavonoids. In J.B. Harborne, T.J. Mabry and H. Mabry (eds), *The Flavonoids*, Chapman and Hall, London, pp. 1056–1095.

Harborne, J.B. (1985) Phenolics and plant defence. In C.F. Van Sumere and P.J. Lea (eds), *Annual Proceedings of the Phytochemical Society of Europe. Vol. 25. The Biochemistry of Plant Phenolics*, Clarendon Press, Oxford, pp. 393–408.

Harborne, J.B. (1986) The natural distribution in angiosperms of anthocyanins acylated with aliphatic dicarboxylic acids. *Phytochemistry*, 25, 1887–1894.

Harborne, J.B. and Mabry, T.J. (1982) *The Flavonoids: Advances in Research*, Chapman and Hall, London.

Harborne, J.B. and Turner, B.L. (1984) *Plant Chemosystematics*, Academic Press, London.

Harborne, J.B. and Williams, C.A. (1971) 6-Hydroxyluteolin and scutellarein as phyletic markers in higher plants. *Phytochemistry*, 10, 367–378.

Harborne, J.B., Mabry, T.J. and Mabry, H. (1975) *The Flavonoids*, Chapman and Hall, London.

Hegnauer, R. (1966) *Chemotaxonomie der Pflanzen. IV.* Birkhäuser Verlag, Basel.

Hegnauer, R. (1989) *Chemotaxonomie der Pflanzen. VIII.* Birkhäuser Verlag, Basel.

Hernández, L.M. (1985) *Dotación de Flavonoides en el Género* Thymus *L. y su Contribución Quimio-taxonómica.* Tesis Doctoral. Universidad de Alicante, Alicante.

Hernández, L.M., Tomás-Barberán, F.A. and Tomás-Lorente, F. (1987) A chemotaxonomic study of free flavone aglycones from some Iberian *Thymus* species. *Biochem. Syst. Ecol.*, 15, 61–67.

Husain, S.Z. and Markham, K.R. (1981) The glycoflavone vicenin-2 and its distribution in related genera within the Labiatae. *Phytochemistry*, 20, 1171–1173.

Jalas, J. (1971) Notes on *Thymus* L. (Labiatae) in Europe. I. Supraspecific classification and nomenclature. *Bot. J. Linn. Soc.*, 64, 199–215.

Jalas, J. (1972) Gen. *Thymus* L. In T.G. Tutin, V.H. Heywood, N.A. Burges, D.M. Moore, D.H. Valentine, S.M. Walters and D.A. Webb (eds), *Flora Europaea*, 3, University Press, Cambridge, pp. 172–182.

Jalas, J. and Kaleva, K. (1970) Supraspezifische Gliederung und Verbreitungstypen in der Gattung *Thymus* L. (Labiatae). *Feddes Repert.*, 81, 93–106.

Khodair, A.I., Hammouda, F.M., Ismail, S.I., El-Missiry, M.M., El-Shahed, F.A. and Abdel-Azim, H. (1993) Phytochemical investigation of *Thymus decassatus*. 1. Flavonoids and volatile oil. *Qatar Univ. Sci. J.*, 13, 211–213.

Kulevanova, S., Stafilov, T., Anastasova, F., Ristic, M. and Brcik, D. (1997) Isolation and identification of flavonoid aglycones from some taxa of Sect. *Marginati* of genus *Thymus*. *Pharmazie*, 52, 886–888.

Kulevanova, S., Stefova, M. and Stafilov, T. (1998) HPLC analyses of the flavonoids in taxa and genus *Thymus* L. *T. tosevii, T. longidens* var. *lanicaulis* and *T. jankae* (var. *jankae*, var. *pantotrichus* and var. *patentipilus*). *Anal. Lab.*, 7, 103–108.

Kümmell, H.J. (1959) Doctoral thesis, Braunschweig. Quoted from G. Stoess (1972) *Phyto-chemische und physiologische Untersuchungen über Polyphenole in* Thymus vulgaris *L. und* Thymus pulegioides *L.* Doctoral thesis, Münster.

Kurkin, V.A., Braslavskii, V.B., Krivenchuk, P.E. and Plaksina, T.I. (1988) Compounds in the aerial parts of *Thymus bashkiriensis*. *Khim. Prir. Soedin.*, p. 758.

Litvinenko, V.I. and Zoz, I.G. (1969) Chemotaxonomic study of *Thymus* species in the Ukraine. *Rast. Resur.*, pp. 481–495.

Marhuenda, E., Alarcón, C., García, M. and Cert, A. (1987a) Sur les flavones isolées de *Thymus carnosus* Boiss. *Ann. Pharm. Fr.*, 45, 467–470.

Marhuenda, E., Alarcón, C. and García, M. (1987b) Demonstration of antibacterial properties of phenolic acids from *Thymus carnosus* Boiss. Isolation of caffeic, vanillic, *p*-coumaric, *p*-hydroxy-benzoic and syringic acids. *Plant. Méd. Phytothér.*, 21, 153–159.

Martínez, F. (1980) *Contribución al Estudio Fitoquímico y Quimiotaxonómico del Género* Thymus L. Tesis Doctoral. Universidad Autónoma de Barcelona. Barcelona.

McClure, J.W. (1975) Physiology and functions of flavonoids. In J.B. Harborne, T.J. Mabry and H. Mabry (eds), *The Flavonoids*, Chapman and Hall, London, pp. 970–1055.

Merghem, R., Jay, M., Viricel, M.R., Bayet, C. and Voirin, B. (1995) Five 8-C-benzylated flavonoids from *Thymus hirtus* (Labiatae). *Phytochemistry*, 38, 637–640.

Miski, L., Ulubelen, A. and Mabry, T.J. (1983) Structural revision of the flavone majoranin from *Majorana hortensis. Phytochemistry*, 22, 2093–2094.

Miura, K. and Nakatani, N. (1989) Antioxidative activity of flavonoids from thyme (*Thymus vulgaris* L.). *Agric. Biol. Chem.*, 53, 3043–3045.

Morales, R. (1985) *Taxonomía del Género* Thymus L., *excluida la Sección Serpyllum (Miller) Bentham en la Península Ibérica*. Tesis Doctoral. Universidad Complutense de Madrid. Madrid.

Morimitsu, Y., Yoshida, K., Esaki, S. and Hirota, A. (1995) Protein glycation inhibitors from thyme (*Thymus vulgaris*). *Biosci. Biotechnol. Biochem.*, 59, 2018–2021

Olechnowicz-Stepien, W. and Lamer-Zarawska, E. (1975) Study of the flavonoid fraction of some plants of the Labiatae family (*Herba serpylli* L., *Herba thymi* L., *Herba marjoranae* L., *Herba origani* L.). *Herba Pol.*, 21, 347–356.

Semrau, R. (1958) *Über die Flavone in der Familie der Labiaten*. Doctoral thesis, München.

Simonyan, A.V. (1972) Flavone glycosides of some species of the *Thymus* genus. *Khim. Prir. Soedin.*, p. 801.

Simonyan, A.V. and Litvinenko, V.I. (1971) Flavone aglycones of some *Thymus* species from the Caucasus. *Rast. Resur.*, pp. 580–582.

Stoess, G. (1972) *Phytochemische und Physiologische Untersuchungen über Polyphenole in* Thymus vulgaris L. *und* Thymus pulegioides L. Doctoral thesis, München.

Swain, T. (1975) Evolution of flavonoid compounds. In J.B. Harborne, T.J. Mabry and H. Mabry (eds), *The Flavonoids*, Chapman and Hall, London, pp. 1096–1129.

Tomás-Barberán, F.A. and Wollenweber, E. (1990) Flavonoid aglycones from the leaf surfaces of some Labiatae species. *Pl. Syst. Evol.*, 173, 109–118.

Tomás, F., Hernández, L., Barberán, F.A.T. and Ferreres, F. (1985) Flavonoid glycosides from *Thymus membranaceus. Z. Naturforsch.*, 40c, 583–584.

Tomás-Barberán, F.A., Grayer-Barkmeijer, R.J., Gil, M.I. and Harborne, J.B. (1988a) Distribution of 6-hydroxy-, 6-methoxy- and 8-hydroxyflavone glycosides in the Labiatae, the Scrophulariaceae and related families. *Phytochemistry*, 27, 2631–2645.

Tomás-Barberán, F.A., Husain, S.Z. and Gil, M.I. (1988b) The distribution of methylated flavones in the Lamiaceae. *Biochem. Syst. Ecol.*, 16, 43–46.

Van den Broucke, C.O., Lemli, J. and Lamy, J. (1982a) Action spasmolytique des flavones de différentes espèces de *Thymus. Plant. Méd. Phytothér.*, 16, 310–317.

Van den Broucke, C.O., Dommisse, R.A., Esmans, E.L. and Lemli, J.A. (1982b) Three methylated flavones from *Thymus vulgaris. Phytochemistry*, 21, 2581–2583.

Vila, R. (1987) *Contribución al Estudio de Polifenoles y Aceites Esenciales en el Género* Thymus L. Doctoral thesis, Universidad de Barcelona. Barcelona.

Voirin, B., Viricel, M.R., Favre-Bonvin, J., Van den Broucke, C.O. and Lemli, J. (1985) 5,6,4′-Trihydroxy-7,3′-dimethoxyflavone and other methoxylated flavonoids isolated from *Thymus satureioides. Planta Med.*, 51, 523–525.

Washington, J.S. and Saxena, V.K. (1983) A new acylated apigenin 4′-O-β-D-glucoside from the stems of *Thymus serpyllum* L. *J. Inst. Chem. (India)*, 57, 153–155.

Washington, J.S. and Saxena, V.K. (1986) Scutellarein-7-O-β-D-glucopyranosyl(1–4)-α-L-rhamnopyranoside from the stems of *Thymus serpyllum* L. *J. Indian Chem. Soc.*, 63, 226–227.

Wollenweber, E. (1985a) Exkret-Flavonoide bei höheren Pflanzen arider Gebiete. *Pl. Syst. Evol.*, 150, 83–88.

Wollenweber, E. (1985b) On the occurrence of acetylated flavonoid aglycones. *Phytochemistry*, 24, 1493–1494.

Wollenweber, E. and Dietz, V.H. (1981) Occurrence and distribution of free aglycones in plants. *Phytochemistry*, 20, 869–932.

Wollenweber, E., Favre-Bonvin, J. and Jay, M. (1978) A novel type of flavonoids: flavonol esters from fern exudates. *Z. Naturforsch.*, 35c, 831–835.

Zinchenko, T.V. and Bandyukova, V.A. (1969). Flavonoids of the Labiatae family. *Farm. Zh.* (Kiev), 25, 49–55.

6 Field culture, *in vitro* culture and selection of *Thymus*

Charles Rey and Francisco Sáez

INTRODUCTION

People have known and used plants from the Labiate family for many centuries, *Thymus* being one of these, due to its medicinal and flavouring properties that have long been recognised. Poetic descriptions of early Persian gardens include *Thymus* among the plants that they cultivated, showing their interest in the plant. The demand has always increased with the growth of the human population, especially in the last decades with the investigation of its pharmacological properties. Lawrence (1992) reports a production of 25 tons of essential oil of *T. zygis* in 1989, mostly provenant from collections of wild material.

A continually growing demand for thyme products is not likely to be supported by natural populations, which are threatened by destructive gathering methods and insufficient/irregular rainfall in traditional source areas. Additionally, the interests of the pharmaceutical/food industries do not focus on all chemotypes available in nature, but only on a few, namely thymol-, carvacrol- and linalool-types. Therefore an increase in the demand of thyme of cultivated origin must be expected with standardised composition and yield of the essential oils and with uniform organoleptic properties of the leaves.

In this chapter the recent efforts made by the different researchers to meet the demands of growers and markets will be reviewed, focusing on the selection and cultivation techniques used both under natural conditions and under a controlled environment in the laboratory. Special attention is given to the organic culture of thyme, which is of great interest in order to produce plant material for the food industry. Switzerland is a well-known centre for the cultivation of thyme, and the experiments performed there will serve to illustrate a number of procedures involved in the development of new varieties with commercial interest in thyme.

ORGANIC CULTURE OF COMMON THYME (*THYMUS VULGARIS* L.) IN SWITZERLAND

Organic culture of common thyme is developed basically in marginal areas (characterised by worse climatic, edaphic and accessibility conditions than traditional agricultural areas) by motivated and conscientious growers. Today it has a small market growing progressively year by year. In Switzerland about 30 tons of dry plants were produced under contract during 1999, in mid-range mountain areas between 600 and 1200 m altitude

(Gammetter, 2000). This production basically feeds local markets related to the food industry and the phytotherapy.

The Swiss organic label by the Association Suisse des Organisations d'Agriculture Biologique (ASOAB) can only be obtained under stricter conditions than in the normal European directives for cultivation (prohibition of synthetic chemical fertilisers, herbicides, insecticides and fungicides). The industries interested in the organic products usually develop their own pricing system taking into consideration the higher production costs, which must be calculated as the sum of a salary about 15–20 Swiss francs (9–12 US$) per hour plus the specific costs for the cultivation.

Today, based on the selection activities developed by the 'Station fédérale de recherches en productions végétales de Changins à Conthey', the growers have developed thyme varieties available that are adapted to the continental climate of mountain ranges. These varieties are thymol chemotypes (Rey, 1993a, 1994a,b). They stand out due to their homogeneity, productivity, and good yield in essential oil (>3.5 per cent). It should be noted that the Swiss Pharmacopoeia VII had fixed the minimum essential oil content at 1.5 per cent, in the dry herbs of thyme. Today the European Pharmacopoeia demands a minimum of 1.2 per cent oil content for thyme herb.

For the success of the culture of this thermophyllous species it is of great importance to choose a convenient place. In mountain ranges between 700 and 1200 m altitude, warm, sheltered places guarantee good productivity and good quality. Under these conditions, avoiding harmful frosts, the cultivation may continue for 3–5 years. These qualities have made the Valais and Poschiavo valleys (Grisons) the preferred places for culture.

Cultivation

Common thyme (*T. vulgaris*) prefers a light and permeable soil, somewhat rich in organic matter and mineral fertilising elements (ITEIPMAI, 1983; Rey, 1990a; Anonymous, 1992, 2000). Its culture is preferably established on land previously cultivated with cereals or legumes. Bovine manure ($0.5–1 m^3$/are) constitutes the basic fertilizer. SRVA's directives for thyme are that the usual fertilisation is 80 nitrogen, 80 phosphorus and 100 potassium units. The establishment of the culture is done by mid-May at the usual plant density of 57 000 plants/ha (70 × 25 cm). Direct sowing by early September is more advantageous. This technique, recently fine-tuned, is only possible in proper soils (Rey, 1993b). It is important to note that plants bearing only few pairs of leaves at the beginning of the winter (2–5 cm height) are quite frost-resistant. Weeds must be removed 4 or 5 times during the season to maintain the culture adequately. Although selective herbicides are properly used for industrial culture, they are strictly prohibited in organic cultures. Thus, in order to reduce this unpleasant work, cultivation with black plastic is possible (My-Pex plastic). Regular watering enhances the response after spring harvest and guarantees a second harvest by mid-August. Concerning parasites, one must expect to find cicadas (*Eupterix decemnotata* Rey) which may cause damage in hot summers. But this damage is usually of little importance, and antiparasitic protection based on natural authorised insecticides (Parexan, Biocide...) is sufficient.

Provided that cultivation starts in spring, the yield for *T. vulgaris* is from 2000 to 2500 kg/ha of dry plant material in the first year after the August harvest. From the second year on, the harvests in June and August provide an annual yield of

Figure 6.1 Culture of common thyme in full bloom before harvest.

3500–5000 kg/ha. A culture started by seeds directly sown on the fields in September allow two harvests the following year. If the spring harvest (Figure 6.1) is made at full bloom (depending on the kind of market), the August harvest is always performed on the branches bearing leaves, since the plants do not produce new flowers, or just a little bloom. The harvest is carried out during dry and sunny weather using motorized shears or a harvester with an adequate cutting bar (Figure 6.2). Cutting at a height of 10–15 (20) cm above the soil level is advisable in order to avoid problems with frost. If harvesting takes place too late at the end of the summer, problems with cold temperatures may arise (Rey, 1991).

Qualitative and economical aspects

There are two aspects that influence the quality: (a) The proportion of leaves and stems, leaves presenting a higher quality, and (b) The content of essential oil. In order to obtain a proportion of leaves higher than 50 per cent, it is important to harvest before blooming. Contrarily, to obtain an optimal yield in essential oil, harvest during full bloom is preferable. The yield in essential oil is dependent on local climatic conditions and on seasonal variations. Thus, in mid-range mountain cultures where two harvests per year are obtained, a 50–100 per cent improvement in the yield of the summer harvest with respect to the spring harvest can be achieved.

The drying of the material must be done carefully in dry places and protected from light. The plants are placed in stratified beds of 1–2 m height, which are ventilated with warm air at 30 and 45 °C. Under these conditions drying takes 2–3 days

Figure 6.2 Harvest of common thyme in a mountain field.

(10–12 per cent moisture as a maximum at the end of the process) depending on the density of the beds, thus keeping all the intrinsic properties as well as a good appearance. With regard to organic cultures the market for the pure leaf is small. If demanded, separation of stems and leaves is done using sieves.

Cultivation in the mountains, developing patches at different altitudes, notoriously increases the production costs because it is not possible to provide a mechanised response to all needs of natural cultivation. The culture of thyme needs about 1 500–2 000 hours of work per hectare, more than half of the time due to the elimination of weeds.

SELECTION OF COMMON THYME (*THYMUS VULGARIS* L.) FOR MARGINAL AREAS

In marginal areas such as the mid-range mountain, only the common thyme from the German race, or 'German thyme' (AL), could be cultivated with success. With respect to the Mediterranean thyme or the French thyme, it is better adapted to lower temperatures. However, it has quite a heterogeneous phenotype, and this results in less regularity of the culture and lower quality of the final product. The yield in essential oil is insufficient and does not reach the minimum of 1.5 per cent required by the Swiss Pharmacopoeia VII or 1.2 per cent by the European Pharmacopoeia. People growing thyme in the mountains have demanded a homogeneous variety better adapted to the peculiarities of a mountainous climate that can be grown from seed. With the aim of fulfilling this demand, a selection program was started at the end of the 1980s at the 'Station fédérale de recherches en productions végétales de Changins' at 'Centre des Fougères de Conthey' (Valais).

Plant material

Common thyme (*T. vulgaris*) is of Mediterranean origin. The natural distribution area is from Italy to Spain. It is mostly erect, with lignified stems of 10–40 cm height. The leaves, with involute margins, are linear to lanceolate and of variable size. They have an acute tip and bear glands. On the upper side they are greyish-green to greyish-blue, while being whittish in the lower face. The flowering stems bear capitated or verticillated inflorescences. The flowers, with pink petals, bloom from April to June depending on the altitude. Hermaphrodite flowers are bigger than female ones. The seeds become mature 1 month after the bloom. One thousand seeds weigh about 0.25–0.29 g. The leaves, flowers and herbaceous stems bear glandular hairs that contain the essential oil. The tector hairs that constitute the hairy characteristic of the leaves and stems protect the plant from evaporation of water. In natural conditions a thyme plant may live for 15 years.

The floral biology of common thyme has been described in detail by Assouad and Valdeyron (1975). To maintain the vigour of the species, the crossed fertilisation between male fertile (hermaphrodites, MF) and male sterile (MS), enhanced by protandry (development of anthers before pistils), is more frequently found, although auto-fertilisation is possible among MF individuals. The use of male-sterility of common thyme for breeding and production of new varieties was first proposed by Rey (1990a). Another aspect to keep in mind during the selection activities is that the European Pharmacopoeia focuses on the thymol chemotype, postulating 30–70 per cent thymol and 3–15 per cent carvacrol. The linalool and carvacrol chemotypes are preferred by the condiment market (ITEIPMAI, 1983).

The commercial varieties/populations of German common thyme and a natural population of common thyme in the Aoste valley in northern Italy have served as starting material for the selection. German thyme is a result of an empiric selection managed by generations of farmers. Its frost-resistance allows it to be cultivated even in northern Europe, e.g. Holland and Finland (Aflatuni *et al.*, 1994; Simojoki *et al.*, 1994) or in Canada (Laflamme *et al.*, 1994) (at least in annual culture), but its quality is not as good as when produced in more southern latitudes. Its foliage is greyish-green, and it blooms later (about 10 days) than the thyme from the Aoste valley. Its habitus is mainly erect. German thyme is generally quite heterogeneous with respect to its vigour and the colour and dimension of the flowers.

The thyme from the Aoste valley or 'Valdôtain thyme' (here abbreviated as VA) corresponds to a more typical Mediterranean thyme. It is more sensitive to frost than German thyme, but its quality with regard to essential oils is superior. Its foliage is mainly greyish-blue and the stems are more lignified, resulting in a more erect habitus. The Valdôtain ecotype represents the most northern occurrence for this species. In this internal valley of the Alps, at subcontinental climate, thyme is placed in the warmest areas and the driest places, at a maximum of altitude 1 600 m observed in this region. As a pioneering species in these limiting conditions common thyme presents a wide variability of the phenotypic characters such as habitus, vigour, colour and size of the leaves and flowers, and its sex, and therefore the plants are interesting elements for selection. In contrast to that the chemical variation is low, only the thymol chemotype being found in this area. This chemotype, preferring the warmer and drier areas of the Mediterranean climate, represents a good bioclimatic indicator of such conditions in the area (Rey, 1989, 1990b). Although being a stable chemotype, a certain variability is observed in the yield of essential oil among the individuals from this population.

Method of selection

A large number of essays for comparison of German thyme and Valdôtain strains were made at the mountain sites of Arbaz valley (920 m altitude, sunny exposure) and Bruson (1 100 m altitude, ombrous exposure) and have preceded and oriented the selection activities (Rey, 1988, 1993a, 1994b). As an example, Table 6.1 shows the differences of yields in dry matter and active matter, as well as the differences in flowering and frost-resistance obtained from German thyme and Valdôtain thyme.

The complementary characters of these two races of common thyme suggested the idea of making crosses between them, in the hope of obtaining more regularity, vigour and quality in the hybrids. A study of the variability within the best provenances has been made. Thus, it was possible to isolate and multiply by classic cutting or *in vitro* culture (Lê, 1989) the elite plants from both sexes MS and MF. After 1989 many crossings have been made each year using the best clones. The cultivation of parental clones with the aim of obtaining hybrids, as well as the method itself, have been described by Rey (1993a). Briefly, the scheme for selection is as follows:

(a) Localisation of elite plants, male-sterile and male-fertile, in the natural populations of the Aoste valley and in the varieties/populations of German thyme.
(b) Vegetative multiplication by spring cuttings or micropropagation of these initial clones to verify their performance under field conditions (agronomic test, laboratory analysis and annotation of flowering period). Only the best are retained to test their crossing value.
(c) Isolation and crossing of the best clones using bees, by pairs MS–MF. Only the seeds from male-sterile parents are collected.
(d) This hybrid seed is sown and the experimental hybrid judged for performance and homogeneity. The parents of the best hybrid are multiplied on a large scale. The production of seeds of the commercial hybrid may thus start under isolation.

More than 120 different crossings have been made up to now. The results presented here concern crossings performed in 1989 up to 1991 (Figure 6.3; see also Rey, 1993a, b) for which we have 3 years of harvests, this being the normal life span for a culture of German thyme. These results are based on a mean of 50 plants.

Productivity

With respect to productivity we recognise four criteria to characterise the yields and quality for each variety. These are:

Yield in dry matter. In the 1989 experiment, with 3 years accumulated and comprising 5 harvests in total, the mean yield in dry matter from hybrids VA×AL was 1.95 times higher than the reference, 1.45 times higher than the polycross VA×AL and 1.72

Table 6.1 Comparison of yields between Valdôtain thyme from Petit-Bruson (population) and German thyme, in two experimental sites, Arbaz and Bruson (VS), in 1987 and 1988 (culture established in the end of May 1987)

Provenance	Cultures	Harvest dates	Yields				Observations			
							Flowering		Ice	
			Dry matter (g/m²)	Leaves/branches %	Essential oil %	Essential oil l/ha	Plants in flower %	Degree	Frozen plants %	Frozen buds %
Petit-Bruson (Aoste)	Arbaz (920 m)	20/08/87	118	62.0	5.15	37.7	20			
		7/06/88	179	62.5	4.80	53.7	100	3/4	11	50
		29/07/88	150	61.5	5.20	48.0	5			
			447*		5.05†	139.4* rank: 1st				
	Bruson (1100 m)	26/08/87	82	60.0	4.85	23.9	7			
		17/06/88	63	61.9	3.45	13.3	50	1/2	21.7	100
		9/08/88	89	58.6	4.75	24.8	0		27.0	62.5
			234*		4.35†	62.0* rank: 4th				
Magnin (German)	Arbaz (920 m)	20/08/87	288	58.0	2.40	40.1	100	1/2		
		7/06/88	389	53.1	2.35	48.6	75	1/2	7	25
		29/07/88	219	61.0	2.65	35.4	100	1/4		
			896*		2.47†	124.1* rank: 2nd				
	Bruson (1100 m)	26/08/87	196	60.0	2.30	27.1	100	1/2		
		17/06/88	197	60.2	1.65	19.6	30	1/4	14	50
		9/08/88	170	60.0	2.65	27.0	80	1/4		
			563*		2.20†	73.7* rank: 3rd				

Notes
* Total.
† Mean.

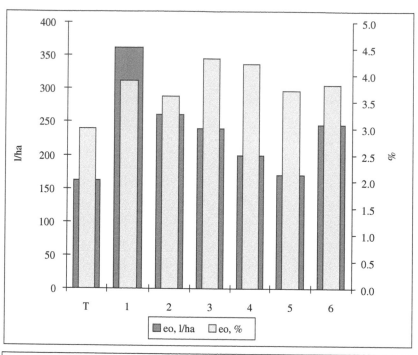

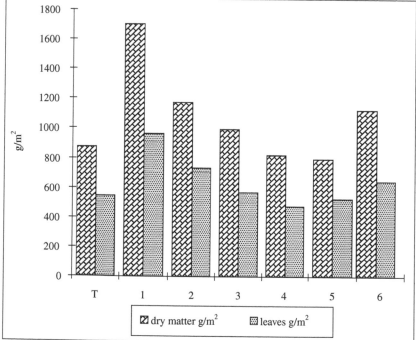

Figure 6.3 Productivity of hybrids of thyme in 1989 up to 1991 (5 harvests in 3 years). Upper graph: cumulative yields of essential oil (EO) in l/ha and mean percentages of EO. Lower graph: cumulative yields of dry matter and leaves in g/m^2. T = control AL; 1 = Hybrids VA×AL; 2 = Polycross VA×4 AL; 3 = Hybrid VA×VA; 4 = Clones VA MS; 5 = Clone AL MF; 6 = Clone VA MF

times higher than the hybrid VA×VA (Figure 6.3). This advantage was observable from the first harvest in the second year of culture. The yields from the Valdôtain and German parents is quite different from those presented by the hybrid itself. In the experiment in 1991, the higher mean yield of the four hybrids VA×AL with respect to the reference was confirmed. The five retrocrossed hybrids (VA×AL) × VA and (VA×AL)×AL lose all advantages with respect to F1 cross. Four autofecundations VA and AL showed the depressing effect attached to consanguinity.

Yield in leaves. Calculated from portions of dry plant, with 100 g, the leaves separated manually, these yields show the same patterns as the dry matter. In the 1989 experiment, the yield in leaves of the hybrids VA×AL was superior (Figure 6.3) to the control, the polycross and the hybrid VA×VA, respectively. The hybrids were also superior to the parents here.

Percentage of essential oil (EO). It was obtained from samples of 100 g of dry leaves, mentioned in Figure 6.3, and are means for the 5 harvests. The hybrids VA×AL have surpassed 1.3 times the value of the control. Their level anyway was 10 per cent lower than the 2 MS Valdôtain parents and also the hybrid VA×VA. In the 1991 experiment, the superiority of hybrids VA×AL with respect to the control was confirmed. The hybrids AL×VA, VA×VA, AL×AL, (VA×AL)×VA and (VA×VA)×AL showed comparable results with a mean of more than 4 per cent. One should note the high level of 4.55 per cent of hybrid AL×AL that results from the careful selection of elite plants within a population of German thyme. Plants obtained from autofecundation VA and AL have proven to be low in essential oil.

Yield in l/ha of essential oil. For this parameter the hybrids VA×AL in the 1989 experiment for example, produced a mean of 362 l/ha, confirming their superiority with respect to the control (Figure 6.3). The polycrosses, with 260 l/ha, showed an intermediate value. The yields of the German and Valdôtain parents are respectively between 170 and 245 l/ha.

During these two crossing trials, the hybrids VA×AL and AL×VA have largely surpassed the AL controls, concerning the yields in dry matter, in leaves and in l/ha of essential oil. Considering the genetical distance between these two races of thyme, the effect of heterosis observed is not surprising. For the first experiment the medium level of hybrids VA×AL represents an appreciable enhancement of 30 per cent with respect to the AL control. The quality of hybrids VA×AL and AL×VA from the second experiment was also very satisfactory.

Homogeneity of the phenotype

In the 1989 experiment, the hybrids of clones VA×AL showed a very good regularity in the morphological characters with respect to the German control and the polycross VA×AL. This homogeneity concerned the habitus and dimension of the plants, the size and colour of the leaves and the flowers (Table 6.2). This was not comparable to the one obtained with the clonal culture. Similar results were reproduced in the second trial with certain hybrids VA×AL as well as some hybrids in the inverted way AL×VA. For the transmission of sex for example, 98–100 per cent MS individuals were obtained if the female parent was homozygous for this character. In the retrocrossed hybrids the sex is usually 50 per cent of MS and 50 per cent of MF, thus diminishing the phenotype regularity.

Table 6.2 Mean percentage for the transmission of floral characteristics and the colour of the leaves depending on the plant type

Plant-type	Flowers				Leaves		
	Sex %		Size %		Colour %		
	MS	MF	Small to medium	Medium to big	Grey-blue	Grey	Grey-green
Control AL	45	50	50	50	0	5	95
1. Hybrid VA 1 × AL	98	2	98	2	97	0	3
4. Clone VA 1 MS	100		100		100		
1. Hybrid VA 2 × AL	99	1	0	100	94	3	3
4. Clone VA 2 MS	100			100	100		
3. Hybrid VA × VA	99	1	98	2	97	0	3
2. Polycross VA × 4 AL	51.5	44	40	60	67	1.5	31.5
5. Clone AL MF		100		100			100
6. Clone VA MF		100		100	100		

Notes
AL, German.
VA, Valdôtain.
MS, Male sterile.
MF, Male fertile.

The homogeneity obtained by the clone hybrids is itself a great advantage with respect to the heterogeneity of the initial varieties/populations. Since the hybrid thyme is produced by crossing two heterozygous parents, its homogeneity is comparable to that of a double hybrid (from 4 parents). Only the simple hybrid with homozygous parents allows a rigorous homogeneity. The simple hybrids are easily obtained from autogamous species, spontaneously homozygous, but more difficult from allogamous species like thyme. For the polycross VA × AL, where the same parents MS VA are crossed with 4 parents MF AL, the sexual characters have the tendency to level off to a 1:1 proportion. This method of selection is thus not the best with respect to homogeneity matters.

Frost-resistance

Table 6.3 with the results obtained in the winter of 1990/1991 presents the damages due to frost sorted according to their severity. Common thyme, when frozen about 20–25 per cent, recovers completely and does not yield significantly less. On the other hand, plants strongly affected, about 30–90 per cent, show a considerably lower yield. The hybrids VA × AL were almost as resistant to winter frost as the German control plants. The polycross VA × AL were also the same. On the contrary, much lower resistance was noted for the hybrid VA × VA and its Valdôtain parents. The plants obtained by autofecundation were found more sensitive to frost than their respective parents. After 3 years of cultivation most of the hybrids showed a very high resistance to winter frost with respect to the German control, with more than 70 per cent frost damage.

Production of seeds

The crossings between Valdôtain and German clones produced a variable quantity of seed depending on the size of the plants, the flowering stadial and the quantity of

Table 6.3 Mean percentage of frost damages per plant, by the plant type in winter 1990/1991

Plant-type	Level of ice damage per plant	
	0–25%	30–90%
control AL	100	0
1. Hybrids VA×AL	98	2
2. Polycross VA×4 AL	100	0
3. Hybrid VA×VA	82	18
4. Clones VA MS	60	40
5. Clone AL MF	95	5
6. Clone VA MF	79	21

Notes
AL, German.
VA, Valdôtain.
MS, Male sterile.
MF, male fertile.

bees around. For the crossings studied in 1989 the production of seeds varied from 1.2 g for the VA×VA hybrid, to 5.5 g for a VA×AL polycross. The autofecundation of MF clone VA produced only 0.3 g of seeds, that is 4 times less than the VA×VA hybrid. The same tendencies were noted during the crossings developed afterwards, with a maximum production of 8.1 g for a VA×VA crossing. The mean production was 2–3 g per plant, or 10–15 g/m^2, and this is 30–100 per cent higher than the reports by Heeger (1956).

Germinative ability

Only allogamy can warrant the quality of the seeds, its germinative ability and strength. A low germination percentage and weak plantlets from autofecundation.

Cultural considerations

The heterogeneity of common thyme has so far been a problem for its rational culture by seeds. That is the reason why the culture of French clones (Mediterranean race) by lignified cuttings is strongly recommended for the Mediterranean countries. For Nordic countries the culture by seeds from German thyme, which is more frost-resistant, is desirable. In mountain areas such as the Valais and Poschiavo valleys, the limiting climatic conditions do not allow a safe production of French thyme. So far only the German thyme is cultivated from seeds grown in nurseries or to a smaller extent directly sown in fields (Rey, 1993c) always keeping in mind their cultural requirements (Rometsch, 1993). A race similar to French thyme but somewhat more rough, the cloned Valdôtain thyme, could be recommended for warm and protected situations. The hybrids between Valdôtain and German clones are more interesting because they are more robust.

VARICO, a new variety of thyme

In order to complement the demands from growers and consumers, a first variety of hybrid thyme named **VARICO** was placed on the market in 1994. It is characterised

Figure 6.4 VARICO thyme hybrid flowering in spring.

by the homogeneity in its phenotype, an erect habitus and great vigour (Figure 6.4). Its greyish-blue foliage differentiates it clearly from varieties/populations of German origin with their greyish-green colour. Its pale pink flowers bloom from May 20 until June 10 in Arbaz (Switzerland, 920 m alt.). The productivity of dry matter reaches more than 1500 g/m^2 after 3 years of culture. Its good quality is reflected in the mean yield in essential oil (3.9 per cent v/w after 5 harvests) and its percentage of thymol of over 50 per cent. Its resistance to frost is good (Rey, 1994a). The company DSP at Delley (Switzerland) produces VARICO thyme seeds. Keeping in mind the selection efforts involved, the price is higher than for other seeds in the market. The first cultures of VARICO thyme established by plantlets and by direct sowing satisfy the growers (Figure 6.5).

IN VITRO CULTURE OF THYMUS

The *in vitro* culture techniques provide a wide array of tools to the breeder that complement the selection activities performed *ex vitro* out of the laboratory. By modifying the conditions to which the plant is exposed the researcher is able to influence and even to determine the metabolic pathways in the plant. These conditions include a multitude of parameters such as temperature, quality and intensity of light, and composition of the substrate or the atmosphere. Mulder-Krieger *et al.* (1988) reviewed the production of essential oils and aromas in cell and tissue cultures of plants from about 70 species, including bryophytes, conifers, monocotyledons and dicotyledons. Frequently significant

Figure 6.5 Culture of VARICO thyme hybrid in Bassins (Switzerland) destinated to the production of
essential oil.

differences between *in vitro* and *in vivo* plant material are found; the desired compounds
may even be absent when grown in the laboratory.

Despite the potential applications of the *in vitro* culture methods few researchers
have used them with *Thymus.* Furmanowa and Olszowska started their research with
thyme using *T. vulgaris* reviewed in 1992. Lê (1989) took another approach using a
different culture medium following the modifications that Collet (1985) made to the
MS culture medium previously developed by Murashige and Skoog (1962). The CMS
medium suggested by Collet has also been applied by Sáez *et al.* (1984) working with
T. piperella, after realising the beneficial effect of these changes compared with the original
formulation of MS salts. The experiments performed with these two species and the
results obtained will be described briefly.

In vitro Culture of *T. vulgaris*

T. vulgaris was cultured *in vitro* for the first time by Furmanowa and Olszowska (1992)
and references therein. They regenerated plants from buds in Nitsch and Nitsch (NN)
culture medium, with varying concentrations of two auxins and three cytokinins, finding optimal results with the use of either (a) 0.1 mg/l Kin + 0.1 mg/l NAA. b) 0.1 mg/l
Kin + 0.3 mg/l IBA or 0.5 mg/l IBA. When using nodal segments, they tested different
concentrations of auxins (NAA, IBA) and cytokinins (Kin, BA, 2-iP). The best rooting
was achieved with 0.5 mg/l IBA, and cytokinins were found to have little or no effect
on plantlet development at 0.05 mg/l. When higher concentrations were used, gradual
inhibition of rooting and shoot growth was observed.

Lê (1989) tested eight treatments, i.e. two different compositions of mineral salts (MS, CMS) and four combinations of growth regulators . Using stem cuttings bearing axillary buds he concludes that the CMS medium without growth regulators performs best when trying to get the best development, and the presence of growth regulators disturbs the growth of axillary shoots, perhaps due to an interaction with these substances from an endogenous origin. This is in agreement with Furmanowa and Olszowska (1992), even though they use different culture media formulations. Lê explains the better performance of CMS medium than MS to be the reduction of concentration of the ammonium ion (NH_4^+) in the CMS medium. These results by Lê characterise mainly the *in vitro* establishment of thyme. Under these conditions he describes an anomalous development of axillary shoots in the presence of growth regulators. This was also observed by Sáez *et al.* (1994), but this behavior is only achieved at the beginning of the experiments. When the plantlets are acclimated to the *in vitro* conditions of growing, there is a more clear response to changes in the type and concentration of growth regulators.

The production of secondary metabolites in *T. vulgaris* was the object of Tamura *et al.* (1993), who focused on the selection of the callus cells and management of cell environments for the production of flavour metabolites. They recorded the relation between the color of the callus and the presence of thymol and carvacrol. They found that on an agar medium, green and yellow calli produced trace amounts of thymol, while white calli produced ethanol and no thymol. The addition of mevalonic acid enhanced the amount of the volatiles produced up to two-fold in comparison to the reference control, showing that enzyme activities of the monoterpene synthesis were latent in the calli. Unfortunately the oil yield was very low, about 1/500 to 1/1000 of the normal plant.

In vitro culture of *T. piperella*

The *in vitro* culture of *T. piperella* was started by Sáez *et al.* (1994) with the aim of testing the behaviour of a species quite different from the already studied *T. vulgaris*. Indeed *T. piperella* presents a more herbaceous habit, sometimes becoming lianoid, with very long internodes. It is much less hairy, with glabrous leaves. Additionally, it is of economic interest as a source of phenols and as a condiment. Furthermore, it is a species with a small dispersal area and therefore it may not be collected intensively.

The procedure involves three major steps: (a) Acquisition of an *in vitro* established population of individuals growing in the same conditions. (b) Determination of the effects by different growth regulators on several characteristics of the plant material, to help to select the culture media adapted to specific needs: multiplication, rooting. (c) Determination of the effects that different concentrations of macronutrients, sucrose and vitamins, have on the production of roots, to help to select the culture media adequate for root production before putting the plants in *ex vitro* conditions. These steps are described below.

Obtaining of plantlets growing in vitro

The initial step for obtaining an *in vitro* plant population was the collection of seeds from natural populations (Figures 6.6 a,b). The bigger ones were selected and stored at a low temperature until the activities in the laboratory started. Previous experiments had shown two problems in these initial steps, one being the difficulties to remove completely

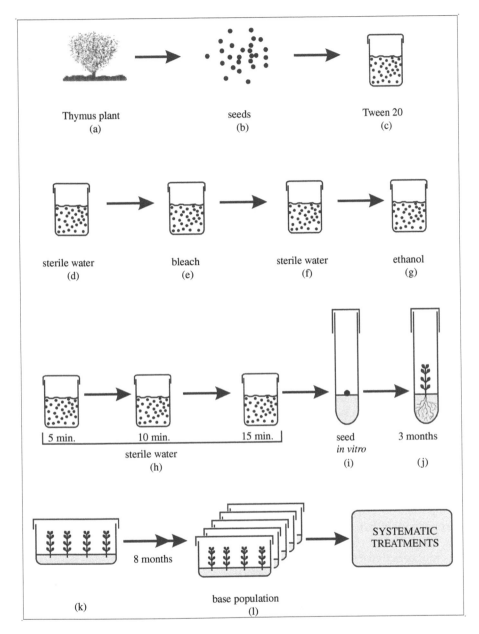

Thymus plant
(a)

seeds
(b)

Tween 20
(c)

sterile water
(d)

bleach
(e)

sterile water
(f)

ethanol
(g)

5 min.

10 min.

15 min.

seed
in vitro
(i)

3 months
(j)

sterile water
(h)

8 months

SYSTEMATIC
TREATMENTS

(k)

base population
(l)

Figure 6.6 Protocol for the establishment of an initial population of *T. piperella* plants in vitro.

the contaminants (bacteria, fungi) from the surfaces, and the other related to the different ability to germinate and/or develop the initial plantlets.

The procedure to remove contaminants from the seed surface is described in Figure 6.6 c–h. Once cleaned, the seeds are placed on the surface of the culture medium inside glass tubes, 3–4 seeds per tube. Some of the tubes containing the seeds have to be removed due to fungal growth (clearly visible after 2 days), bacterial growth (after 5–6 days) or poor growth of the young plants, sometimes affected by vitrification (characterised by an excess of water in their tissues).

After about two months the plants are 5–6 cm long and present several internodes (Figure 6.6 j). They are cut into portions with 2–3 internodes and transferred to larger glass vessels (Figure 6.6 k) with translucent caps, the roots being removed. They produce axillary shoots that are periodically removed and transferred to new vessels, thus producing a population of plantlets growing under the same conditions (Figure 6.6 l). During this period of time, a change in the morphology of the leaves occurs, from smaller, deeper green and tough, to wider, lighter green and herbaceous consistency can be noticed.

The systematic treatments used portions of plants from this *in vitro* population containing an apical bud and three nodes. They were transferred to new vessels with a culture medium adequate to test the different combinations of growth regulators (2 cytokinins and 2 auxins), sucrose, macronutrients and vitamins. Two groups of treatments were performed. The first one tested five levels in the concentration of each growth regulator in the products IAA×BA, IAA×Kin, NAA×BA and NAA×Kin (100 combinations). The second one tested three concentrations of sucrose, four of macronutrients and the presence/absence of vitamins (24 combinations). These two experiments were run independently.

Influence of growth regulators

The effects of the addition of an auxin and a cytokinin to the growth media were measured by (a) the number of shoots greater than 5 mm long, (b) the number of shoots less than 5 mm long, (c) the quantity of roots produced per explant, (d) the number of explants that showed abnormal growth, either by the presence of callogenic structures in the base or by vitrification. The axillary shoots produced were classified into two categories due to the higher ability of shoots greater than 5 mm to produce new well-developed plantlets. Figure 6.9 represents the results (mean, standard error of the mean and 95 per cent confidence intervals) obtained for the number of shoots >5 mm, comparing the different growth regulators and the concentrations used for each one. Stronger activity of BA than Kin, when promoting the development of axillary shoots, and when inhibiting the formation of roots was found. Similarly, NAA as an auxin presents higher activity than IAA, expressed as stronger inhibition of shoot development and promotion of root development. Furthermore, NAA produced calli at the base of the explants more readily than IAA. Considering to all the variables studied, the most suitable combinations of growth regulators were found to be 1.0 mg/l BA or 1.5 mg/l BA without auxin to promote shoot growth, and 0.5 mg/l IAA without cytokinin to promote root growth. Figures 6.7 and 6.8 show the effects of different combinations of growth regulators.

Influence of macronutrients, sucrose and vitamins

The role of macronutrients, sucrose and vitamins added to the culture medium in relation to the root development of the plantlets was tested in 24 experiments containing different variations of them plus the addition of 0.5 mg/l IAA (proved to enhance this metabolism when testing the different growth regulators). The results are shown in

Figure 6.7 T. piperella growing *in vitro* with high concentration of cytokinins.

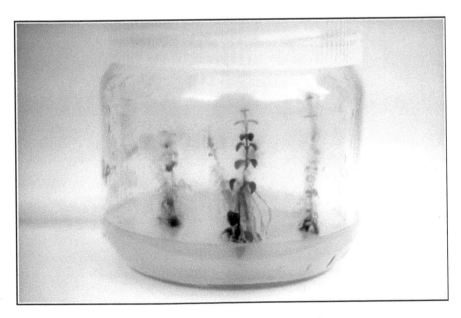

Figure 6.8 T. piperella growing *in vitro* with high concentration of auxins.

Figure 6.10. The best combination seems to be a culture medium with 25 per cent sucrose, and 50 per cent or 25 per cent macronutrients plus the addition of vitamins. It is advisable to reduce the concentration of sucrose to promote rhizogenesis, but in the absence of vitamins, low levels of sucrose gave the lowest yields. Also a reduction to

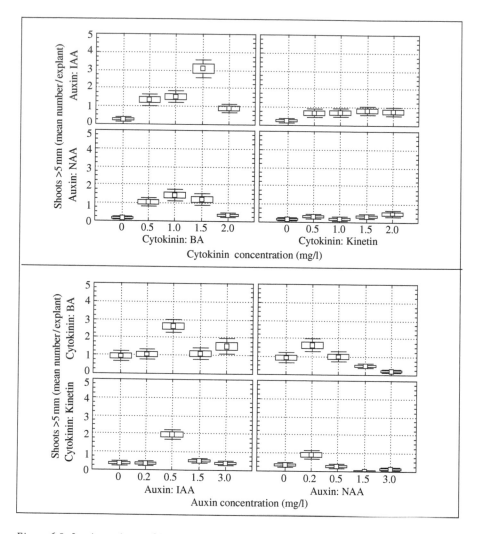

Figure 6.9 In vitro culture of *T. piperella*. Production of axillary shoots greater than 5 mm long. Comparison of results by cytokinin (up) and auxin (down) concentrations.

50 per cent or 25 per cent of the macronutrients seems to promote rhizogenesis. There is a very high variability in the results obtained, suggesting a certain influence from endogenous growth regulators.

Résumé

Obtaining different varieties of thyme suitable for field culture under different climatic conditions is highly desirable, as previously stated. Renewed efforts have to be made in the localisation of ecological/chemical variants in nature, as well as in the characterisation of their properties under controlled environments, both *ex vitro* and *in vitro*. New

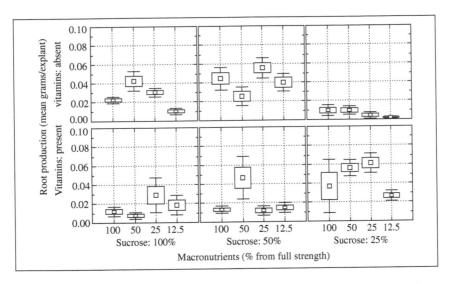

Figure 6.10 In vitro culture of *T. piperella.* Root production in different concentrations of macronutrients, sucrose, and vitamins, in CMS culture medium supplemented with 0.5 mg/l of IAA.

techniques such as production of haploid and doubled-haploid plants from microspores may help to obtain elite cultivars in the near future.

ACKNOWLEDGEMENTS

The collaboration of the following persons is sincerely appreciated: P. Achermann, P. Bruchez, J.-P. Bouverat-Bernier, J. Burri, C.-A. Carron, F. Fournier, A. Fossati, B. Galambosi, J. Grétillat, P. Imhof, L. Laflamme, G. Marguerettaz-Gaetani, B. Nendaz, I. Kapetanis, A. Schori, I. Slacanin and N. Verlet. Thanks are given also to 'Fundación Séneca', Murcia, Spain.

REFERENCES

Aflatuni A., Pessala R., Hupila I., Simojoki P., Huhta H., Virri K., Kemppainen R., Järvi A. and Galambosi G. (1994) Yield of *Thymus vulgaris, Melissa officinalis* and *Origanum vulgare* grown between 601 and 681 latitudes in Finland. Abstract of seminar n1. 240. *Production of herbs, spices and medicinal plants in the nordic countries.* Mikkeli, Finland, August 2–3.

Anonymous (1992) ITEIPMAI. Technical dossier for balm *(Melissa officinalis,* Lamiaceae). ITEIPMAI, Chemillé, Angers.

Anonymous (2000) SRVA. Technical dossier for common thyme. In *File for Medicinal and Aromatic Plants.* Working group coordinated by P. Amsler, SRVA, Lausanne.

Assouad, W. and Valdeyron, G. (1975) Remarques sur la biologie du thym, *Thymus vulgaris* L. *Bull. Soc. Bot. Fr.,* **122,** 21–34.

Collet, G.F. (1985) Enracinement amélioré lors de la production *in vitro* de rosiers. *Rev. Suisse Viticult. Arboric. Hort.,* **17,** 259–263.

Furmanowa, M. and Olszowska, O. (1992) Micropropagation of Thyme (*Thymus vulgaris* L.). In Y.P.S. Bajaj (ed.), *Biotechnology in Agriculture and Forestry*, 19, pp. 230–243.

Gammetter, M. (2000) Annual report Plantamont 1999.

Heeger E. F. (1956) *Handbuch des Arznei- und Gewürzpflanzenbaues. Drogengewinnung.* Deutscher Bauernverlag, Berlin. pp. 775.

ITEIPMAI (1983) Domestication de la production, conditionnement et définition du thym (*Thymus vulgaris* L.). Bull. d'inform. série Monographie.

Laflamme L., Tremblay N. and Martel C. (1994) Winter survival of medicinals plants in Québec (Canada). Compte-rendu du séminaire no. 240. *Production of herbs, spices and medicinal plants in the nordic countries*. Mikkeli, Finland, August 2–3.

Lawrence, B.M. (1992) Chemical components of *Labiatae* oils and their exploitation. In R.M. Harley and T. Reynolds (eds), *Advances in Labiate Science*, Royal Botanic Gardens, Kew, pp. 399–436.

Lê, C. L. (1989) Microbouturage *in vitro* du thym (*Thymus vulgaris* L.). *Revue suisse Vitic. Arboric. Hortic.*, 21, 355–358.

Mulder-Krieger, T., Verpoorte, R., Baerheim Svendsen, A. and Scheffer, J.J.C. (1988) Production of essential oils and flavors in plant cell and tissue cultures. A review. *Plant Cell, Tissue and Organ Culture*, 13, 85–154.

Murashige, T. and Skoog, F. (1962) A revised medium for rapid growth and bioassays with tobacco tissue cultures. *Physiol. Plant*, 15, 473–497.

Rey, Ch. (1988) Comparaison de provenances de thym vulgaire. Internal report RAC.

Rey, Ch. (1989) Le thym vulgaire (*Thymus vulgaris* L.) du Val d'Aoste: une particularité botanique de haut intérêt. *Rev. Valdôtaine d'Hist. Naturelle*, 43 , 79–97.

Rey, Ch. (1990a) La culture du thym en Suisse. *Revue horticole suisse*, 63, 20–22.

Rey, Ch. (1990b) *Thymus vulgaris* L. du Val d'Aoste (Italie): un écotype intéressant pour les zones marginales. *Revue suisse Vitic., Arboric., Hortic.*, 22, 313–324.

Rey, Ch. (1991) Incidence de la date et de hauteur de coupe en première année de culture sur la productivité de la sauge officinale et du thym vulgaire. *Revue suisse Vitic. Arboric. Hortic.* 23, 137–143.

Rey, Ch. (1993a) Hybrides de thym prometteurs pour la montagne. *Revue suisse Vitic. Arboric. Hortic.*, 25, 269–275.

Rey, Ch. (1993b) Selection of tyme (*Thymus vulgaris* L.). *Acta Hort.*, 344, 404–410.

Rey, Ch. (1993c) Semis direct au champ du thym (*Thymus vulgaris* L.). *Revue suisse Vitic. Arboric. Hortic.*, 25, 401–403.

Rey, Ch. (1994a) Une variété de thym vulgaire 'Varico'. *Revue suisse Vitic. Arboric. Hortic.*, 26, 249–250.

Rey, Ch. (1994b) La sélection du thym (*Thymus vulgaris* L.). Actes du 3e Colloque Médiplant sur le thème 'Ressources et potentiels de la Flore médicinale des AlpesA', 20 october 1994, Domaine de Bruson (RAC) Bruson-Bagnes (Valais/Suisse).

Rometsch, S. (1993) Ecology and cultivation assessment of Thyme (*Thymus vulgaris* L.) in the Canton Valais, Switzerland. *Acta Hort.*, 344, 411–415.

Sáez, F., Sánchez, P. and Piqueras, A. (1994) Micropropagation of *Thymus piperella. Plant Cell, Tissue and Organ Culture*, 39, 269–272.

Simojoki P., Hupila I., Pessala R., Galambosi B. and Aflatuni A. (1994) Yield potential of thyme, lemon balm and anyse hyssop grown in different latitudes of Finland. Abstract of seminaire no. 240. *Production of herbs, spices and medicinal plants in the nordic countries.* Mikkeli, Finland, August 2–3.

Tamura, H., Takebayashi, T. and Sugisawa, H. (1993) *Thymus vulgaris* L. (thyme): *In vitro* culture and the production of secondary metabolites. In Y.P.S. Bajaj. (ed.), *Biotechnology in Agriculture and Forestry*, 21, pp. 413–425.

7 Harvesting and post-harvest handling in the genus *Thymus*

Petras R. Venskutonis

INTRODUCTION

There are many *Thymus* species, however only few of them are of commercial significance (Reineccius, 1994; Clarke, 1994), namely *T. vulgaris* L., *T. zygis* L. ssp. *gracilis* Boiss. (red thyme), *T. satureioides* Cosson, *T. serpyllum* L. (wild thyme), *T. capitatus* Hoffmanns. and Link (syn. *Thymbra capitata* (L.) Cav., Spanish "origanum"). The first two are the most widely used *Thymus* species and surveyed literature sources deal mainly with these two herbs. *T. vulgaris* is the only species, which is cultivated commercially in reasonable amounts. Other *Thymus* species are collected in wild-growing sites mainly as sources of dried medicinal herbs.

In order to obtain the product of the best quality, harvesting as well as growing and all other processing steps of thyme should be optimised considering several important factors. These factors, which in general are very common to many aromatic and medicinal plants, shall be briefly discussed in this chapter. Two flow diagrams of manual and mechanised herb harvesting and processing are presented in Figure 7.1 (Heindl and Müller, 1997). The cutting and processing steps shown in the diagram represent most traditional procedures, which have been generally used in the preparation of dried aromatic and medicinal herbs.

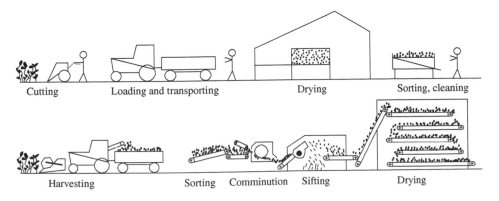

Figure 7.1 Flow diagrams of harvesting and processing of herbs (Heindl and Müller, 1997).

HARVESTING METHODS AND THEIR INFLUENCE ON THE QUALITY OF THE PRODUCT

Harvesting date, time of day and weather conditions

T. vulgaris is indigenous to the Mediterranean region, however it is also widely cultivated in different parts of the world including North America, Europe and North Africa. In general, thyme is most aromatic during the period of blooming (or at the beginning of full bloom); therefore this period is considered as the best time for harvesting. However, the period of vegetation and blooming can be different in various geographical zones depending on their climatic conditions. Spain is one of the major producers of thyme in the Mediterranean region, and harvest takes place during the blooming period from February to August, depending on the species (Tainter and Grenis, 1993). Collection of plant material for Moroccan thyme oil (*T. satureioides*) takes place in the beginning of April with plants (flowering tops) growing at 1000 m above sea level, the harvest ends in August with plants growing up to 2000 m. In France, thyme can be harvested twice a year, once in May and then again in September.

In the central regions of Russia thyme is harvested during the second year of plant vegetation (Poludennij and Zhuravlev, 1989). Usually, the first cutting is performed in June during flowering, the second one in September–October, i.e. 1.5–2 months before the end of vegetation. Aerial parts are cut at 10–15 cm height from ground. *T. serpyllum* is harvested starting from the first year of vegetation in June–July at the beginning of full blooming (Kudinov *et al.*, 1986; Mashanov and Pokrovskij, 1991). It has been recommended to harvest *T. serpyllum* during the flowering period. The best time is considered to be the middle of August, because after cutting there will be enough time for new stems to grow and this helps plants to prepare for the winter (Jaskonis *et al.*, 1983).

The quality of the herb depends on the site and climatic conditions of growing. Accumulation of essential oil (EO) in herbs directly or indirectly depends on light. Li *et al.* (1996) studied the essential oil production in thyme and sage grown at 15 per cent, 27 per cent, 45 per cent and 100 per cent of full sunlight. Their results showed that the highest yield of essential oils and percentage of thymol and myrcene in thyme occurred in full sunlight.

Weather conditions during the day of harvest are also important. In general, sunny days should be preferred, after morning dew has disappeared. The plants harvested after rain or with dew moisture are difficult to dry; they deteriorate much faster, their colour and useful properties become inferior (Kauniene and Kaunas, 1991).

Harvesting techniques and machinery

Most wild growing medicinal plants are collected by hand. The main objective of such harvesting is to collect the most valuable anatomical parts of the plants. In case of thyme (*T. serpyllum*, *T. pulegioides*, *T. vulgaris* and other species) these are the leaves and flowering parts. Woody stems, which are of minor value, must be avoided as far as possible. For instance, recently Guillén and Manzanos (1998) studied the composition of a Spanish *T. vulgaris* and found that the yield obtained by the extraction with dichloromethane from leaves and flowers was much higher than that obtained from stems; the chemical composition of the extracts obtained from different anatomical parts was also different. It is also important to avoid damaging plant growing sites to ensure renewable crops during the following years of harvesting. Simple means are suggested in the numerous manuals for medicinal herb gatherers, e.g. to use scissors or knives to cut the herb instead of tearing it out with roots (Börngen, 1979). *T. vulgaris* as a commercially cultivated herb can be harvested more effectively by common harvesting machinery, which cuts the plant at 10–15 cm height from the ground (Mashanov and Pokrovskij, 1991).

Effect of harvesting time on the essential oil content and composition

The essential oil yield and chemical composition are the most important characteristics of aromatic herbs. Senatore (1996) investigated the oil of a wild Italian thyme, *Thymus pulegioides* L., at different growth times (Table 7.1). From April to September the essential oils consisted mainly of α-terpinene, *p*-cymene, thymol, and carvacrol, which varied from 57.3 to 62.5 per cent of the total oil content. From their results it was concluded that the best time to harvest this species of thyme, for both essential oil yield and phenol content, is during or immediately after full bloom.

Cabo *et al.* (1987) collected the aerial parts of *Thymus hyemalis* Lange throughout its complete vegetative cycle (April 1981 to March 1982) and determined the content of essential oil and its composition. The results of this study are summarised in Table 7.2, showing that the oil yield varied from 0.15 per cent (December 23) to 0.58 per cent (July 4). The percentages of the main constituents were quite stable during the period of investigation with few exceptions. For instance, the content of monoterpene hydrocarbons (MH) was significantly lower during the period of September–November and February–March. Thymol and carvacrol were determined in reasonable amounts only during October–December. However, it should be noted that recently Sáez (1998) reported the results of a comprehensive study of the variability in essential oils from populations of *T. hyemalis*, which supported the concept that the linalool chemotype is intrinsic to this species, while borneol and 1,8-cineole types are not. The latter compound was one of the major constituents in *T. hyemalis* investigated by Cabo *et al.* (1987).

Karawya and Hifnawy (1974) examined the oil of *T. vulgaris* grown in Egypt and collected at different stages of growth, before flowering, during flowering, and at fruiting stages. They found the highest thymol and carvacrol concentrations during the beginning of the flowering stage.

Arrebola *et al.* (1994) investigated *Thymus serpylloides* Bory ssp. *serpylloides* for 3 years. They found that the oil content in stems obtained by steam distillation was lower than 0.5 per cent v/w. Independent of its sexual characteristics, the highest oil yield had been obtained from plants collected during the full flowering period (average measures of the 3 years: 1.56 per cent v/w from hermaphrodite, and 1.05 per cent v/w from females). The authors also observed that the highest content in carvacrol was found in a sample of hermaphrodites collected during full flowering in 1991, while the lower content was found during fruiting stage. In previous years, the amount of carvacrol during the full flowering period was lower (1989) or quite similar (1990) to the fruiting stage. The annual variation of carvacrol may be due to variations in the climate experienced during the 3 years of harvesting.

The essential oil composition of *T. vulgaris* grown in Lithuania has been investigated at different vegetation phases (Venskutonis, unpublished data). The results on the total oil content and some of its constituents in leaves (L) and flowers (F) are tabulated

Table 7.1 Essential oil yield[a] and composition[b] of Italian thyme (*Thymus pulegioides* L.) at different dates of harvest

Characteristic	Apr 18[c]	May 2[c]	May 12[c]	May 12[d]	May 24[c]	May 31[c]	May 31[d]	June 10[f]	June 21[c]	July 7[c]	July 20[f]	Aug 2[f]	Aug 19[f]	Sept 13[c]
Yield of oil	0.38	0.71	0.84	0.50	1.11	1.01	0.87	0.90	0.93	0.87	0.78	0.70	0.68	0.59
MH	50.3	41.3	39.6	37.4	32.6	29.2	29.9	33.6	36.8	39.3	37.8	36.3	42.2	44.8
OCM	32.4	39.3	43.0	46.8	45.0	45.9	52.4	43.2	40.9	39.4	40.3	39.4	35.6	34.9
p-Cymene+γ-terpinene	43.3	33.7	32.3	30.4	28.1	23.1	25.1	27.9	30.0	32.5	33.2	33.5	37.5	38.7
Thymol+carvacrol	19.2	23.6	26.5	33.9	32.5	36.2	43.3	33.9	31.6	29.3	27.4	25.2	24.0	22.4
TPC	26.8	31.7	34.3	43.4	39.4	40.8	49.4	38.4	35.7	33.4	31.2	29.7	27.9	26.8
SH	7.6	9.9	8.2	7.5	8.8	8.4	8.8	7.9	8.6	8.1	8.3	8.0	8.8	7.4
OCS	0.6	1.6	1.6	0.8	2.3	1.1	0.6	1.3	1.5	1.9	0.9	0.7	1.3	0.7
Others	4.2	4.5	3.7	4.9	6.8	12.9	5.2	9.3	6.4	7.9	9.6	11.1	10.1	10.0
Undetermined	5.0	3.5	4.1	2.7	4.9	2.8	3.3	4.9	6.2	3.6	3.5	4.8	2.6	3.0

Notes
a g/100 g of fresh material.
b GC peak areas percentages.
c Leaves.
d Flowers
MH, Monoterpene hydrocarbons.
OCM, Oxygen-containing monoterpenes.
TPC, Total phenol content.
SH, Sesquiterpene hydrocarbons.
OCS, Oxygen-containing sesquiterpenes.
Source: Senatore, 1996.

Table 7.2 *Thymus hyemalis* Lange – essential oil yield and its composition (%) at different dates of harvest

Characteristics	Date of harvest											
	Apr 22	May 26	July 4	Aug 3	Sept 4	Oct 10	Nov 23	Dec 23	Jan 29	Feb 27	March 30
Yield of oil (% v/w)	0.52	0.42	0.58	0.45	0.48	0.45	0.35	0.15	0.31	0.35	0.40
MH*	50.9	51.2	52.5	47.7	42.5	39.4	30.4	48.1	48.2	43.3	38.9
Myrcene	31.3	28.3	24.5	18.9	14.7	8.9	9.1	11.2	16.5	18.4	20.9
1,8-Cineole	17.1	19.8	20.0	26.8	22.4	13.3	22.1	16.6	18.8	20.7	22.5
Camphor	12.0	12.9	10.9	12.5	20.9	16.9	17.1	12.2	17.0	17.9	19.0
Alcohols	11.6	10.6	11.1	8.4	9.6	15.5	11.8	7.9	9.5	11.4	14.4
Acetates	11.5	6.6	5.2	4.3	4.4	10.4	8.4	6.0	5.7	5.4	7.5
Thymol+carvacrol	tr	tr	tr	tr	tr	2.5	3.5	4.4	tr	tr	tr

Notes
* MH, monoterpene hydrocarbons (including myrcene).
tr content below 1%.
Source: Cabo *et al.*, 1987.

Table 7.3 Composition (%) of the essential oil from thyme (*Thymus vulgaris*) at different dates of harvest

Compound	L1 May 25	L2 June 6	LF3 June 16	LF4 June 28	LF5 July 7	LF6 July 19	LF7 August 7
α-Thujene	0.59	0.62	0.67	0.72	0.50	0.97	0.99
α-Pinene	0.67	0.50	0.55	0.50	0.43	0.64	0.63
Camphene	0.52	0.26	0.25	0.22	0.22	0.32	0.34
Oct-1-en-3-ol	0.61	0.56	0.51	0.71	0.73	0.52	0.76
Myrcene	1.17	1.07	1.19	1.14	0.84	1.52	1.66
α-Terpinene	1.37	1.00	1.23	1.03	0.69	1.82	1.89
p-Cymene	14.43	10.21	11.19	8.27	10.83	7.36	6.63
Limonene	0.29	0.25	0.28	0.24	0.35	0.33	0.39
β-Phellandrene	0.76	0.45	0.48	0.50	0.72	0.49	0.37
1,8-Cineole	0.29	0.25	0.28	0.24	0.36	0.33	0.42
γ-Terpinene	13.40	6.52	7.50	7.58	5.23	16.39	17.21
tr-Sabinene hydrate	0.27	0.64	0.60	0.71	0.79	0.86	0.82
Linalool	1.60	2.04	1.91	2.23	2.51	1.79	1.70
Camphor	0.25	0.19	0.08	0.12	0.23	0.17	0.19
Borneol	1.30	0.78	0.65	0.63	0.73	0.78	0.83
Terpinen-4-ol	0.83	0.67	0.67	0.62	0.72	0.57	0.54
α-Terpineol	0.27	0.17	0.22	0.23	0.21	0.21	0.18
Thymol methyl ether	2.61	0.59	0.34	1.33	0.52	0.73	0.46
Carvacrol methyl ether	1.58	0.44	0.23	0.89	0.32	0.52	0.13
Thymol	49.12	65.98	63.11	65.20	66.94	57.13	57.91
Carvacrol	1.66	3.35	3.33	2.95	3.79	2.55	2.83
β-Caryophyllene	1.87	2.74	2.44	2.41	1.90	1.76	1.75
γ-Cadinene	0.40	0.31	0.30	0.37	0.26	0.24	0.17
δ-Cadinene	0.22	0.40	0.28	0.33	0.25	0.16	0.15
Caryophyllene oxide	0.44	0.41	0.37	0.32	0.60	0.23	0.13
T-Cadinol	0.78	0.28	0.38	0.48	0.32	0.46	0.26
Essential oil (%)	1.75	1.40	2.90	2.86	2.14	3.24	4.29

Notes
L, Leaves.
F, Flowers.

in Table 7.3. The highest amount of essential oil was distilled from the flowering parts harvested at the later phases. The percentage of *p*-cymene was highest in May, while the content of the major phenolic constituent thymol at the same phase was the lowest. Further it increased and was quite stable during the period of vegetation. It is interesting to note that the percentage of α-terpinene in June – beginning of July was rather low, however it considerably increased in the end of June and beginning of August, when some reduction in the content of thymol was determined. Mohamed (1997) also reported the effect of time of harvest on the composition of essential oil from *T. vulgaris*.

Moldão-Martins *et al.* (1999) investigated seasonal variations in yield and composition of *T. zygis* ssp. *sylvestris* (Hoffmanns. et Link) Brot. essential oil. The authors determined that the yield in essential oil peaked at the flowering stage (0.9–1.4 per cent) and was lowest during the dormancy period (about 0.15 per cent). The composition also showed different patterns at different phases of the vegetative cycle. At the flowering stage, the essential oil was rich in thymol and geraniol while *p*-cymene was highest when thymol was at a minimum (post-flowering period). Concerning the use of the essential oil as a food ingredient, it is suggested that the most interesting stage is the post-flowering period, the essential oil at this time being rich in thymol (about 21 per cent), geranyl acetate (about 17 per cent) and geraniol (about 13 per cent). Başer *et al.* (1999) also reported the variations in chemical composition of the essential oils of *T. pectinatus* Fisch. et Mey. var. *pectinatus* at different stages of vegetation.

PROCESSING OF FRESH AND DRIED PRODUCTS

Only a small part of harvested thyme can be consumed as fresh plants. Processing tech-nologies in general and their parameters in particular, which are and/or could be, applied to *Thymus* genus are similar to the processing technologies commonly applied for many other labiates. These technologies and processing parameters have been comprehensively described in several internationally recognised manuals, handbooks, edited books and monographs (Tainter[8] and Grenis, 1993; Heath and[5] Reineccius, 1986; Farrell[2], 1985; Heath[4], 1981; Reineccius[6], 1994; Underriner and Hume, 1994; Richard[7], 1992; Ashurst[1], 1991; Gerhardt[3], 1994). The content of this chapter is based on the materials provided in the above mentioned literature sources. It should be mentioned that specific inform-ations on the processing of thyme is rather scanty, therefore, the description of processing treatments provided in this chapter is mostly of a general character.

Processing of fresh products

Shelf life of fresh herbs is usually very short and therefore, traditionally, herbs have been used as dehydrated products. However, processes to prolong shelf life of herbs and spices have been developed and used. Storage of freshly harvested herbs at a temperature close to 0 °C is the simplest method to prolong shelf life of the fresh herbs. Such storage can delay deterioration only a few days. Some processes for producing frozen herbs, which retain their flavour and appearance for a considerably longer time, have been recently developed and tested. For instance, LaBell (1991) describes a process during which herbs are cleaned, chopped and coated lightly with canola oil within a few hours of harvesting. A small amount of acid is also added to the herbs to inhibit browning. The herbs are then stored frozen and this way they are stable for 1 year. Such processing enables more of the volatile top notes to be retained than by drying.

Also chemical agents were used to prolong shelf life of the fresh herbs. Mohammed and Wickham (1995) used this method for the plants of shado benni (*Eryngium foetidum* L.), which were harvested with intact roots, dipped in Gibberellic Acid (GA3) and stored in perforated and non-perforated low density polyethylene (LDPE) bags up to 22 days at 20–22 °C and 28–30 °C. It was shown that GA3 effectively retarded plant senescence up to 22 days at both temperatures when stored in non-perforated LDPE bags. Despite the external maintenance of marketable quality, flavour life was 17 days as development of off-flavours and reduction in pungency occurred after this period. Thus, the combin-ation of polyethylene packaging, GA3 dip treatment and reduced temperature storage extended the shelf life of the plants in a fresh, turgid and decay-free condition for an excess of 2 weeks. Most likely, such procedure could be applied to other plants including *Thymus* species.

The processing method of fresh vegetable products and particularly fresh herbs and spices invented by Hsieh and Albrecht (1988) is summarised in Figure 7.2. The flow diagram shows that modern technologies, such as fluidised bed drying (the process

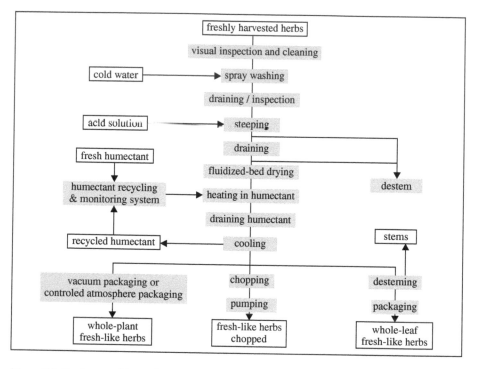

Figure 7.2 Fresh vegetable products treating flow chart (Hsieh and Albrecht, 1991).

when the solid particles are suspended in a rising stream of air) and using humectants (substances having distinct hygroscopic properties and retarding moisture changes) provide several different treatments for preserving the quality of fresh herbs.

Processing of dried products

Drying is undoubtedly the most ancient and still most widely used method of the fresh herb processing. At the time of harvest, most herbs and spices contain 60–85 per cent of water. In order to obtain stable products, which will withstand long periods of storage without deterioration this must be reduced to 8–10 per cent. Drying is the most critical process due to the volatility and susceptibility to chemical changes of the contained volatile oil (Heath, 1982).

Cleaning

In 1975, the Food and Drug Administration (FDA) initiated a three-year study, including more than 1000 samples, to develop data on insect, bird, rodent and other animal contamination levels in selected retail market, ground and unground spices and mould in ground paprika (Gecan *et al.*, 1986). The sampling and analytical details on nine spice products (including thyme) from that program were presented in the report of that study. Frequency distribution of insect fragment counts in ground and unground thyme are presented in Table 7.4, while a statistical summary of defects found in thyme

Table 7.4 Frequency distribution of insect fragment counts; 1267 of 25 g samples of unground thyme and 1332 of 10 g samples of ground thyme

Insect fragments	Ground	Unground	Rodent hair	Ground	Unground	Feather barbule	Ground	Unground
0	5	43	0	1164	1112	0	1162	995
1–100	339	760	1	109	97	1	124	153
101–200	302	299	2	38	34	2	28	58
201–300	212	95	3	11	11	3	7	31
301–400	160	47	4	5	4	4	7	11
401–500	97	10	5	–	3	5		3
501–1000	183	9	6	3	3	6		2
1001–1500	28	2	8	2	2	7		4
1501–2000	5	1	11–50	1	1	8		1
2001–2500		1				9	1	2
						10		1
						11–25	3	4
						26–50		2

Mite	Ground	Unground	Thrips	Ground	Unground	Aphid	Ground	Unground
0	1293	588	0	1284	582	0	1291	694
1	26	196	1	33	164	1	21	130
2	6	115	2	7	107	2	8	102
3	2	74	3	2	74	3	3	64
4	2	52	4		71	4	2	42
5	1	24	5	1	32	5	3	30
6		23	6	1	27	6	1	21
7		9	7	1	22	7		19
8		14	8		33	8	3	26
9		7	9		19	9		11
10		2	10	2	19	10		20
11–20		34	11–20	1	70	11–20		71
21–30		15	21–30		24	21–30		24
31–40		6	31–40		11	31–40		5
41–50		4	41–50		2	41–50		4
51–100	1	22	51–100		10	51–100		4
101–200	1	18						
201–300		20						
301–400		3						
401–500		7						
501–999		34						

Source: Gecan *et al.*, 1986.

is presented in Table 7.5. The investigations show that thyme was a heavily contaminated herb, particularly with insect fragments, mites, thrips, and aphids.

Count means of insect fragments varied from 7.8 for 10 g of ground allspice to 287.7 for 10 g of ground thyme; samples containing insect fragments ranged from 70.8 per cent for ground allspice to 98.6 per cent for ground thyme. Mite counts ranged from 0–2 for 25 g of ground paprika to 0–999 for 25 g of unground thyme; count means varied from 0.0 for 25 g of ground paprika to 35.4 for 25 g of unground thyme; samples containing mites ranged from 2.8 per cent for ground thyme to 53.6 per cent for

Table 7.5 Statistical summary of defects found in thyme

Defects	Ground thyme*			Unground thyme**		
	Counts		Samples with defects (%)	Counts		Samples with defects (%)
	Mean	Range		Mean	Range	
Insect fragments	287.7	0–3625	99.6	100.7	0–2257	96.6
Rodent hairs	0.2	0–32	12.6	0.2	0–28	12.2
Feather burbules	0.2	0–12	12.8	0.5	0–50	21.5
Mites	0.2	0–179	2.9	35.4	0–999	53.6
Thrips	0.1	0–12	3.6	3.8	0–99	54.1
Aphids	0.1	0–8	3.1	3.1	0–83	45.2

Notes
* In total 1332 of 10 g samples were examined.
** In total 1267 of 25 g samples were examined.
Source: Gecan *et al.*, 1986.

unground thyme. Thrip counts ranged from 0–1 for 10 g of ground allspice and 25 g of ground paprika to 0–99 for 25 g of unground thyme; counts means varied from 0.0 for 10 g of ground allspice and 25 g of ground paprika to 3.8 for 25 g of unground thyme; samples containing thrips ranged from 0.2 per cent for ground paprika to 54.1 per cent for unground thyme. Samples containing aphids ranged from 0.0 per cent for ground paprika to 45.2 per cent for unground thyme.

Tainter and Grenis (1993) have described the general principles of cleaning methods and equipment. The principles of all cleaning equipment are based on physical difference (i.e. shape, density) between the spice and/or herb and the foreign material being removed. The equipment can consist of magnets, sifters, air tables, destoners, air separators, indent separators and spiral separators. The choice of cleaning methods and equipment depends on the physical characteristics of the herb, the cost of the machinery and the process effectiveness of the removal of foreign objects, and the loss of the main material during cleaning. Anyway, it is impossible to perform a cleaning operation at reasonable production rates that result in a pile of foreign material completely free of thyme and a pile of thyme completely free of foreign material. Therefore in optimising the process it is necessary to define the limits of foreign material in the cleaned products. Some of these characteristics are usually provided in the regulations for a particular spice or herb.

Magnets are widely used for eliminating ferromagnetic particles from herbs and spices. Modern quality assurance systems, e.g. HACCP (Hazard Analysis Critical Control Points), require that every herb cleaning system should include magnets or other metal detecting devices. The main purpose of removing metals from the products is to protect the end-user from physical hazards. Remaining pieces of metals can also damage other processing equipment, e.g. milling machinery.

Dehydrated thyme can already be considered as a prepared product for utilisation. However, when leaves and/or blossoms are used as a spice, a separation process is necessary to remove stems, especially woody parts, which usually are of inferior quality as compared to green plant parts. Several techniques can be used for that purpose. The simplest method is to dry the product in such a way that the flowering parts are sufficiently dried to detach them, whereas the stems remain humid. The other technique deals with passing dehydrated plants through modified threshers.

The most basic cleaning operation is the utilisation of sifters. By running the herb over a set of screens (Table 7.6), it is possible to remove larger and smaller particles from the product that is being cleaned. Although the principle of sifting is quite simple, it is rather difficult in operation, because dried herbs are random pieces of leaves. Therefore, sifters are generally not used for cleaning, but for sizing.

An air table or a gravity separator is the most versatile piece of cleaning equipment for herbs and spices. Actually, an air table is a wire mesh screen with a stream of air blowing up through it. The lighter pieces on the screen are suspended higher than the heavier ones. The air stream blows the very lightest pieces out of the system. During operation the screen is tilted and all thyme particles move to the bottom end of the screen. Rotational vibration is imparted to the screen, which is adjusted so as to just touch the heavier particles and tap them, pushing up the screen, while the light filth migrates to the bottom of the screen. In practice, the tilt of the screen, the rotational vibration of it, and the airflow through the screen are adjusted so that the cleaned thyme migrates to the middle of the screen, the heavy filth to the top of the screen.

The main disadvantage of an air table is that it may or may not be able to separate particles of different sizes and different densities if the air stream floats a large surface area particle of relatively heavier weight at the same height as a small surface area particle of lighter weight. However, it can efficiently separate particles of the same density and different size and the particles of the same size and of different densities.

There are some other types of cleaning equipment, e.g. indent separators which try to make use of the difference in shape between the spice and the foreign material, spiral separators, which work well separating round seeds from nonround foreign material. However, physical characteristics of dried thyme are not convenient for such techniques. Very fine dusty particles can be removed from dried raw herb materials by using an apparatus called a cyclone.

Harvesting of thyme should be performed in a way to minimise the contamination with stones and rocks. However, depending on the excellence of the whole process some

Table 7.6 Comparison of various screen size measurement systems*

USS screen	Tyler screen	Mill screen	Stainless-steel screen	USS screen	Tyler screen	Mill screen	Stainless-steel screen
4	4	4		25	24	26	28
6	6	6		30	28	30	32
8	8	9		35	32	36	38
10	9	10		40	35	40	48
12	10	12	14	45	42	45	56
14	12	14	16	50	48	55	66
16	14	16	18	60	60		74
18	16	18	22	80	80		94
20	20	22	24				

Note
* This information shows the closest match of screen sizes for various measurement systems. The actual aperture for each is not necessarily identical and some tolerance is needed to build into specifications.
Source: Tainter and Grenis, 1993.

amount of stones will remain in the dried product. Destoners work on the same principle as the air tables but are generally much smaller in size. Where an air table is able to separate the product stream into as many divisions as is desired, a destoner is generally set up to remove only heavier stones and rock from thyme. Once again, by varying the airflow, the inclination, vibration and the type of screen, it is possible to make the stones "walk" up the screen and thus affect a separation from the lighter leaves of thyme (Tainter and Grenis, 1993).

Dried thyme can be imported from various countries or sources, which sometimes do not ensure good sanitary processing conditions. Therefore, its reconditioning is used to remove contaminants and bring the product into conformity with specifications. Reconditioning involves the same cleaning steps, which are briefly outlined above. Some legislation institutions, e.g. FDA in the US, want to know if the spice is planned to be reconditioned prior to performing the work. They may want to supervise the operation to ensure adequate removal of the contaminant. Under American Spice Trade Association (ASTA) procedures, supervision is not necessary, but the lot must be resampled and tested by an independent laboratory.

Comminution and particle size selection

Thyme can be used as a whole, leaves and flowering parts, and as a ground product. The use of whole herbs in food processing is limited as they are not ready sources of flavour but they do improve the appearance of certain products. With few exceptions, herbs and spices are normally milled to powder before use. Size reduction may also be an essential primary stage for extraction or distillation, for incorporation into food product mixes or for use as they are at table or in cooking.

There are many different ways to cut, grind, break or otherwise reduce the size, however the basics of herb comminution are very simple. There is a variety of equipment used to grind thyme, and they are generally designed to crush, or to shatter the plant particles. To minimise negative grinding effects such equipment ranges from high-speed disintegrators, hammer mills, pin mills, plate mills, roller mills, ball mills to slow speed, vertical and horizontal buhrstone mills in a whole range of sizes and capacities. The machinery can be roughly classified into the following groups:

(a) course cutters, crushers or breakers designed to give an intermediate product further reduction from the large plant parts received in bulk;
(b) slow-speed attrition mills, which reduce the product to a fine powder;
(c) high-speed impact mills, whereby a wide range of sizes may be produced, and throughput is rapid;
(d) micromills for extremely fine powders.

There are two main problems encountered in the grinding procedure: (i) breaking of herb oil bearing particles and (ii) formation of heat. In natural herbs the aromatic components are retained within a protective cell wall. This gives to the whole herb a long shelf life so long as certain basic storage conditions are observed. By grinding all the oil structures containing the volatile oil, the oil becomes available for reaction (e.g. oxidation) or evaporation. Grinding also generates some heat, which will tend to vaporise the volatile oil leading to a reduction in flavour strength. The strong odour of thyme, which penetrates environment during grinding, indicates some degree of flavour loss. Therefore it is necessary to keep the temperatures during the grinding process as low as possible to minimise the loss of volatile oil.

Various mills can be configured in various manners by changing internal screens, speed and internal clearances to control the heat build-up. As a rule, finer particle sizes develop higher temperatures during grinding. To limit the formation of heat the product can be prechilled before feeding. Usually the user defines particle size. There are sophisticated grinding methods that avoid heat release during grinding. For instance, Cohodas in 1969 patented a process for cryogenic milling in which the spice stock and a gaseous refrigerant (e.g. nitrogen) are fed into the milling head at a constant rate. Such a technique considerably minimises losses of volatiles and discoloration of herb. By freezing the herb and solidifying the volatile oils, these herbs grind and sift a lot easier. Cryogenic grinding will also minimise oxidative deterioration of the flavours due to the nitrogen blanket during grinding. Therefore, a spice ground cryogenically may have a different flavour profile, usually retaining the top notes and giving a fuller flavoured product.

Comminution operations often include sieving and sifting procedures. The mills may have internal screens that in part dictate the final particle size, or the sifting operation may be a separate operation where the oversized particles are returned to the mill for further processing. In either case, the set-up of the mill or sifters determine the particle size of the finished spice. Since nearly all spice and herb specifications contain a granulation parameter, e.g. granulation of ground thyme should be 95 per cent minimum through a United States Standard (USS) #30 (ASTA), it is important to look at the particle size control of the ground herb. The manufacturers may also adopt their own empirical classification.

POSTPROCESSING TREATMENTS

Control or reduction of microbiological populations found in herbs and spices as they are harvested is an important task demanding some postprocessing treatments of the product. Kneifel and Berger (1994) screened a total of 160 samples of 55 different spices and herbs originating from six different suppliers and retailed at outlets in Vienna for their microbial quality. Arithmetical mean values and a range of viable counts of thyme are presented in Table 7.7. In the other source the contamination of rubbed thyme herb is estimated as 3.5×10^4 counts/g (Heath, 1982). High microbiological loads can lead to a significantly reduced shelf life and may produce problems if there is not a significant heat processing step in the finished food. The process of sterilisation adds to the cost of the herb or spice but it is well worth it to the ultimate user (Farrell, 1985).

Pathogens (microorganisms which cause human diseases) can be a problem, especially for herbs and spices processed in countries with lower sanitary standards compared to those countries that may have well-defined levels. As the presence of microorganisms in herbs can cause food spoilage and illness, a case can be made for sterilisation. However, herbs and spices generally constitute only 0.1–2.0 per cent of most food products and are most often used in cooked form. As a result, the risk of spoilage and illness is greatly reduced.

A number of different sterilising procedures have been applied to spices and herbs in an attempt to reduce high microbial levels. Both physical and chemical methods have

Table 7.7 Microbial contamination of random samples of thyme

Microorganisms	Mean values	Viable count range log CFU/g
Total aerobic mesophilic bacteria	5.1×10^6	4.1–7.6
Enterobacteria	4.5×10^5	
Pseudomonades and Aeromonades	3.0×10^5	3.4–6.4
Bacilli	6.6×10^4	4.1–6.4
Lactobacilli	<10	
Enterococci	8.8×10^2	
Viable counts		2.8–5.0

Source: Kneifel and Berger, 1994.

been used with varying degrees of success. These include steam and heat sterilisation, ultra-violet (UV) irradiation, ionising irradiation, filtration, infrared irradiation, and fumigation. Mention may also be made of methods such as drying, freezing and salting which are generally used in a supplementary role. A common industry practice is to refer to these treatments as sterilisation or "bacteria treatment". Neither of these terms is accurate. The treated spice is not commercially sterile nor has it been treated with bacteria (ICMSF, 1980; see Table 7.8). All microbial decontamination methods can be classified into physical and chemical treatments. The principles of these methods are comprehensively described elsewhere (Gerhardt, 1994; Tainter and Grenis, 1993).

Ethylene oxide (ETO) or methyl bromide treatment

Many commercial food processors fumigate herbs and spices with methyl bromide to eliminate insects or with ETO or its mixture with carbon dioxide (90 : 10) to eliminate bacteria and mould. Both methyl bromide and ETO are extremely toxic, and methyl bromide is potentially capable of depleting the atmospheric ozone layer (Thayer *et al.*, 1996). The US Environmental Protection Agency (EPA) (1982) places a maximum tolerance of 50 ppm for ETO in or on ground spices; after 25 years of use in Germany ETO was recently banned as well as in the other European Union (EU) member states. Propylene oxide is also approved as a microbiological treatment process for spices, but it is not nearly as effective as ETO because its penetrating ability is weaker.

Blends of spices can be treated with ETO as long as no salt is present. Salt will react with ETO to form chlorohydrins that are toxic. Gustafsson (1981) examined the residue levels of ethylene chlorhydrin (2-chloroethanol) in various products and found that 5 samples of thyme contained 105, 230, 390, 450 and 1290 mg/kg of chlorhydrin.

Irradiation

Most of the chemical microbial decontamination treatments are insufficient or have a serious disadvantage (Eiss, 1984). Irradiation for the sterilisation of spices has been studied for several years and the treatment has been shown to be a very effective tool of reducing microbial populations. Alongside traditional methods of processing and pre-serving food, the technology of food irradiation is gaining more and more attention around the world. Herbs and spices are being irradiated in many countries including Argentina, Brazil, Denmark, Finland, France, Hungary, India, Indonesia, Israel, Norway, the United States, and Yugoslavia.

The Joint Expert Committee on Food Irradiation (JECFI) in 1980 concluded that "the irradiation of any food commodity" up to an overall average dose of 10 kGy "presents no toxicological hazard" and requires no further testing. Dried herbs and spices are allowed the highest radiation treatment level (up to 30 kGy) because (i) they are considered as a minor part of our diets; (ii) they are safer to irradiate due to the low moisture content; and (iii) they are irradiated for the purpose of controlling both insects and microorganisms, which necessitates high doses of radiation.

The International Atomic Energy Agency (IAEA) and the International Consultative Group on Food Irradiation (ICGFI) concerning herb and spice irradiation issued the following important documents:

- IAEA/TECDOC-688 (Technical Document 688): Irradiation of Spices, Herbs and Other Vegetable Seasonings – A Compilation of Technical Data for its Authorisation and Control;
- ICGFI (International Consultative Group in Food Irradiation) Document No. 4: Code of Good Irradiation Practice for the Control of Pathogens and Other Microflora in Spices, Herbs and Other Vegetable Seasonings.

Ozone and other methods of treatment

Ozone is an extremely reactive substance and a very effective antimicrobial agent. Decontamination can be performed in storage tanks or in the milling machinery during the comminution procedure. Ozone treatment can reduce microbial counts down to 2–3 per cent of the initial load (Gerhardt, 1994). However, the main disadvantage of ozone application deals with the safety issues due to the possibility of explosion.

UV-irradiation, microwaving, treatment with high frequency electric currency, high pressure treatment are rarely used methods for herb decontamination, some of them are in the stage of development, laboratory and pilot-plant testing. Some other methods, such as a focused microwave system (Anonymous, 1991a) and pulsed electric fields (PEF) (Keith *et al.*, 1997) were also tested for disinfestation of herbs and spices.

Steam sterilisation is usually applied to the whole herb or spice before grinding and decontamination can be performed by pressurised, atmospheric and sub-atmospheric steam (Anonymous, 1999). Although described in the literature, steam sterilisation methods and apparatus only in some cases were tested on thyme, it is reasonable to briefly overview these methods, so far as they can be applicable for the sterilisation of *Thymus* herbs as well.

Hosokawa Micron Europe BV developed a process for sterilisation of herbs and spices, based on rapid heating to the required temperature with saturated steam, holding for the required period, and rapid vacuum cooling/drying (Spook, 1993; Anonymous, 1993). Dudek (1996) patented the process involving exposure to elevated pressures and temperature for a predetermined time in a series of chambers; Leife (1992) described a method for sterilisation of spices and herbs with rapid pulses of steam. Herbs and spices can be pasteurised in Torbed equipment (from Torftech, UK) described by Dodson (1996). Particles are forced into a toroidal motion, exposing all surfaces to heat and breaking down the insulating boundary layer in seconds. A natural steam pasteurisation process for herbs and spices was developed by McCormick as a superior alternative to UV or Infra-Red (IR) radiation, microwave, cryogenic or ultrasonic methods of microbial control (Anonymous, 1991b). Mercati (1992) proposed a multi-step processing method, which is characterised by a stage in which the herbs stay in hot air chambers at an adjustable temperature and for a period of time ranging from 1 to 60 h sufficient to cause the death of the bacteria.

DRYING TEMPERATURES, CONDITIONS AND DURATION OF STORAGE

Dehydration (drying) is undoubtedly the most ancient process for the preservation of spices and aromatic herbs. As medicinal and aromatic plants are usually harvested at 80 per cent moisture content and stored at 11 per cent, drying of this crop requires high energy equivalent to 1–2 l of fuel oil per kg of crude drug. Depending on the herb or spice the ratio of the weight of fresh raw material with the weight of dry product can be from 1:1 to 10:1; for thyme this ratio can be from 3:1 to 5:1 (Gerhardt, 1994). The process of drying for many medicinal and aromatic plants is the crucial one in deter-mining the quality of the dried material and products produced from it. Heindl and Müller (1997) comprehensively described drying of medicinal plants and spices.

Traditionally herbs have been cut and dried in the sun or in the shade. Dried herbs then were sold to a processor who beats the leaves from the stalk by hand or machine and then sieves them to remove stalks and stones. The process of removing the leaves from the stalk is called "rubbing" and the sieving called "sifting".

Methods of drying

In general, all drying methods can be divided into thermal and non-thermal drying. Thermal methods can be classified into (1) natural direct drying (air drying with the aid of sun energy), (2) solar indirect drying, and (3) artificial drying (with the aid of heat, cold or IR). Non-thermal drying can be performed by using (1) moisture-absorbing materials, (2) drying agents and (3) electrolytes (Gerhardt, 1994). There are different modifications of these methods, which are used depending on the economical and quality requirements.

Natural drying is the simplest way to prepare herbs for storage and further processing. So far as the method is cheap and does not require costly equipment it is still widely used in many countries. There are several methods of drying of raw material such as sun-drying (SD), drying in the shade, solar drying, hot air drying, practised in com-mercial processing in the different countries.

The method of sun-drying is the most usual as well as the one most widely employed. In most cases the material is first comminuted and spread on the ground on mats and exposed to direct sunlight. This method is simple and quite cheap; however, it possesses some disadvantages: (a) the possibility for the introduction of contaminants in many different ways (rodents, insects, and their resultant contamination), (b) the loss of volatile matter, (c) the degradation of heat or light-sensitive constituents, (d) dif-ficulties in controlling the process, (e) usually long time of drying. Natural drying of the whole *T. vulgaris* herb is particularly problematic because the shrub consists of comparatively fast drying leaves and slowly drying rather hard stems. Therefore the process can take up to 120 h causing severe loss of the volatile oil (Raghavan *et al.*, 1995).

A more sophisticated method of solar drying is also employed, but this is only used in cases of small volume high value products. This method is excellent for leafy material as it maintains the rich green colour making the product look attractive. It is, however, a method, which involves much initial outlay and will be uninteresting for most producers. Different types of solar dryers have been developed. They have been successfully tested on different Labiatae aromatic plants (mainly mint and sage) and can be used for drying *Thymus* spp. In 1991 a solar drying device, based on the use of solar-energy, started to operate in Krka – Drug factory, p.o. (Novo mesto, Yugoslavia, Program of Green Drugs). In 1993 Müller *et al.* developed a solar-heated dryer in modular design at the Institute for Agricultural Engineering in the Tropics and Subtropics of Hohenheim University (Figure 7.3). The roof of a standard plastic film covered greenhouse was used as a solar air heater area. A batch dryer was installed inside the greenhouse. A comparative economic evaluation showed that the plastic-house type solar dryer is economically more efficient than the conventional drying system, as soon as supplementary heating is used.

Hot air drying allows for more rigid control of the process, it is rapid and clean, however requires high capital and operational costs and can cause overheating. This method is useful if the end product is of high value and the quantity to be handled is of reasonable magnitude (Wijesekera, 1993). Some companies provide high output, automatically controlled dehydration lines specially designed for leafy plant material. The scheme of one such line produced by Heindl GmbH, Maschinen- und Anlagenbau, Germany, is provided in Figure 7.4.

Although traditional hot air drying is a simple method in herb processing, however, its main parameters can be varied and should be tailored to every particular herb mainly to minimise flavour loss and to perform the process at reasonable time and energy costs. Raghavan *et al.* (1995) compared cross flow and through flow drying methods on Indian thyme at 40, 50 and 60 °C and found that through flow drying at

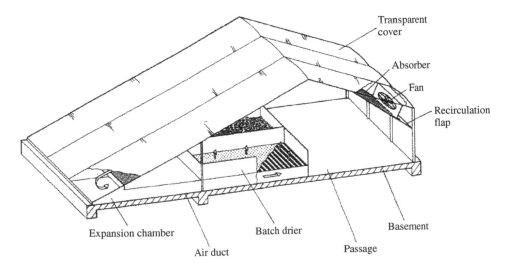

Figure 7.3 A solar heated dryer in modular design (Müller *et al.*, 1993).

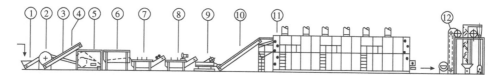

Figure 7.4 Dehydration line for leaf vegetables and officinal herbs (Heindl GmbH, Maschinen-und Anlagenbau). (1) feed hopper, (2) leaf cutter, (3) conveyor belt, (4) conveyor belt, (5) stalk separator, (6) dosing hopper, (7) paddle washer (optional), (8) fan washer, (9) vibratory conveyor for removal of surface water (optional), (10) oscillating conveyor belt, (11) five-band-dryer, (12) milling and air separating unit.

40 °C gave the best results. Kakis (1986) invented the method by which foodstuff is firstly pressed to remove a substantial amount of its less tightly bound water, then contacted with an absorbent to remove a substantial amount of its more tightly bound water and provide a low moisture foodstuff. Subjecting this low moisture foodstuff to ambient temperature air drying in a low moisture atmosphere provides a dry-to-the-touch dehydrated foodstuff which retains its flavourful and aromatic volatiles and is resistant to spoilage. Ground fresh thyme (63.2 g) was subjected to the pressing at 8000–10 000 psi and absorption steps by using 5 g of Syloid 244. After the initial pressing step the sample lost 32 per cent and after the absorption step an additional 25 per cent of its original weight. After air drying at 26 °C for 18 hours, 18 g of dry material were obtained. Thus, the total weight loss for the three-step process was 72 per cent of the original.

Freeze drying is based on evaporation of water directly from ice under a high vacuum. Herbs are rapidly frozen to less than −18 °C temperature to form microcrystalline ice structure, which does not damage plant cell structure. The products obtained by this method are usually of a better appearance (colour) and aroma quality. High cost is the main disadvantage of freeze drying limiting wider use on a commercial scale. The effect of freeze drying on the chemical composition of thyme will be discussed in the Chapter 8.

Drying conditions

It is well established that the higher the drying temperature is, the shorter the time of the process is; however, bigger losses of volatile oil occur. Drying temperature and conditions should be tailored for every aromatic plant to achieve optimal quality characteristics.

Poludennij and Zhuravlev (1989) recommend drying harvested thyme in the shade in ventilated premises. When oven-drying is used, the temperature should not exceed 43–45 °C. *T. serpyllum* is also dried in the shade or in special oven-dryers at 40 °C. After drying, leaves and flowers are ground and sifted. Wooden stems and branches are removed (Kudinov *et al.*, 1986; Mashanov and Pokrovskij, 1991).

So far traditional drying conditions require long drying time. There are some interesting reports about using different techniques intended to improve the process of herb processing. Sometimes very simple alterations can be implemented to improve natural drying conditions, e.g. Aliev and Kuliev (1989) suggest to hang the herbs up to dry in 0.5–4 m plaits to reduce product loss, and facilitate subsequent storage and transport. Some authors describe more sophisticated new combination techniques, e.g. including such procedures as blanching, treatment with osmotic agents and surfactants. So far as the application of these techniques can be considered in the processing of thyme they will be briefly described.

Mastrocola *et al.* (1988) applied four different sets of drying conditions to retail samples of basil and found the importance of blanching in retaining the colour. Rocha *et al.* (1993) used steam blanching and surfactant pretreatment to increase the drying rate of basil and found that drying rates were increased by a factor of 10 and 14 for steam blanching and surfactant pretreatments, respectively. Blanching leaves of Indian spearmint (*Mentha spicata* L.) prior to drying yielded products, which were unattractive with respect to colour and appearance and were also bland and odourless due to loss of volatile oil during drying whereas shade-drying leaves resulted in a product with a good green colour and small loss of volatile oil (Raghavan *et al.*, 1994). Blanching of *Eryngium foetidum* L. herb in hot water at 96 °C using a quick dip step followed by drying in the indirect drier reduced the loss of green colour normally observed on direct drying without pretreatment (Sankat and Vashti-Maharaj, 1994).

Investigations with parsley leaves showed that compared with convection drying, when leaves where dried in a microwave oven the loss of the aroma fractions was approximately five times less, sensory scores were considerably higher, and drying times were three times shorter (Zarebski and Mroczkowski, 1995). Aung and Fulger (1993) describe a method for drying fresh herbs whilst preserving their colour, aroma, flavour and overall appearance involving treatment with an osmotic agent which, on completion of the drying process, forms a solid amorphous mass. This mass coats and infuses the treated herb.

Effect of SD, solar cabinet drying (SCD) and tray drying (TD) on the colour of dehydrated fenugreek (*Trigonella foenum-graecum*) and mustard (*Brassica campestris* var. *sarson*) leaves was investigated and significant differences were observed. It was recommended that SCD could be used in tropical regions (Ramana *et al.*, 1988).

A process for preservation of herbs or other aromatic plants is based on mixing the edible part of the herb plant with a substance which reduces water activity; freezing the mixture; partial freeze drying to yield a mixture of dried and frozen or non-dried herb material; and homogeneous mixing to achieve equilibration of moisture content between the dried and partially dried material (Darbonne, 1996). Various spices were dried by a dehumidifying dryer, in which trays of products are superimposed and air is circulated by a fan. After passing through the product, the humid air is cooled in order to remove the water vapour, and then heated before re-entering the products (Rattanapant and Phongpipatpong, 1990).

A process for osmotic partial drying of edible plant material (e.g. vegetables, herbs) has been described. The plant material is treated with an aqueous solution (refractometric dry matter (DM) content 25–40 per cent) containing one ingredient selected from the group gum arabic, carboxy methyl cellulose (CMC), modified starch and ethyl maltol, together with a polyol (sorbitol, mannitol, xylitol or glycerol) and/or a sugar (sucrose, glucose, lactose, or maltodextrin with dextrose (DE) less than 30). The solution should have pH in the range 4.5–6.5. Contact time between the plant material and the solution is 15–50 min, at 10–50 °C. The plant material is dried to a residual moisture content of 10–15 per cent, corresponding water activity (a_w) of 0.3–0.6. Optionally, the plant material may be blanched before the osmotic drying process (Darbonne and Bain, 1991).

Bousser (1990) describes a method for improvement of natural drying of biological materials (including fruit, vegetables, herbs and medicinal plants) which is based on addition of glucose to the raw material, then comminution of the mixture to a homogeneous mass which is then dried. Optionally, sorbitol and/or citric acid and/or salt may be added to the glucose; a typical formulation is 95 per cent glucose, 2.5 per cent sorbitol, two per cent NaCl and 0.5 per cent citric acid. Metabisulphite may optionally be added. The glucose-based mixture is added to the raw material at a level approximately three times higher than the quantity of water in the product to be dried. The mixture is dried in a thin layer at 25–30 °C with gentle ventilation. The products may also contain other ingredients, e.g. dried milk, chocolate, starch or breadcrumbs.

A drying system, utilising conductive and convective heating to dehydrate a variety of foods (including vegetables, herbs, fruits, cheese, meat and dairy products) with low-temperature air, is described. Feed material enters the top of the dryer and is uniformly distributed across a moving bed of balls made of stainless steel, ceramic, aluminium or food-grade plastics. Heated air enters at the base of the dryer. Products can be dried to 3–7 per cent moisture, and have flavour and colour characteristics comparable to freeze-drying (Swientek, 1988).

A method and apparatus for treating (washing/steeping/drying/cooling) freshly harvested vegetable products, particularly herbs and spices, are described (Hsieh and Albrecht, 1988).

Herbs are prepared by stabilising them in the freshly harvested state (or, if frozen products are used, during or immediately upon thawing), and drying them in the presence of a suitable carrier under mild conditions. As carriers, salts (electrolytes), proteins, carbohydrates, or mixtures can be used. The stabilisation step consists of either heating to 50–150 °C or mixing with an electrolyte or both (Bezner *et al.*, 1987).

PACKAGING AND STORAGE

Dried thyme is a long shelf life product. As a rule dried herbs are considered stable until the development of some noticeable off-flavour (unusual taste and/or odour). The period of minimal stability is a time during which the herb is fully suitable for use and its essential specific characteristics such as aroma, pungency, colour, etc. are maintained. The quality and minimal period of stability can be assessed by the sensory evaluation of colour, odour and taste and by the determination of essential oil (Gerhardt, 1994).

The stability of thyme depends on the following aspects:

- moisture content;
- comminution method (finer grinding means lower stability);
- quantity and package size (bigger package brings higher stability);
- packaging material (lower permeability to water and air results in higher stability);
- penetration of air into the package;
- effect of light and humidity (higher humidity and light access causes lower stability);
- storage temperature (lower temperature means higher stability).

In general storage below −18 °C is a guarantee for unlimited storage time; when the product is stored at 5–7 °C dried herb can be stored more than 12 months, whereas at room temperature stability considerably decreases.

After all processing steps have been completed thyme herbs are packaged. It is an important procedure, because during comminution the structure of the cells is usually more or less damaged and volatile constituents can be easily released. There are two main tasks for packaging (Niebergall *et al.*, 1978): (a) to protect against exterior effects, (b) to increase the stability against negative internal changes (enzymatic, non-enzymatic, chemical reactions, etc.).

Optimal packaging materials are glass and metals: both are completely impermeable and provide the best protection of aroma. Due to economical and some other reasons different new materials, mostly synthetic plastics, have increasingly substituted those traditional packaging materials. However, among them there is no ideal material for the packaging of herbs. For instance, polyethylene efficiently protects against water and gives resistance, however it is permeable to fat and aroma compounds and is difficult for machine treatment; polyamide is impermeable to gas, whereas its waterproof qualities are less efficient (Gerhardt, 1994). Modified atmosphere packaging can be used to protect packaged herbs against oxidation. The oxygen can be removed by vacuuming or by replacing air with inert gas. Oxygen absorbents can also be used for this purpose.

The quality of thyme decreases during storage. The changes depend on several factors, drying method and parameters, moisture content, cleaning procedures, grinding technique (e.g. perfection of milling equipment minimising heat build-up during grinding) and particle size, sterilisation treatment, storage conditions, packaging type, etc. In general, dried thyme should be stored in cool, dry conditions away from light. Ideally, it should be in airtight packaging to reduce oxidation.

Finely milled thyme is often advantageous in use, however, it does tend to lose volatiles more rapidly than medium or coarsely ground material and must be stored in well-closed containers. Storage in multi-layered paper sacks having an impervious lining is also satisfactory but not so good once the sack has been opened (Heath, 1981).

The content of essential oil in herbs and spices reduces during storage. The loss of volatiles depends on various factors, e.g. botanical plant characteristics (structure and distribution of oil-bearing particles), essential oil composition, processing (mainly drying and grinding procedures), packaging and storage conditions. For instance, in one of the early studies, thyme herb, which was packaged in paper bags, was stored in dark premises at ambient temperature and after 6 years of storage it was found that the loss of essential oil constituted 71.4 per cent, whereas essential oil reduction in sage after 7 years of storage was only 20.6 per cent (Stamm and Willner, 1934). Some changes in volatile oil composition also take place during storage.

Fehr and Stenzhorn (1979) studied the change of essential oils in relation with the storage time (up to 38 months). The essential oil content in thyme decreased at a rate of 0.002–0.022 ml/month. The authors determined significant differences in the composition of essential oils (Table 7.8) and proposed mathematical description of the long-time storage stability of dried thyme.

The concentrations of some quantitatively important thyme flavour compounds after a storage period of 0, 1, 5 and 10 months were also studied by Venskutonis *et al.* (1996) and are presented in Table 7.9. It was found that the changes during storage are highly significant, but the differences vary between compounds and it is possible to divide the compounds into groups of relatively small changes and relatively large changes. The compounds belonging to the two groups are as follows: (a) small differences: α-pinene, *p*-cymene, linalool, borneol, thymol, carvacrol, and β-caryophyllene; (b) large differences: α-thujene, myrcene, α-terpinene, γ-terpinene, *trans*-sabinene hydrate and caryophyllene oxide. Some of the compounds did not show systematic changes and generally were not reduced during the storage period, that is the case with thymol, while others like

Table 7.8 Contamination of some untreated Labiatae herbs with bacteria and moulds (ICMSF)

Spice	Cumulative percentage incidence of					
	Aerobic plate count/g			Mould count/g		
	10^5	10^6	10^7	10^4	10^5	10^6
Thyme	85	53	0	87	6	0
Basil	86	38	0	6	0	0
Marjoram	78	33	5	29	0	0
Oregano	32	9	0	9	0	0
Sage	47	6	0	50	0	0
Savory	10	0	0	0	0	0

Table 7.9 Composition of thyme essential oils (%) depending on storage time

Compound	1/75	2/76	3/76	4/76	5/77	6/77	7/78	8/78	9/78	Range	Mean
α-Pinene	1.1	1.4	1.4	1.7	1.5	1.3	1.9	1.3	1.9	1.1–1.9	1.5
Camphene	0.5	0.6	0.6	0.7	0.6	0.6	0.7	0.5	1.0	0.5–1.0	0.6
Myrcene	0.7	0.9	0.8	0.9	1.3	0.8	1.3	0.8	1.0	0.7–1.3	0.9
α-Terpinene	0.8	1.0	1.0	1.0	1.3	0.8	1.4	0.9	1.0	0.8–1.4	1.0
Limonene	0.3	0.4	0.4	0.5	0.5	0.4	0.4	0.4	0.4	0.3–0.5	0.4
1,8-Cineole	0.6	0.6	0.7	0.8	0.8	0.6	0.7	0.6	0.8	0.6–0.8	0.7
γ-Terpinene	3.2	4.1	3.9	3.8	6.9	3.2	7.6	3.2	6.0	3.2–7.6	4.7
p-Cymene	30.6	31.4	34.1	42.8	29.3	34.8	27.3	33.2	34.6	27.3–42.8	33.1
Linalool	2.2	2.3	2.6	2.5	2.9	2.7	3.0	2.7	3.0	2.2–3.0	2.7
Terpinen-4-ol	2.0	2.2	2.2	2.5	2.3	2.3	1.4	1.5	1.2	1.2–2.5	2.0
β-Caryophyllene	0.2	0.3	0.3	0.3	0.2	0.3	2.1	1.1	2.2	0.2–2.2	0.8
Borneol	1.1	1.2	1.2	1.2	1.3	1.2	1.5	1.1	1.8	1.1–1.8	1.3
Thymol	43.6	42.7	39.1	30.0	39.5	39.3	39.6	41.1	33.2	30.0–43.6	38.7
Carvacrol	4.3	4.1	4.0	3.5	4.1	4.3	3.6	5.4	3.5	3.5–5.4	4.1
Hydrocarbons	37.4	40.1	42.5	51.7	41.5	42.2	42.7	41.4	48.1	37.5–51.7	43.1
Oxygenated compounds	53.8	53.1	49.8	40.5	50.9	50.4	49.8	52.4	43.5	40.5–53.8	49.4

Source: Fehr and Stenzhorn, 1979.

α-thujene, myrcene and sabinene hydrate are reduced by 21–40 per cent after 10 months of storage.

Besides the most important compounds quantitatively caryophyllene oxide is also included in Table 7.10 as the content of this compound increased during storage. This could indicate some oxidation during storage. It is interesting to note that the reduction of the content of β-caryophyllene between 1 and 10 months of storage was by 69 mg/kg, whereas the increase of the concentration of its oxide during the same period was by 56 mg/kg, i.e. nearly equal. However, this tendency was not found during the first month of storage, when according to the statistical assessment there were no significant differences in the content of β-caryophyllene. Tressl *et al.* (1978) found 12-fold increase in the concentration of caryophyllene oxide in dried quince after a storage period of

Table 7.10 Changes of the average content of volatile compounds in thyme during storage (mg/kg)

Constituent	Time of storage, months			
	0	1	5	10
α-Thujene	255[a]	179[b]	92[c]	54[d]
α-Pinene	257[a]	222[c]	241[b]	254[ab]
Myrcene	299[a]	268[b]	166[c]	117[d]
α-Terpinene	220[a]	198[b]	151[c]	106[d]
p-Cymene	3863[a]	3694[b]	3360[c]	3538[bc]
γ-Terpinene	1296[a]	1173[a]	805[b]	410[c]
tr-Sabinene hydrate	260[a]	233[b]	97[c]	81[c]
Linalool	345[a]	355[a]	323[b]	324[b]
Borneol	189[a]	160[c]	169[bc]	183[ab]
Thymol	6912[b]	8153[a]	6441[b]	6426[b]
Carvacrol	827[b]	1027[a]	880[b]	979[a]
β-Caryophyllene	283[a]	311[a]	303[a]	242[b]
Caryophyllene oxide	34[c]	58[b]	68[b]	114[a]

Note
a–d Values with the same letter within same row are not significantly different.
Source: Venskutonis *et al.*, 1996.

3 years, whereas the content of all identified non-oxygenated terpenes severely reduced during the same period.

The changes in the amount of the aroma compounds during storage may be explained by oxidation and other chemical changes, as indicated by the increase of caryophyllene oxide. However, as the polyethylene bags were not aroma tight some losses may be due to evaporation of volatile compounds (most of non-oxygenated monoterpenes) through the PE bags. In this view fairly stable content of α-pinene during storage seems rather contradictory.

Bendl *et al.* (1988) studied the effect of freeze-drying of sage and thyme on the stability-properties following storage in comparison to drying at increased temperature. The authors determined that by using freeze-drying there could be found greater amounts of the characteristic flavour components particularly immediately after preparation but also after 80 days of storage (Table 7.11). These studies indicate that by freeze-drying sage and thyme there could be obtained a storable product with high spice and flavour quality. Some special measures to improve storability were applied. For instance, storing fresh leaves of Indian spearmint (*Mentha spicata* L.) for 12 h after spraying with water increased volatile oil content by about six per cent (Raghavan *et al.*, 1994). Anokhina *et al.* (1990) performed tests on the effect of freeze-drying on dill and parsley from a retail perspective and reported that storage in laminated polyethylene containers is recommended in order to minimise vitamin C loss.

Koller (1988) gives examples of inadequate treatment changing the characteristic aroma of herbs and spices, so that they no longer fulfil their function. Processes such as drying (temperature, method, pH) and storage conditions (air access, light, temperature, pH, packaging material), as affecting colour and aroma of thyme, sage, cloves and marjoram were investigated. Storage temperature has been found to be most decisive on the changes of headspace volatiles and consequently aroma.

Table 7.11 Effect of storage on the content of the essential oil, thymol and carvacrol in thyme

| Storage, days | Amount in ground herb | | | | | |
| | Essential oil (µl/g) | | Thymol (mg/g) | | Carvacrol (mg/g) | |
	FD	OD	FD	OD	FD	OD
0	15.2	13.3	9.6	7.9	0.5	0.4
20	12.5	11.8	8.7	7.7	0.4	0.3
40	10.5	10.6	7.9	7.1	0.4	0.3
80	10.0	10.0	7.1	6.5	0.4	0.3

Notes
FD, freeze dried.
OD, oven dried.
Source: Bendl *et al.*, 1988.

Other constituents of herbs also can change during storage. Bakowski and Michalik (1988) investigated suitability of some plants for drying and observed high losses of vitamin C during dehydration and storage (after 6 months by 90 per cent). After dehydration the carotene content decreased by 10 per cent, after 6 months storage, by 20 per cent. Chlorophyll content in leaves also decreased during dehydration and storage period but did not change the colour. The dehydrated leaves had high contents of calcium, phosphorus, potassium, magnesium and iron.

REFERENCES

Anonymous (1991a) To focus on the microwave field. *Food Technol. N. Z.*, 26, 16–17, 19.
Anonymous (1991b) McCormick masters the microbes. *European Food & Drink Review*, Autumn, 64, 66–67.
Anonymous (1993) Turning up the heat. *Food Manuf.*, August.
Anonymous (1999) Pressurised steam decontamination. Atmospheric steam decontamination. Sub-atmospheric steam decontamination. *Newsletter.* Food Refrigeration and Process Engineering Centre, 22, 3–5.
Anokhina, V.I., Ovchinnikova, I.F., Prokudina, V.E., Kolotilova, L.A. and Kioseva, L.V. (1990) Retail evaluation of freeze-dried herbs. *Tovarovedenie*, 23, 9–11 (Russian).
Aliev, Sh.A. and Kuliev, G.Yu. (1989) Drying herbs. *USSR-Patent*, SU 1 528 417 (Russian).
Arrebola, M.L., Navarro, M.C., Jiménez, J. and Ocaña, F.A. (1994) Yield and composition of the essential oil of *Thymus serpylloides* subsp. *serpylloides. Phytochemistry*, 36, 67–72.
Ashurst, P.R. (ed.) (1991) *Food Flavourings*, Blackie Academic & Professional, Glasgow and London.
Aung, T. and Fulger, C.V. (1993) Process for preparing dehydrated aromatic plant products and the resulting products, *US Patent* US 5 227 183.
Bakowski, J. and Michalik, K. (1988) Suitability of several vegetable species for drying. Przydatnosc niektorych gatunkow warzyw do produkcji suszu. *Biuletyn-Warzywniczy (Poland). Bulletin of Vegetable Crops Research Work*, 29, 191–211.
Başer, K.H.C., Demirci, B., Kürkçüoglu, M. and Tümen, G. (1999) Composition of the essential oils of *Thymus pectinatus* Fisch. et Mey. var. *pectinatus* at different stages of vegetation, *J. Essent. Oil Res.*, 11, 333–334.

Bendl, E., Kroyer, G., Washüttl, J. and Steiner I., (1988) Untersuchungen über die Gefriertrocknung von Thymian und Salbei. *Ernährung/Nutrition*, 12, 793–795.

Bezner, K., Biller, F., Kellermann, R. and Bohrmann, H. (1987) Flowable dried aromatic plant product and process for making the same. Manila (Philippines). Philippine Patents Office. July 1987. 14 p. *Philippine patent document* 21074-C.

Börngen, S. (1979) *Pflanzen helfen heilen*, VEB Verlag Volk und Gesundheit, Berlin.

Bousser, R. (1990) Procédé pour améliorer le séchage naturel des produits biologiques, et produits biologiques ainsi traités. *French Patent Application Demande de Brevet d'Invention* FR 2 642 617 A1.

Cabo, J., Crespo, M.E., Jiménez, J., Navarro, C. and Risco, S. (1987) Seasonal variation of essential oil yield and composition of *Thymus hyemalis*. *Planta Med.*, 53, 380–382.

Clarke, M.W. (1994) Herbs and spices. In E.W. Underriner and I.R. Hume (eds), *Handbook of Industrial Seasonings*, Blackie Academic & Professional, London, pp. 43–61.

Cohodas, A.M. (1969) Spice grinding process, *Can. Pat.* 808, 644, Mar. 18.

Darbonne, L.F.M. (1996) Vegetaux stockables à temperature basse positive et negative et procede de traitement de vegetaux frais en vue de leur obtention. *French Patent* FR 2 725 111 A1.

Darbonne, L. and Bain, J. (1991) Process for dehydration of edible plants. *French Patent* FR 2 649 297 A1.

Dodson, C. (1996) Fruitful new opportunities in the food processing industry. Food-Tech-Europe, 3, 52, 54, 56.

Dudek, D.H. (1996) Sterilization method and apparatus for spices and herbs. *US Patent* 5 523 053.

Eiss, M.I. (1984) Irradiation of spices and herbs. *Food Technol. in Australia*, 36, 362–363, 366.

Farrell, K.T. (1985) *Spices, Condiments and Seasonings*, The AVI Publ. Co. Inc., Westport, Connecticut.

Fehr, D. and Stenzhorn, G. (1979) Untersuchungen zur Lagerstabilität von Pfefferminzblättern, Rosmarinblättern und Thymian. *Pharm. Ztg*, 124, 2342–2349.

Gecan, J. S., Bandler, R., Glaze, L. E. and Atkinson, J. C. (1986) Microanalytical quality of ground and unground marjoram, sage and thyme, ground allspice, black pepper and paprika. *J. Food Prot.*, 49, 216–221.

Gerhardt, U. (1994) *Gewürze in der Lebensmittelindustrie: Eigenschaften, Technologien, Verwendung*, B. Behr's Verlag, 2. Auflage, Hamburg.

Guillén, M.D. and Manzanos, M.J. (1998) Study of the composition of the different parts of a Spanish *Thymus vulgaris* L. plant. *Food Chem.*, 63, 373–383.

Gustafsson, K.H. (1981) Rester etylenklorhydrin I vissa importerade industrikrydor. *Vår Föda*, 33, 15–21 (Swedish).

Heath, H.B. (1981) *Source Book of Flavors*. The AVI Publishing Company, Inc. Westport, Connecticut, (USA).

Heath, H.B. (1982) Spices and aromatic extracts, influence of technological parameters on quality. In J. Adda and H. Richard (Coord. Scient.), *Int. Symp. on Food Flavors*, Tec. Doc.-Lavoisier, A.P.R.I.A., Paris, pp. 138–175.

Heath, H.B. and Reineccius, G. (1986) *Flavor Chemistry and Technology*, Macmillan Publishers Ltd.

Heindl, A. and Müller, J. (1997) Trocknung von Arznei- und Gewürzpflanzen. *Z. Arzn. Gew. pfl.*, 2, 90–98.

Hsieh, R.C. and Albrecht, J.J. (1988) Method and apparatus for treating fresh vegetable products. *European Patent Application* EP 0 285 235 A1.

ICMSF (1980) Spices. In J.H. Siliker, R.P. Elliot, A.C. Baird-Parker, F.L. Bryan, J.H.B. Christian, D.S. Clark, J.H. Olson and T.A. Roberts, Jr. (eds), *Microbial Ecology of Foods. International Commission on Microbiological Specifications for Foods*, Vol. 2., Academic Press, New York, pp. 731.

Jaskonis, J. (ed.) (1983) *Growing of Medicinal Plants*. Mokslas, Vilnius, Lithuania (Lithuanian).

Kakis, F.J. (1986) Food dehydration process, *USA Patent* No. 4707370.

Karawya, M.S. and Hifnawy, M.S. (1974) Analytical study of the volatile oil of *Thymus vulgaris* L. growing in Egypt. *J. AOAC*, 57, 997–1001.

Kauniene, V. and Kaunas, E. (1991) *Medicinal Plants*, Varpas, Kaunas, Lithuania (Lithuanian).

Keith, W.D., Harris, L.J., Hudson, L. and Griffiths, M.W. (1997) Pulsed electric fields as a processing alternative for microbial reduction in spice. *Food Res. Int.*, 30, 185–191.

Kneifel, W. and Berger, E. (1994) Microbial criteria of random samples of spices and herbs retailed on the Austrian market. *J. Food Prot.*, 57, 893–901.

Koller, W.D. (1988) Problems with the flavour of herbs and spices. In G. Charalambous (ed.), *Developments in Food Science "Frontiers of Flavor"*, Elsevier Science Publishers BV., Amsterdam, The Netherlands, 17, pp. 123–132.

Kudinov, M. A., Kuchareva, L. V., Pashina, G. V. and Ivanova, E. V. (1986) *Spicy and Aromatic Plants*, Uradzhai, Minsk (Russian).

LaBell, F. (1991) Frozen fresh chopped herbs. *Food Processing*, USA, 52, 106–108.

Leife, Å. (1992) Steriliserar kryddor med pulserande ånga. *Livsmedelsteknik*, 34, 24–25 (Swedish).

Li, Y.-L., Craker, L.E. and Potter, T. (1996) Effect of light level on essential oil production of sage (*Salvia officinalis*) and thyme (*Thymus vulgaris*). *Acta Hortic.*, 426, 419–426.

Mashanov, V.I. and Pokrovskij, A.A. (1991) *Spicy and Aromatic Plants*. Agropromizdat, Moscow (Russian).

Mastrocola, D., Barbanti, D. and Armagno, R. (1988) Ricerche sull' essiccamento in corrente d' aria del basilico (*Ocimum basilicum*). 1: Influenza sul colore. *Industrie Alimentari (Italy)*, 27, 341–344.

Mercati, V. (1992) Process for preservation of vegetable products. *European Patent Specification*, publ. no. 0 243 567 B1.

Mohamed, M.A. (1997) Effect of plant density and date of cutting on *Thymus vulgaris* L. plants, *Egypt. J. Hort.*, 24, 1–6 (Arabic).

Mohammed, M. and Wickham, L. D. (1995) Postharvest retardation of senescence in shado benni (*Eryngium foetidum* L.) plants. *J. Food Qual.*, 18, 325–334.

Moldão-Martins, M., Bernardo-Gil, M.G., Beirão da Costa, M.L. and Rouzet, M. (1999) Seasonal variation in yield and composition of *Thymus zygis* L. subsp. *sylvestris* essential oil, *Flavour Fragr. J.*, 14, 177–182.

Müller, J., Conrad, T., Tešic, M. and Sabo, J. (1993) Drying of medicinal plants in a plastic-house type solar dryer. *Acta Hortic.*, 344, 79–85.

Niebergall, H., Humeid, A. and Blöchl, W. (1978) Die Aromadurchlässigkeit von Verpack-ungsfolien und ihre Bestimmung mittels einer neu entwickelten Meßapparatur. *Lebensm. Wiss. Technol.*, 11, 1–4.

Poludennij, L.V. and Zhuravlev, Ju.P. (1989) *Medicinal Plants in the Home Garden*, Moskovskij Rabotchij, Moscow (Russian).

Raghavan, B., Abraham, K.O., Jaganmohan, R.L. and Shankaranarayana, M.L. (1994) Effect of drying on flavour quality of Indian spearmint (*Mentha spicata* L.). *J. Spices Arom. Crops*, 3, 142–151.

Raghavan, B., Abraham, K.O. and Koller, W.D. (1995) Flavour quality of fresh and dried Indian thyme (*Thymus vulgaris* L.). *Pafai Journal*, 17, 9–14.

Ramana, S.V., Jayaraman K.S. and Mohan-Kumar, B.L. (1988) Studies on the colour of some dehydrated green leafy vegetables. *Indian Food Packer*, 42, 19–23.

Rattanapant, O. and Phongpipatpong, M. (1990) Drying of spices by using a dehumidifying dryer. *Food*, 20, 253–263.

Reineccius, G. (1994) *Source Book of Flavors. 2nd Edition*. Chapman and Hall, New York.

Richard, H. (coord.) (1992) *Epices et Aromates*, Tec. Doc.-Lavoisier, A.P.R.I.A., Paris.

Rocha, T., Lebert, A. and Marty-Audouin, C. (1993) Effect of pretreatments and drying conditions on drying rate and colour retention of basil (*Ocimum basilicum*), *Lebensm. Wiss. Technol.*, 26, 456–463.

Sáez, F. (1998) Variability in essential oils from populations of *Thymus hyemalis* Lange in south-eastern Spain. *Journal of Herbs Spices & Medicinal Plants*, 5, 65–76.

Sankat, C.K. and Vashti-Maharaj (1994) Drying the green herb shado beni (*Eryngium foetidum* L.) in a natural convection cabinet and solar driers. *ASEAN Food J.*, 9, 17–23.

Senatore, F. (1996) Influence of harvesting time on yield and composition of the essential oil of thyme (*Thymus pulegioides* L.) growing wild in Campania (Southern Italy). *J. Agric. Food Chem.*, 44, 1327–1332.

Spook, W.J.A. (1993) Stoomsterilisatie van kruiden en specerijen. *Voedingsmiddelen-technologie*, 26, 75–76 (Dutch).

Stamm, I. and Willner, E. (1934) Gehaltsminderung an ätherischem Öl durch längere Aufbewahrung von Drogen. *Farmacia*, 14, 296.

Swientek, R.J. (1988) Low-temperature drying saves energy. *Food Processing*, USA, 49, 45–46.

Tainter, D.R. and Grenis, A.T. (1993) *Spices and Seasonings. A Food Technology Handbook*, VCH Publishers, New York.

Thayer, D.W., Josephson, E.S., Brynjolfsson, A. and Giddings, G.G. (1996) Radiation pasteurization of food. *Council for Agricultural Science and Technology. Issue Paper*, 7, 1–10.

Tressl, R., Friese, L., Fendesack, F. and Köppler, H. (1978) Studies of the volatile composition of hops during storage. *J. Agric. Food Chem.*, 26, 1426–1430.

Underriner, E.W. and Hume, I.R. (eds) (1994) *Handbook of Industrial Seasonings*, Blackie Academic & Professional, London.

Venskutonis, P.R., Poll, L. and Larsen, M. (1996) Influence of drying and irradiation on the composition of the volatile compounds of thyme (*Thymus vulgaris* L.). *Flavour Fragr. J.*, 11, 123–128.

Wijesekera, R.O.B. (1993) Processing of medicinal plant derived preparations in developing countries – prospects & perspectives. *Acta Hortic.*, 332, 63–71.

Zarebski, A. and Mroczkowski, A. (1995) Microwave drying of parsley leaves. *Przemysl-Spozywczy*, 49, 168–169.

8   Thyme – processing of raw plant material

Petras R. Venskutonis

INTRODUCTION

Traditionally dried ground spices and herbs, although being widely used products, possess several serious disadvantages. The most common disadvantages of dried thyme are:

- variable flavour strength and profile;
- unhygienic;
- often contaminated by filth;
- easy adulteration with less valuable materials;
- presence of lipase enzymes;
- flavour loss and degradation on storage;
- undesirable appearance characteristics in end products;
- poor flavour distribution (particularly in thin liquid products such as sauces);
- discolouration due to tannins;
- unacceptable hay-like aroma;
- dusty and unpleasant to handle in bulk.

Therefore, manufacturers are increasingly recognising the advantages of seasoning based on herb extractives. In general, the methods of extraction depend on the desired properties of a final product, characteristics of plant material, economical and technical issues. The most important extraction products that are obtained from thyme are essential oils, herbs, oleoresins and extracts. Therefore these three processes will be discussed more thoroughly in this chapter.

ESSENTIAL OIL: PRODUCTION AND CHARACTERISTICS

The definitions of the most applicable terms to thyme products can be seen as follows (Lawrence, 1995, modified):

Essential oil: The isolated volatile aromatic portion of a plant, produced within distinctive secretory structures. The essential oils generally constitute the odorous principles of the plants. They are either distilled or expressed. In exceptional cases, they may be formed during processing when the plant tissue is brought into contact with water.

Extract: A concentrated product obtained by treating a natural raw material with a solvent. True extracts do not contain significant amounts of the solvent. Depending on the polarity of the solvent extracts consist of polar or less polar compounds.

Oleoresin: Liquid preparations extracted from herbs or spices with solvents which can extract oil and resinous matter from the botanical drugs yielding the oleoresin as evaporation residue. Oleoresins are often used in food and pharmaceutical industries as a replacement of ground spices and spice tinctures. Prepared oleoresins may also contain fixed oils. Natural oleoresins are exudations from tree-trunks, barks, etc.

It can be noticed that the differences in the definitions between extract and oleoresin are not very strict. The content of volatile oil can be considered as the most important characteristic in distinguishing these two products. Oleoresins usually contain significant amounts of volatile oil whereas its content in the extract is much smaller or absent. Artificial combinations of the essential oil and an extract of the same plant are also called oleoresins.

Principles of essential oil isolation

Essential oils are accumulated in different types of secretory structures of the plants, and they can be categorised into superficial and subcutaneous oils, and Labiate's oils belong to the first group. The main methods to obtain essential oils from plant material are water distillation, steam and water distillation, steam distillation, maceration distillation, empyreumatic (or destructive) distillation, and expression. With the exception of the latter process, all others need heat to release the oil.

Water distillation is the simplest method to obtain volatile oils; therefore this method is usually performed in rural areas where no access to a steam boiler is possible. The plant material is loaded into a still fitted with a slow-speed paddle stirrer (to avoid agglomeration) and is always in direct contact with water. The water can be boiled by direct fire and by submerged stem coils. It is very important to maintain a sufficient level of water in the still to avoid overheating and/or charring of plant material and consequently undesired essential oil off-flavours. During the process of boiling, volatile oil evaporates together with water and the vapour afterwards is condensed. Due to its highly hydrophobic property the essential oil can be easily separated from the water in the so-called florentine flask. Thyme essential oil, being lighter than water, separates on its surface. The scheme of the apparatus for water distillation is shown in Figure 8.1 (Heath and Reineccius, 1986).

Steam distillation is performed with the aid of steam, which is generated outside the still, in steam generators generally referred to as boilers. Plant material is loaded into a suitable still on a perforated grid through which steam may be injected from the base. The process of steam distillation is more effective and the most widely accepted process for the production of essential oils on a large scale. Steam distillation units can be stationary or mobile. In case of mobile distillation the process is performed in the harvesting fields, and therefore such time and labour consuming operations as loading of herbs into a cart, transporting to the still and unloading can be avoided. This may result in reducing the time for the whole process, from chopping the wilted plant material to disposing of the spent material, from 6–8 h to less than 3 h, the size of the labour force could be reduced by at least 50 per cent. The mobile steam distillation process is used for the production of oils from such important commercial plants as mints, clary sage and dill.

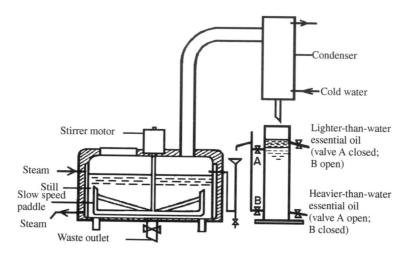

Figure 8.1 Diagrammatic representation of a water distillation unit (Heath and Reineccius, 1986).

The only difference between water distillation and steam and water distillation is that during the latter process the plant material is separated from the water. It can be loaded onto a frame within the still body, fixed above a layer of water. To increase the effectiveness of water and steam and water distillation, cohobation is commonly used, consisting in the return of distilled water to the still after the oil has been separated from it so that it can be re-boiled. This procedure is very important for thyme, because its oil is rich in phenols, which to some extent may dissolve in distilled water. Cohobation on the one hand minimises the loss of oxygenated compounds, on the other hand however, it increases the risk of hydrolysis and degradation of constantly re-vaporised and condensed oxygenated compounds. Therefore, it is not recommended unless the temperature to which the oxygenated compounds dissolved in the distillate exposed is maintained not higher than 100 °C.

Continuous distillation possesses many advantages in comparison with conventional distillation procedures. Short time and high output, reduced energy and water needs, reduced disposal costs for spent material, reduced labour costs, possibility of automation, improved process reproducibility and consequently quality of oil are the most important ones. However, such a process can be efficiently used only when large quantities of essential oils are required.

The principles of essential oil distillation are common to many oil-bearing plants, however, to obtain the highest yield and the best quality product the process has to be tailored for every particular herb depending on its characteristics. The quality of an essential oil is adversely affected by heat, light, air and moisture and since these are inherent parameters of distillation it is small wonder that many commercial oils differ markedly in sensory character (Heath, 1982). The quality of the oil is also affected by the method of distillation. Water-distilled oils are commonly darker in colour and have stronger still notes than oils produced by other methods (Lawrence, 1995). Distillation can cause chemical changes of natural constituents, e.g. formation of p-cymene from γ-terpinene, both compounds being important for thyme (Moyler, 1991).

Characterisation of different thyme oils

In commerce, the designation "thyme oil" is occasionally applied indiscriminately, and erroneously, to oils distilled from plants belonging to species other than *Thymus vulgaris* L. or *Thymus zygis* L. In Fenaroli's Handbook of Flavour Ingredients (Burdock, 1994) the essential oil obtained from *T. vulgaris* and *T. zygis* is described as a brownish-red liquid exhibiting a strong, aromatic odour and a warm, somewhat sharp flavour (red thyme oil). White thyme oil is a pale-yellow liquid obtained by rectification of the distilled red thyme oil, exhibiting similar but milder odour and flavour characteristics.

The main constituents of thyme oil are thymol and carvacrol (up to 70 per cent). Other chemotypes of *T. vulgaris* are limited to specific areas and yield, e.g. oils that contain geraniol, linalool, α-terpineol, and 1,8-cineole; these oils are of minor importance. Some chemotypes of *T. zygis* produce an essential oil with other dominant constituents (linalool, carvacrol, geraniol/geranyl acetate, 1,8-cineole/linalool, linalool/thymol, 1,8-cineole/linalool/thymol). For instance, Sáez (1995) who comprehensively reviewed recent investigations on *T. zygis* in one of the samples grown in south-eastern Spain (ssp. *gracilis*) determined 91.40 per cent of linalool, and only 0.31 per cent of thymol.

The essential oil of *Thymus capitatus* (today: *Thymbra capitata*) is a clear, pinkish to reddish-brown oily liquid with odour reminiscent of origanum (Spanish origanum). Usually the oil from *Thymus capitatus* is richer in carvacrol than the oil from *T. vulgaris* and *T. zygis*.

EXTRACTION: METHODS AND EXTRACT CHARACTERISTICS

Numerous companies all over the world produce different extracted thyme products. The examples of such products are provided in Table 8.1. Usually standardised thyme oleoresins

Table 8.1 Examples of standardised thyme products

Product	Producer	Characteristics
Standardised oleoresin Thyme FD0718	Bush Boake Allen Limited, London, England	Volatile oil content (%, v/w) 54–60
Standardised oleoresins Thyme HX2089	Lionel Hitchen Essential Oil Company Limited, Barton Stacey, Hants, UK	Volatile oil content (%, v/w) 50 Dispersion rate kgs = 100 kg of spice 1
Dispersed spices – salt Thyme	Bush Boake Allen Limited, London, England	Volatile oil content 0.3–0.4% (v/w)
Dispersed spices – dextrose Thyme	Bush Boake Allen Limited, London, England	Volatile oil content 0.3–0.4% (v/w)
Dispersed spices – rusk Thyme FD5781		Volatile oil content 0.6–0.8% (v/w)
Standardised emulsion oleoresins Thyme HF107	Felton Worldwide SARL, Versailles, France	Strength compared to ground spice 4×
Standardised emulsion oleoresins Thyme FD6136	Bush Boake Allen Limited, London, England	Strength compared to ground spice 5×
Encapsulated standardised oleoresins Thyme FD4040	Bush Boake Allen, "Saronseal Encapsulated spices"	Strength compared to ground spice 10×

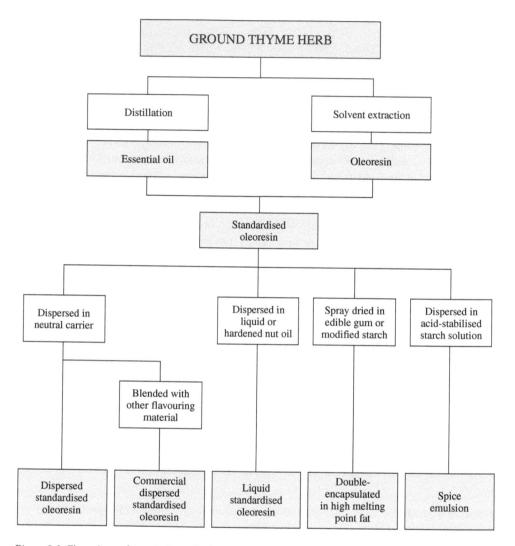

Figure 8.2 Flow chart of a typical standardised oleoresin range of thyme (Moyler, 1991).

are produced by adding to the extract some distilled essential oil. Such oleoresins can be further used in preparation of different thyme products, which are shown in Figure 8.2.

Extraction methods

Extraction solvents

General principles of the extraction and equipment used for this purpose are comprehensively described in various handbooks and manuals on herb and spice processing (Heath, 1981; Moyler, 1991; Peyron and Richard, 1992; Richard and Loo, 1992; Reineccius, 1994; Lawrence, 1995). There are several solvents, which are legally approved for the extraction of aromatic materials. These are tabulated in Table 8.2.

Table 8.2 Extraction solvents (IOFI, Europe)

Butane	Ethyl acetate
Propane	Diethyl ether
Isobutane	Dibutyl ether
Toluene	Dichloromethane
Cyclohexane	Trichloroethylene
Petrol ether	Dichlorofluoromethane
Methanol	Dichlorodifluoromethane
Butanol-1	Trichlorofluoromethane
Acetone	Dichlorotetrafluoromethane
Ethyl methyl ketone	Carbon dioxide

The quality of the extracts and often their composition depends on the solvent nature, particularly its polarity: the polar solvents better extract polar constituents. Therefore, possessing the knowledge of the constituents of a plant, it is possible to predict which components will be extracted under a given set of extraction conditions. For instance, the main constituents in the essential oil of *T. vulgaris* are both polar (thymol, carvacrol) and non-polar (*p*-cymene, γ-terpinene) compounds. Another important characteristic of the solvent influencing the profile of an extract is its boiling temperature. So far as the solvent has to be removed from the extract some natural volatiles can be lost. To avoid losses of volatiles, the choice of the solvent has to be a compromise between extractive potency and boiling temperature.

Solvent viscosity and its latent heat of evaporation are also important characteristics. The former affects solvent penetrability into the extracted material, the latter is directly related to the extraction energy costs. The removal of solvent from the extract is of ultimate importance due to the following reasons: (a) the residues of most of the solvents are strictly limited by laws; in the US maximal permitted residue of acetone is 30 ppm, methanol 50 ppm, isopropanol 50 ppm, hexane 25 ppm, and chlorinated solvents 30 ppm (Reineccius, 1994); (b) the residue of the solvent affects the quality of the extract and has to be minimal; (c) removal of the solvent can cause the loss of more volatile constituents, which are usually extracted together with less volatile substances.

Extraction at atmospheric pressure

The most widely used extraction process at atmospheric pressure involves the following unit operations:

- preparation of the raw material;
- exposure of the material to the solvent;
- separation of miscella from the extracted material;
- removal of the solvent.

Comminution is the main and most important preparation procedure for herb extraction. It is necessary to obtain the optimal particle size, sufficient to enable the solvent to penetrate the mass completely, but not too fine to reduce the rate of penetration. Exposure of the material to the solvent involves three phases: (i) the addition of the solvent and its penetration into the dry mass; (ii) the achievement of equilibrium; and (iii) the replacement of the solute with new solvent (Reineccius, 1994). The process can be carried out on a batch basis (e.g. when small quantities of different materials are handled) or continuously, when large amounts of unique raw materials are processed. Separation of miscella from extracted material is a process during which the ground material acts as its own filter and consequently the clear miscella passes directly to a still or evaporator.

To meet all extract concentration requirements the miscella is usually processed in two stages: (1) the removal of the main part of the solvent (approximately 95 per cent of the solvent can be removed in a standard falling-film, raising-film or other type of evaporator); (2) removal of the rest of the solvent, e.g. by using vacuum treatment.

Besides the conventional extraction procedures some authors proposed to use an optional measure. For instance, Honerlagen and Steiner (1990) in their patent proposed to add a drying agent before or after separation of the extract from the exhausted solids, to eliminate water from the extract. The solvent is then distilled off, to leave the extracted lipophilic material. Recently some interesting experiments were carried out on the use of microwaves in the development of extraction processes. The so-called microwave transparent solvents, which allow all the energy to be absorbed by the plant material have been used in such experiments with mint, cedar leaves and garlic (Paré *et al.*, 1991). The principle of microwave use is that the sudden increase in temperature causes the cell walls of the essential oil glands to rupture and release their oil to the solvent. Spiro and Chen (1995) examined the kinetics of this process and found that under the severe thermal stress the oil glands of peppermint not only ruptured but also totally disintegrated into aggregates of powdery fragments.

High-pressure extraction

Using modern high-pressure extraction techniques can successfully solve the main problems of the conventional extraction. A great number of low boiling temperature solvents can be used for this purpose, however, carbon dioxide (CO_2), is the most suitable material in various food applications. All dry botanicals with an oil or resin content can be extracted with CO_2. This pressured solvent behaves during extraction in a similar way to any of the other solvents. As a solvent it has some significant advantages compared to alternatives.

Commercially, CO_2 can be used in two distinct modes of extraction, which are dependent on its operation above or below the critical point in the phase diagram for CO_2. From this point of view, CO_2 as a solvent can be used for the extraction in sub-critical and supercritical states (Pc > 73.8 bar, Tc > 31.3 °C). The main advantages of CO_2 are the following (Moyler, 1991; Lawrence, 1995):

- odourless, colourless, tasteless and non-toxic;
- non-combustible;
- inexpensive and readily available;
- easily removed leaving no solvent residue;
- because of a low viscosity it can readily penetrate comminuted dry plant material;
- by varying the temperature and pressure it can be used in a more selective manner.

However, wider commercial application of the supercritical CO_2-extraction is limited by economical reasons. The capital costs of the equipment are still rather high and the process can be used only for the high-added value products, e.g. hops and coffee.

Table 8.3 Solubility of botanical components in liquid CO_2

Very soluble	Sparingly soluble	Almost insoluble
Low MW aliphatic hydrocarbons, carbonyls, esters, ethers (e.g. 1,8-cineole), alcohols monoterpenes, sesquiterpenes	Higher MW aliphatic hydrocarbons, esters, etc.; substituted terpenes and sesquiterpenes; carboxylic acids and polar N and SH compounds; saturated lipids up to C12	Sugars, protein, polyphenols, waxes, inorganic salts; high MW compounds, e.g. chlorophyl, carotenoids, unsaturated and higher than C12 lipids
MW up to 250	MW up to 400	MW above 400

Note
MW; Molecular weight.
Source: Moyler, 1987.

A comparison with traditional forms of extraction shows that CO_2 is a versatile solvent. By using traditional isolation procedures we can obtain either essential oil (distillation) or oleoresin (solvent extraction). By using CO_2-extraction we obtain oleoresin which can be fractionated into essential oil and resin.

Using extraction conditions of 50–80 bar pressure and 0 to $+10\,^{\circ}C$, it is commercially viable to extract essential oils as an alternative to steam distillation. The energy savings of CO_2 offset some of the capital expenditure of the extraction equipment. The solubility of natural compounds in liquid CO_2 are provided in Table 8.3 (Moyler, 1987). In order to increase the yield of the extracts, sometimes CO_2 is used together with some entraining solvent. For *T. vulgaris*, Calame and Steiner (1987) used supercritical conditions with hexane as entrainer at 150 bar and $40\,^{\circ}C$ with subsequent subcritical fractionation at 50 bar and $9\,^{\circ}C$ to obtain a similar yield of 2 per cent to that of steam distillation. Thyme leaf was extracted by supercritical CO_2 and ethanol and the reported yield was 2.1 per cent (Moyler, 1993). Oszagyan *et al.* (1996) carried out supercritical fluid extraction (SFE) of *T. vulgaris* under different extraction conditions. A stepwise increase of the extraction pressure resulted in the fractionation of the extracts into liquid and pasty products. SFE of thyme gave a product which contained 10–15 per cent thymol and 30–35 per cent carvacrol, while steam distillation produced an oil containing 48–50 per cent thymol and only 8–10 per cent carvacrol.

Characterisation of different thyme extracts

The quality of the extracts depends on various factors as previously pointed out. This can be illustrated by some experiments targeted on the investigation of different properties of thyme extracts. In early 1952 Chipault *et al.* assessed the antioxidant activity of different herb and spice products and found that the antioxidant index of a thyme fraction soluble in petroleum ether was two times lower than the antioxidant index of the fraction soluble in alcohol (Table 8.4). Somewhat similar results were obtained with other herbs, excluding savoury.

T. vulgaris was extracted by CO_2 in several studies with different purposes. Cardoso *et al.* (1993) compared supercritical fluid carbon dioxide (SFC) extraction to steam and hydrodistillation in a Clevenger type apparatus. In thyme extracts, higher production yields were always obtained by the conventional distillation method. Using SFC extraction high yields were obtained but the extracts included other kinds of compounds. With

Table 8.4 Antioxidant properties of ground spices and of petroleum ether and alcohol-soluble fractions

Spice	Antioxidant index determined by active oxygen method at 98.6 °C, employing as substrate prime steam lard with a stability of 6.5 h		
	Ground spice	*Petroleum ether-soluble fraction*	*Alcohol-soluble fraction*
Thyme	3.0	1.5	3.0
Rosemary	17.6	2.2	5.5
Sage	16.5	2.2	5.2
Oregano	3.8	1.4	2.7
Savory	1.6	1.3	1.3

Source: Chipault *et al.*, 1952.

Table 8.5 Yields of the isolates obtained from dried thyme and their antioxidant activity as evaluated by the β-Carotene Bleaching Test

Type of the extract	Yield, g/kg dry matter	Antioxidant activity, 0 (low)/5 (high = BHT)
Essential oil	62.7	2
Deodorised acetone extract	4.4	3
Deodorised water extract	73.2	0
Acetone oleoresin	35.0	4
Methanol-water extract	73.9	3
CO$_2$ extract (300 bar, 40 °C, 5 min)	54.6	4

Source: Dapkevičius *et al.*, 1998.

respect to chemical composition, steam and hydrodistilled extracts showed similar profiles. Supercritical extracts presented besides the same components as that of steam distillation, non-volatile and non-aromatic compounds. Thymol was extracted at similar levels by all the tested methods. However, *p*-cymene, which was present in a high content on the other extraction procedures (11 per cent for Clevenger, 5.6 per cent for steam distillation and 7.2% for 20 MPa) was almost absent in the SFC extract obtained at 10 MPa (0.07 per cent) and was very low in the product obtained at 15 MPa (1.5 per cent). In general, the choice of the extraction method depends on the composition and characteristics of the required products.

In another study *T. vulgaris* extracts were prepared by different methods to test their antioxidant properties (Dapkevičius *et al.*, 1998). The results obtained are summarised in Table 8.5. The yield of the extract obtained by CO$_2$ at 300 bar, 40 °C, 5 min was 54.6 g/kg and it possessed significant antioxidant activity. The experiments were further expanded separately with thyme stems and leaves by using 120 and 450 bar pressure. The antioxidant activity of the extracts obtained at different pressures was similar; however, the concentration of active substances in the stems was considerably lower than in the leaves (Figure 8.3).

A significant effect of the extraction conditions on the quality of extracts is clearly demonstrated in Figure 8.4. By selecting the solvent and optimising the extraction procedures, it is possible to isolate the largest amount of the substances of interest. For instance, by

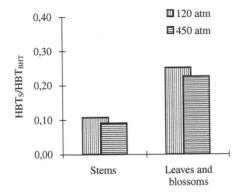

Figure 8.3 Antioxidant activity of CO$_2$ extracts from different parts of thyme.

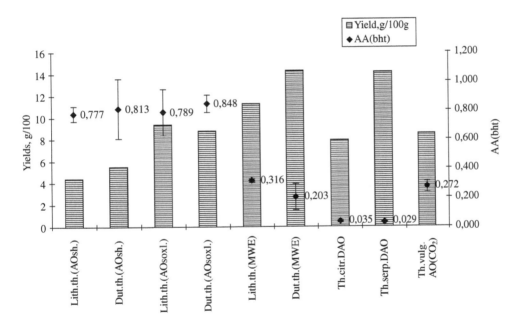

Figure 8.4 Antioxidant activity and yields of thyme extracts.
Lith.th., *Thymus vulgaris* grown in Lithuania; Dut.th., *Thymus vulgaris* grown in The Netherlands; AO, acetone oleoresin; sh., extracted by shaking dried plant material with solvent; soxl., extracted in the Soxhlet apparatus; MWE, methanol-water extract; DAO, deodorised acetone oleoresin; AO(CO$_2$), AO after CO$_2$ extraction.

using methanol-water it was possible to obtain the highest extract yield although the anti-oxidant activity of the extract was very low (about five times lower than the antioxidant activity of butylated hydroxytoluene (BHT)). The yields of acetone extracts obtained in a Soxhlet apparatus were lower, however their antioxidant activity was comparable to that of BHT.

Nguyen *et al.* (1991) patented a method for extracting antioxidants from Labiate herbs, which settles the following extraction and fractionation parameters for *T. vulgaris*: extractor

500 bar/95 °C; 1st separator 120 bar/80 °C; 2nd separator 35 bar/15 °C. These conditions being applied, the yield of the essential oil fraction was 0.7 per cent while the yield of the antioxidant fraction was 2.0 per cent.

INFLUENCE OF PROCESSING PROCEDURES ON THE QUALITY OF THYME AND THYME PRODUCTS

Processing procedures are known to change some characteristics of herbs and spices. This problem must be mainly focused on

(a) the essential oil content and its chemical composition and consequently the sensory profile of the product aroma;
(b) the composition of other (non-volatile) constituents and consequently the properties as regards taste and nutritional value;
(c) the structure of natural pigments and consequently the colour of the product;
(d) microbial contamination.

Usually the flavour impact of the freshly cut herb is appreciably higher and of a different character from that of its dried counterpart. This is due to the loss or modification of the low boiling fractions of the oil. The clean characteristic top notes associated with the freshly cut green herb are, in the dried material, overlaid with a dull hay-like aroma (Heath and Reineccius, 1986). Drying can be carried out by different methods, which were in general described in Chapter 7. Literature data indicate that the changes of aroma compounds during drying depend on the drying method as well as on the character of the herbs and spices.

The treatments of foods with ionising radiation to reduce bacterial counts and thus to prevent the spread of food-borne diseases and to improve the shelf life of the food itself or the processed products has become a matter of much greater importance in the past years. Results on the consequences of irradiation are available for about 50 different spices. Some of these have been examined several times by different authors (Schüttler *et al.*, 1991). Most of the papers concerning irradiation of herbs and spices deal with bacteriological decontamination, shelf life and detection of irradiation. So far as irradiation is a concern of food legislation, the methods of the detection of irradiated herbs and spices is also briefly outlined in the present section.

Effect of drying on the composition of volatile compounds

Dehydration is still a very important process for the preservation of spices and aromatic herbs. A great number of studies have been carried out on the effect of drying methods and parameters on the quality of spices and aromatic plants, particularly focused on Labiatae family herbs. The publications on the quality of thyme itself are not very numerous, at least in the international journals and other available sources. Therefore, some results on the effect of drying on the quality of other, particularly Labiatae family herbs have also been included in this section. Considering biological similarities between Labiatae plants this seems to be reasonable.

In general, two main approaches can be used in the assessment of volatile constituents in herbs and spices as well as in other foods. These are: (1) the analysis of the total

Table 8.6 The content of essential oil, thymol and carvacrol in dried thyme

Characteristics	Freeze-dried	Oven-dried
Essential oil (µl/g)	15.2	13.3
Thymol (mg/g)	9.6	7.9
Carvacrol (mg/g)	0.5	0.4

Source: Bendl *et al.*, 1988.

concentration of aroma compounds in the product matrix (essential oil in case of thyme); (2) analysis of the headspace volatiles, which are above the matrix and which are in close relation with the dynamics of release of volatiles (i.e. equilibrium between aroma compounds in the matrix and above it) and consequently with the sensory profile of the product. Therefore, the method of the analysis of volatile constituents is quite important. Comparing several methods for the isolation of volatile compounds from aqueous model systems Leathy and Reineccius (1984) concluded that headspace is very dependent on the volatility of aroma compounds, whereas simultaneous distillation/ solvent extraction (SDE) exhibited reproducible recoveries.

Different methods of isolation for the investigation of the effect of drying on the aroma changes in herbs and vegetables were applied: extraction (Huopalahti *et al.*, 1985, Nykänen and Nykänen, 1987); SDE (Kaminski *et al.*, 1986; Kirsi *et al.*, 1989); headspace method (Koller, 1988). The latter enables one to determine the changes in vapour phase, which can be much more related to the sensory aroma profile of the product. The differences in the percentage composition of volatile compounds in the samples obtained by different methods could be significative as it is demonstrated by Huopalahti *et al.* (1988) in the case of dill.

In general, headspace samples are dominated by the more volatile components, steam distilled concentrates additionally contain some higher boiling compounds, whereas extracts consist of both volatile and non-volatile fractions. Chialva *et al.* (1982), Venskutonis and Dapkevičius (1995) compared the composition of the headspace over fresh herbs with that of steam distilled essential oil and observed significant differences, e.g. some very volatile components were present in headspace and absent in the essential oil. Jennings and Filsoof (1977) conclude that no single sampling system can be regarded as uniformly satisfactory, but that, depending on the sample and what the investigator wishes to study, one or another system may be superior.

Oil yield and oil composition

Oil yield and composition are the most important parameters defining flavour properties of a particular oil-bearing plant. Drying is the most critical process due to the volatility and susceptibility to chemical change of the contained volatile oil. Several important studies have been carried out to determine the effects of drying methods and parameters on the volatile oil content of thyme.

Bendl *et al.* (1988) studied the effect of freeze-drying of sage and thyme on the content of essential oils and their characteristic flavour components in comparison to drying at increased temperature (40 °C). The examination showed (Table 8.6) that the content of essential oil in the spice samples was higher in the freeze-dried products

Table 8.7 Changes in volatile oil yield* with increasing drying temperature

Plant	Temperature (°C)						
	40	50	60	70	80	90	100
Thyme	1.0	1.0	1.0	0.3	0.3	0.2	0.0
Savory	1.3	1.0	0.8	0.7	0.5	0.03	0.0
Basil	1.0	0.7	0.7	0.7	0.3	0.3	0.1
Marjoram	1.7	1.3	1.3	1.0	1.0	0.7	0.5
Rosemary	2.3	2.0	2.0	1.7	1.7	1.0	0.3
Sage	2.0	1.7	1.0	0.5	0.5	0.4	0.3
Tarragon	0.5	0.3	0.3	0.2	0.2	0.2	0.1

Note
* Percentage of volatile oil expressed as yield (v/w) dried plant matter.
Source: Deans *et al.*, 1991.

Table 8.8 Comparison of the major peaks from GC analysis after warm air oven and microwave-drying (%)

Constituent	Oven-dried		Microwave-dried	
	Thyme	Savory	Thyme	Savory
Thymol	47.77	–	18.42	–
γ-Terpinene	16.77	10.40	12.06	19.90
p-Cymene	11.91	22.05	3.04	0.00
Carvacrol	–	37.61	–	66.66
New peaks	–	–	20.25; 6.66; 6.62	–

Source: Deans *et al.*, 1991.

compared to those dried at increased temperature. The concentration of the most important components for aroma of thyme, thymol and carvacrol was also higher in the freeze-dried herb.

Deans *et al.* (1991) studied *T. vulgaris* and six other culinary herbs dried by warm-air and microwave ovens. The volatile oil content (Table 8.7) of seven plant species was determined by Gas Chromatography (GC) following drying at temperatures from 40–100 °C, revealing that at temperatures >60 °C, most of the volatile constituents were lost. The yield of volatile oils was substantially decreased in all microwaved herbs. The qualitative and quantitative changes in the volatile oil profiles were profound. The results on the main constituents of thyme and savoury, which are similar herbs in terms of chemical composition, are presented in Table 8.8.

It is interesting to note that the percentage of thymol in thyme significantly decreased after microwaving (most likely, due to the formation of new compounds), whereas the content of carvacrol, a thymol isomer, in savoury considerably increased (most likely due to the loss of more volatile constituents). Rather controversial results were obtained with two other common compounds both for thyme and savoury, *p*-cymene and γ-terpinene. For instance, contrary to thyme, *p*-cymene was not detected in microwaved savoury (in the oven-dried herb it constituted 22.05 per cent), whereas the percentage of γ-terpinene was almost two times higher in the essential oil distilled

Table 8.9 Effect of drying on the Indian thyme essential oil content (% on moisture free basis) and its relative concentration (%)

Constituents	Fresh	Freeze	Cross flow	Through flow	Shade
Monoterpene hydrocarbons	29.3	29.6	29.6	23.9	0.1
Oxygenated compounds	2.9	2.4	2.4	2.7	0.0
Thymol	60.1	59.5	60.5	64.9	83.5
Sesquiterpenes	3.5	2.8	2.0	3.2	9.5
Essential oil	1.54	0.95	0.90	1.10	0.37

Source: Raghavan *et al.*, 1995.

Table 8.10 Content of some volatile compounds in fresh, air-dried and freeze-dried thyme (mg/kg)

Constituent	Fresh	Air-dried	Freeze-dried
Myrcene	343[a]	314[b]	285[b]
α-Terpinene	254[a]	226[b]	215[b]
γ-Terpinene	1817[a]	1298[b]	1293[b]
Thymol	6465[a]	6917[a]	6907[a]
β-Caryophyllene	213[b]	288[a]	282[a]

Note
a–b values with same letter within same row are not significantly different.
Source: Venskutonis *et al.*, 1996.

from microwaved savoury than from the oven-dried herb. The results obtained by Deans and co-authors clearly show that the behaviour of particular plants even belonging to the same family (Labiatae in this case) can be significantly different.

Raghavan *et al.* (1995) compared the effect of cross flow drying, through flow drying, freeze-drying and shade drying on the Indian thyme (*T. vulgaris*) essential oil content and its composition. At temperatures of 50 and 60 °C losses from 50–75 per cent were registered, therefore these temperatures proved not to be suitable. The results obtained in this study, which are summarised in Table 8.9, also show that drying in the shade was very ineffective and long (120 h). Other methods were comparable and, considering the time of drying and the flavour quality of the dried herb, the authors concluded that flow drying (40 °C, 8.5 h) should be the method of choice.

Venskutonis *et al.* (1996) studied the effect of air and freeze-drying on the content of volatiles and their composition in thyme. Air-drying was carried out at 30 °C and air velocity of approx. 3.3 m/s for 25 h. The final moisture content of the air-dried herb was 8.5 per cent. Freeze-drying was completed in 40 h with the final moisture content of 5.5 per cent. Volatile constituents were isolated by simultaneous SDE procedure in a Likens-Nickerson apparatus. The reduction in the total content of volatile constituents after drying was of approx. 1–3 per cent and no differences between the two drying methods were found. This is less than for basil and marjoram and approximately the same for wild marjoram (air-drying, room temperature, Nykänen and Nykänen, 1987). In Table 8.10 the content of some compounds are expressed that underwent more considerable changes, from fresh to air-dried and freeze-dried thyme. In general, the levels slightly decreased during drying, except for β-caryophyllene and thymol, which

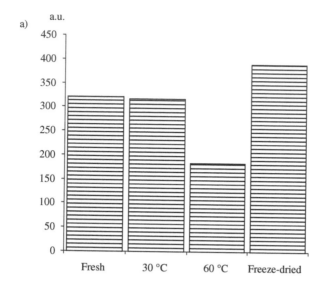

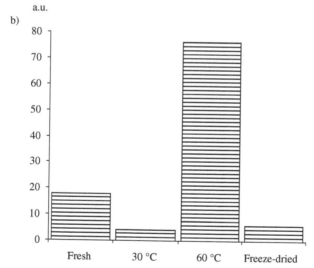

Figure 8.5 Changes of the total content of aroma constituents in thyme during drying, in arbitrary units (a.u.). (a) Simultaneous distillation/extraction (SDE), (b) headspace (HS); (Venskutonis, 1997).

increased (although statistical evaluation did not give significant differences for the latter compound in fresh and dried samples).

In another study, thyme was dried in the oven at temperatures of 30 °C and 60 °C and in the freeze-dryer (Venskutonis, 1995, 1997; Venskutonis *et al.*, 1996). The changes of the total amount of SDE volatiles are demonstrated in Figure 8.5a. Very close concentrations of volatiles were determined in fresh and oven-dried at 30 °C herb. However, the reduction of the total amount of SDE compounds in oven dried at 60 °C herb was 43 per cent. It is worth mentioning that the weight of thyme during 4 h

Table 8.11 Composition of thyme essential oil extract (SDE) and headspace volatiles (HS), arbitrary units

Compound or retention time	SDE				HS			
	Fresh herb	Dried herb		Freeze	Fresh herb	Dried herb		Freeze
		30 °C	60 °C			30 °C	60 °C	
11:50	0.56	0.56	0.28	0.56	0.06	0.02	0.47	0.04
14:46	0.91	0.90	0.48	0.96	0.13	0.02	0.96	0.04
E-2-hexenal		0.06		0.07		0.13	0.03	0.51
α-Thujene	5.05	4.78	2.40	5.32	0.68	0.09	2.16	0.10
α-Pinene	3.70	3.68	2.11	3.99	0.46	0.06	1.46	0.07
Camphene	1.96	1.90	1.10	2.12	0.26	0.04	1.07	0.06
1-Octen-3-ol	3.89	3.79	2.25	4.15	0.08	0.04	1.46	0.06
β-Pinene	1.26	1.20	0.71	1.31	0.13	0.02	0.51	0.02
Myrcene	6.61	6.12	2.76	7.01	0.61	0.14	2.85	0.16
α-Terpinene	3.40	3.25	1.68	3.80	0.31	0.08	1.58	0.10
p-Cymene	82.70	80.50	37.00	87.96	10.66	1.87	38.88	2.04
Limonene	1.74	1.65	0.60	1.69	0.17	0.03	0.74	0.03
1,8-Cineole	3.21	3.04	2.17	3.24	0.26	0.02	1.54	0.04
γ-Terpinene	21.20	20.20	8.24	22.03	1.53	0.59	9.40	0.78
tr-Sabinene hydrate	2.47	3.42	2.49	3.17	0.08	0.03	1.42	0.06
Linalool	9.47	9.08	6.10	10.41	0.14	0.05	2.40	0.10
Isoborneol	3.20	3.12	2.32	3.68	0.03	0.02	0.62	0.05
4-Terpineol	2.35	1.67	1.23	2.42	0.02		0.11	0.01
Thymol	137.20	134.00	87.70	182.00	0.45	0.11	3.66	0.25
Carvacrol	8.43	8.60	5.33	11.29			0.17	
β-Caryophyllene	5.87	7.57	4.66	8.06	0.39	0.08	0.76	0.13
Total	321.72	317.01	183.16	388.33	17.98	3.92	76.14	5.90

Source: Venskutonis, 1997.

drying at 60 °C reduced 3.4 times. Koller and Raghavan (1995) obtained very close results with rosemary: 30 per cent of the essential oil was lost during air convection drying at 50 °C. The total amount of SDE volatiles in the freeze-dried thyme even increased approximately by 20 per cent. Considerable increase in the content of the major compound thymol (by 33 per cent) was the main contribution to the total increase.

Some interesting observations were made concerning the changes of individual flavour constituents (Table 8.11). Most of the thyme SDE volatiles during oven-drying at 30 °C and freeze-drying did not undergo significant changes. Their reduction during oven-drying at 60 °C depended on the volatility and chemical origin of the constituent. For instance, the concentration of the quantitatively major compounds p-cymene, γ-terpinene and thymol were reduced by 2.24, 2.57 and 1.56 times respectively. The amount of β-caryophyllene in the oven-dried at 30 °C and freeze-dried thyme was found to have increased by 29 and 37 per cent respectively. Very close results were obtained in a previous study (Venskutonis *et al.*, 1996). One more tendency in the changes of volatiles is evident from the results obtained: the losses of non-oxygenated terpenes during oven-drying at 60 °C were considerably higher than that of oxygenated compounds, particularly terpene alcohols. Most likely, two reasons could be responsible for this tendency: the differences in the volatility and the formation of oxygenated compounds during drying.

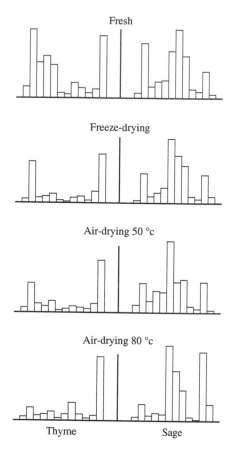

Figure 8.6 Influence of drying temperature on the composition of the headspace gas over thyme and sage (Koller, 1988).

Headspace (HS) constituents

Influences of drying methods and temperatures on the composition of the HS of thyme and sage were analysed by Koller (1988). The so-called histograms (the result of gas-chromatographic "odour" tests) presented in Figure 8.6 showed that the profile of volatile compounds changed even by freeze-drying. However, freeze-drying and air-drying at 50 °C were comparable, while air-drying at 80 °C was not suitable for both herbs due to loss of volatile constituents.

Later some new results of HS analysis were obtained, which also revealed some interesting peculiarities of the influence of drying on the aroma constituents in thyme, Table 8.11, Figure 8.5b (Venskutonis, 1997). The total amount of the collected volatiles in thyme herb HS was the highest in the case of thyme oven-dried at 60 °C: it was 4.2 times higher than in the fresh one, 19.4 times higher than that oven-dried at 30 °C, and 12.9 times higher than the freeze-dried one. It was obvious that the intensity of sensory odour perception of the thyme herb oven-dried at 60 °C was evidently stronger than that at 30 °C.

It is interesting to note that similar experiments were carried out with another Labiate herb, sage, and it was found that the total content of the absorbed compounds on Porapack during dynamic HS purging of the samples was the largest in the case of fresh herb, middle in freeze-dried, and lowest in oven-dried. The results were comparable for 30 °C and 60 °C drying temperatures, when the total content of absorbed HS volatiles was lower in comparison with fresh herb 4.6 and 3.7 times, respectively. It means that the influence of the drying method on the rate of the release of flavour compounds can be very particular for each given herb. For instance, the amount of volatiles in HS of fresh, oven-dried and freeze-dried sage significantly exceeds that of thyme, however, when the temperature of 60 °C was applied, the intensity of aroma release from thyme was "activated" 4.2 times (as compared with fresh herb) in terms of total increase of the absorbed HS volatiles. It could be supposed that thyme leaves undergo significant changes in their botanical structure during drying at higher temperatures. From this point of view, sage leaves could be considered to be more resistant against the effect of drying at higher temperatures.

Assessment of drying effects on thyme aroma

The comparison of data obtained by simultaneous SDE and HS allows one to assess every volatile constituent of herb in terms of its total content (m) and the rate of its release (v). The latter can be related to the concentration of a particular constituent in HS, which depends on the morphological and anatomical characteristics of the structures containing the essential oil in the plant. In this case a certain aroma potential (AP) of every particular volatile constituent can be considered as a function of these two parameters and odour threshold value (c) (Venskutonis, 1997):

$$AP = f(m, v, c)$$

Certainly, such a function is rather conditional and depends on the parameters of SDE and HS analysis. However, by using standardised conditions it is possible to have some mathematical tool, representing a certain aroma potential of a particular volatile constituent in aromatic herb.

In the case of thyme it is demonstrated that the major constituent thymol in SDE concentrate constitutes 42–48 per cent, while in HS only 2.5–4.8 per cent (Figure 8.7). Such volatile compounds as p-cymene, γ-terpinene, and myrcene prevail in HS vapours of thyme. The ratio of the percentages of SDE/HS characterises as a certain coefficient of efficiency (C_e) of a particular constituent in aromatic herb. To some extent, it represents the activity of the participation of such a compound in the creation of the odour. In Table 8.12, percentage concentrations of some major thyme volatile constituents and their C_e coefficients are tabulated. It is interesting to notice that for some similar compounds these coefficients are different in thyme and sage. For instance, C_e of β-caryophyllene in fresh thyme is 2.8 times higher than in fresh sage.

C_e coefficients were also calculated for dried herb. The figures obtained can be informative for the evaluation of the degree of disbalance of the fresh aroma during drying. For some constituents of thyme the changes of C_e after drying are represented in Figure 8.8. The diagrams show that the changes of C_e depend on the chemical origin of the constituent. For instance, C_e of linalool significantly increased after oven-drying at 60 °C, while that of β-caryophyllene was reduced several fold.

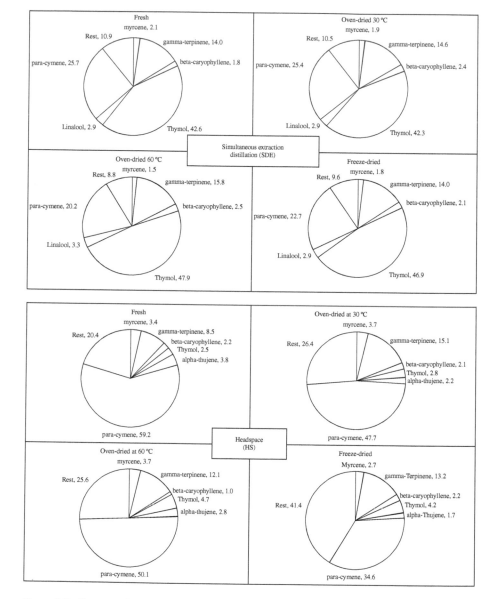

Figure 8.7 Changes of the percentage content of the main SDE and HS constituents in thyme during drying (Venskutonis, 1997).

Effect of drying on colour

The phenomenon of colour changes during drying is quite common for most of the leafy plants, because the chlorophylls define the colour of green herbs and vegetables. They are sensitive to many factors, which cause the shift of maximum absorbance and therefore their natural green colour changes to less desirable colours. Therefore when

Table 8.12 Percentage content of volatile compounds and their coefficients of efficiency (C_e) in fresh thyme

Compound	SDE	HS	C_e
α-Pinene	1.15	2.55	2.22
Camphene	0.61	1.42	2.33
β-Pinene	0.39	0.74	1.90
Myrcene	2.05	3.40	1.66
1,8-Cineole	1.00	1.46	1.46
γ-Terpinene	6.59	8.53	1.29
p-Cymene	25.71	59.28	2.31
Linalool	2.94	0.79	0.27
Thymol	42.65	2.50	0.06
β-Caryophyllene	1.82	2.15	1.18

Source: Venskutonis, 1997.

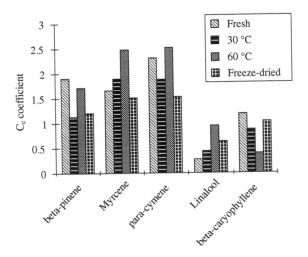

Figure 8.8 Changes of the coefficients of aroma efficiency (C_e) of some thyme volatile constituents during drying (Venskutonis, 1997).

optimising the drying process the parameters should be kept to minimise both the losses of volatile compounds and the change of a natural green colour.

Analyses concerning colour in *Thymus* are very scanty, however, there are some studies dealing with colour changes in Labiate herbs. So far as the process of discolouration and/or colour changes during drying can possess similarities between various aromatic plants, it was considered reasonable to provide a few examples of the relevant investigations.

Takruri and Daqqaq (1984, 1986) studied the effects of storage and the methods of drying on the carotene values in mint, Jew's mallow, thyme, and parsley. They found that a range of 47–92 per cent of the carotenoid content was retained in these plants when dried by the traditional methods of sun-drying and shadow-drying. In addition,

Table 8.13 Effect of drying on weight loss and microbiological quality of thyme

Treatment*	Weight loss	PCA	Media VRBA	BCSA	B-PA
Initial microflora		1.62×10^5	4.0×10^2	2.0×10^1	<10
After 48 h of drying oven at 38 °C	68.2	1.05×10^4	<1	<10	<10
After 5 min microwave-drying	46.5	7.35×10^4	<1	1.0×10^1	<10
After 7.5 min microwave-drying	52.9	6.10×10^2	<1	1.0×10^1	<10

Notes
* Counts per gram fresh material;
PCA, Plate Count Agar.
VRBA, Violet Red Bile agar for coliforms.
BCSA, *Bacillus cereus* selective agar for *Bacillus cereus*.
B-PA, Baird-Parker agar for *Staphylococcus aureus*.
Source: Deans *et al.*, 1991.

44–69 per cent of the carotenoid values in the dried plants were detected after storage for 1 year. The drying method and the drying temperature had varying effects on the carotenoid value. Thus in mint, the percentage recoveries using shadow-drying, sun-drying, and oven-drying at 40 °C, 60 °C and 100 °C were respectively 79 per cent, 76 per cent, 71 per cent, 74 per cent and 54 per cent. For the oven-dried samples, increasing the drying time or the drying temperature over 100 °C resulted in greater losses in the carotenoid content. It is apparent, therefore, that these plants dried by traditional means, or in ovens at temperatures not exceeding 100 °C, remain good sources of carotenes even after one year storage.

Müller *et al.* (1989) studied drying of medicinal and spice plants with solar energy in a plastic film covered greenhouse. In their experiments with mint, sage and hops, they found that solar drying was much superior to conventional drying with regard to colour, texture and contents of active ingredients.

The process of rapid heating and cooling of herbs and spices developed by Hosokawa Micron Europe BV minimise loss of colour and essential oil. Special equipment for this process has been developed (Spook, 1993; Anonymous, 1993).

Rocha *et al.* (1993) used steam blanching and surfactant pretreatment to increase drying rate of basil and found that both pretreatments resulted in better retention of the green coloration of the leaves. Steam blanching was shown to enhance chlorophyll retention. Low air-drying temperatures were needed for samples that were not pretreated, while high air temperatures were acceptable for pretreated samples.

Effect of drying on microbial contamination

Usually herbs and spices are heavily contaminated with microorganisms. Microbiological quality also depends on drying method and conditions. Deans *et al.* (1991) studied the microbiological quality of the raw and dried material of *T. vulgaris* and six other culinary herbs dried by warm air and microwave ovens. The authors determined total bacterial counts, *Staphylococcus aureus*, *Bacillus cereus* and coliforms. Exposure of herbs to microwaves was evaluated as a method of both drying and reducing the microbial load present on the plants. The microflora was reduced by two or three logarithmic cycles (Table 8.13).

Influence of irradiation processes

The use of ionising radiation has been extensively studied over the past 30 years as a mean of sterilising spices. In all cases, microbial populations have been reduced (Eiss, 1984). The reduction of the microbial population both after irradiation and during storage of irradiated herbs and spices is the most important task in post-processing treatment. The studies have resulted in the establishment of optimal irradiation doses for various herbs and spices. Ten kGy has been found to be effective in destroying bacterial spores, while 5 kGy has been sufficient to eliminate mould contamination (Munasiri *et al.*, 1987). The treatment with 5–7.5 kGy was proposed as a sufficient dose for decontaminating thyme (Farkas, 1988). Comparing irradiation with other sterilisation methods, e.g. Vajdi and Pereira (1973), Josimović and Jovanović (1982) found out that γ-irradiation was more effective than ethylene oxide in reducing the bacterial population of spices (pepper, paprika, oregano, allspice, celery seeds, garlic, marjoram, cardamon, caraway and parsley).

The microbiological data of some non-irradiated and irradiated herbs and spices are provided in Table 8.14. The dose of 10 kGy reduced standard plate count in thyme by 3750 times, moulds more than 30 times and completely destroyed maximum probable number (MPN) coliforms. Very good effects were achieved with other treated herbs and spices as well.

In foods irradiation can result in the creation of new chemicals, called unique radiolytic products. These include known carcinogens like benzene, formaldehyde and certain peroxides. However, the FDA has concluded that the amount of any toxic

Table 8.14 Microbiological data: irradiated spices

Spice	Dose kGy	Standard plate count	Yeast	Mould	MPN coliforms
Thyme	0	150.000	0	300	30
	10	40	0	<10	0
Allspice	0	2.278.000	<10	0	0
	10	<10	<10	0	0
Greek oregano	0	1214400	40000	9000	2636
	10	<10	<10	<10	0
Black pepper	0	32.000.000	0	0	25
	10	60	0	0	0
Garlic powder	0	414.000	<10	7800	0
	10	700	<10	<10	0
Egyptian basil	0	3.000.000	>30.000	400	>11.000
	10	1000	<10	<10	0
Mexican oregano	0	1.500.000	>30.000	10.000	5000
	10	30	<10	<10	0
Domestic paprika	0	1.000.000	0	0	0
	10	100	0	0	0
Spanish paprika	0	2.200.000	0	0	410
	10	260	0	0	0
Celery seed	0	440.000	1500	200	25.000
	10	<10	<10	<10	0
Crushed red pepper	0	130.800	<10	0	0
	6.5	<10	<10	0	0

Source: Eiss, 1984.

chemical created by allowable levels of irradiation is too small to be significant; foods cannot be tested to determine that the proper amount of radiation was used. Strictly from the scientific point of view, no ceiling should be set for food irradiated with doses greater than the currently recommended upper level of 10 kGy by the Codex Alimentarius Commission. The food irradiation technology itself is safe to such a degree that as long as sensory qualities of food are retained and harmful microorganisms are destroyed, the actual amount of ionising radiation applied is of secondary consideration.

In the case of irradiation of herbs and spices, this need for a greater average dose has already been recognised in several countries. France permits an average dose of 11 kGy for the irradiation of spices and dry aromatic substances, whereas Argentina and the US permit a maximum dose of 30 kGy for this purpose.

The radiation effects on biological material are ascribed to the sum of two processes, direct (chemical events occurring in the target molecule as a result of energy deposition) and indirect (consequence of reactive, diffusible, free radicals formed from the radiolysis of water: OH^-, e_{aq}, H^+, H_2, H_2O_2) action. Experiments indicate that herbs and spices with water contents of 4.5–12 per cent are very resistant to physical or chemical change when irradiated. Sensory and food applications analyses indicate no significant difference between irradiated samples and controls for all spices tested (Eiss, 1984).

Another important point is that sterilisation by ionising radiation is a cold treatment. Irradiation as high as 50 kGy will only increase the temperature of irradiated food by 12 °C. Therefore, there is little danger of the loss of volatile components of the spice. Irradiation does not require the addition of any chemicals, liquid or gas. In addition, no other methods of processing, e.g. heating, freezing or drying need to be employed (Eiss, 1984).

The effect of irradiation on the sensory profile and chemical composition of herb and spice essential oils has been thoroughly examined. Thyme was also selected as a testing material in some studies. Most of them show that for the microbial decontamination required, irradiation doses do not affect total essential oil content and the sensory profile of the herb. For instance, marked reduction in bacterial count with no deterioration in organoleptic quality of thyme, coriander and paprika was determined after irradiation with 0–10 kGy by Van Dijk in 1970. Some workers compared the effect of irradiation with that of ethylene oxide. The investigations on the effect of irradiation on some widely used Labiate herbs, including thyme, are provided in Table 8.15 (Farkas, 1988).

The effect of irradiation on the chemical composition of herbs and spices has been investigated by using different methods. Venskutonis (Venskutonis, 1994; Venskutonis *et al.*, 1996) studied the effect of 3, 10 and 30 kGy γ- and β-irradiation on the chemical constituents of the oven and freeze-dried thyme isolated by a simultaneous SDE procedure in a Likens-Nickerson apparatus and analysed by capillary GC and GC/mass-spectrometric (MS) methods. The quantitative content of the main constituents in air-dried herb before and after irradiation at different doses is tabulated in Table 8.16. It was shown that the concentration of γ-terpinene has increased after β-irradiation. The tendency for reduction of *p*-cymene after 30 kGy for both types of irradiation was also observed. This tendency was also found for the freeze-dried samples. Statistical analysis of the results showed that irradiation did not have any significant effect on the concentrations of the thyme compounds, except for γ-terpinene.

The lack or small effect of irradiation on the thyme aroma compounds is in agreement with other works. For instance, irradiation of spices including Spanish thyme with a

Table 8.15 Dose requirements for radiation decontamination of thyme and some other Labiatae herbs as compared to the retention of their volatile oil content, and threshold doses of organoleptic changes

Herb	Dose requirement (kGy)	Relative yield of volatile oils* at 8–10 kGy (%)	Threshold dose of organoleptic changes (kGy)
Thyme	5–7.5	101	≥10
Basil	4–10	99	~12.5
Marjoram	7.5–10	100–103	5–10 but also >16
Oregano	≤4	99–100	>10
Sage	4		10
Savory	<5		

Note
* As percentage of the yield of untreated sample.
Source: Farkas, 1988.

Table 8.16 Concentrations of the main volatile constituents in air dried thyme before and after irradiation (mg/kg), none of the results were significant, except for γ-terpinene

Constituent		γ 3 kGy	γ 10 kGy	γ 30 kGy	β 3 kGy	β 10 kGy	β 30 kGy
α-Thujene	169	190	188	171	189	185	182
α-Pinene	222	230	208	210	233	231	232
Myrcene	254	268	272	251	279	280	280
α-Terpinene	179	184	201	192	212	202	223
p-Cymene	3996	4036	3904	3520	4002	3764	3513
γ-Terpinene	987	1060	1035	1110	1181	1392	1406
tr-Sabinene hydrate	219	235	233	228	246	239	248
Linalool	337	357	361	344	373	367	381
Borneol	151	150	151	153	165	169	172
Thymol	8509	7705	7799	7752	8672	8538	8846
Carvacrol	994	1108	980	957	1103	1014	1143
β-Caryophyllene	309	291	312	294	348	333	336
Caryophyllene oxide	76	68	55	44	75	53	65

Source: Venskutonis *et al.*, 1996.

dose up to 50 kGy provided no evidence of noticeable changes of volatile compounds (Eiss, 1984). Bug (1989) applied 29 kGy doses to different herbs including thyme and also did not find any effect on the quantitative and qualitative composition of its essential oil.

Detection of irradiated herbs and spices

As far as the treatment of food with ionising energy is finally becoming reality there is a need of a reliable detection method of irradiated foodstuffs. Usually, before the radiation treatment of spices and herbs can be undertaken on a commercial basis, the respective health authority or regulatory body must give written approval. As the process of food irradiation produces practically no change in appearance, shape or temperature of products, it is controlled mainly by administrative means through requirements for documentary records and labelling. Therefore, there is an interest to supplement administrative control by developing identification methods for irradiated foods (Farkas, 1992). The regulations concerning irradiated foods are very different even between European countries (IAEA, 1988). This has led to increasing interest in methods for detecting prior irradiation of foods. A current problem with the use of irradiation is the lack of a specific method for identifying foods that have been irradiated.

Delincée (1998) recently reviewed significant progress, with the development of analytical detection methods using changes in food with an origin as the radiation treatment. Five detection methods (electron spin resonance, thermoluminescence, lipid hydrocarbons, O-tyrosine and microbiological analysis) have been developed to detect a large variety of irradiated foods and their reliability has been confirmed through a series of collaborative trials.

Some studies have been carried out in establishing special detection methods for irradiated herbs and spices. Göksu (Göksu and Regulla, 1989; Göksu *et al.*, 1990) showed that most of the natural products contain minute amounts of wind blown or intruding dust, which can be separated and used to identify irradiated spices by measuring its thermoluminescence (TL).

Thermoluminescent dust obtained from herbs and spices, including thyme, were investigated and it was concluded that for reliable identification of irradiated spices by TL, it is essential that adhering inorganic dust is used for measurements. It has advantages over the use of whole spices. Dust which is inorganic gives no TL due to incandescence and can be heated up to at least 400 °C. At this temperature more stable TL peaks are accessible. Thus, the identification and quantitative absorbed dose assessment of irradiated spices is possible even after some years.

Sjöberg *et al.* (1990) tested three types of methods for the identification of irradiated spices as potential control methods: a) a microbiological, combining a direct epifluorescent filter technique (DEFT) with a total aerobic plate count (APC), b) a chemoluminescence method and c) chemical GC and GC/MS methods for the analysis of volatile oils. The best methods for control purposes were the microbiological (DEFT+APC) methods combined with chemoluminescence measurements. No differences were detected between the irradiated and non-irradiated samples with the chemical methods.

REFERENCES

Anonymous (1993) Turning up the heat. *Food Manufacture*, August.
Bendl, E., Kroyer, G., Washüttl, J. and Steiner I., (1988) Untersuchungen über die Gefriertrocknung von Thymian und Salbei. *Ernährung/Nutrition*, 12, 793–795 (German).
Bug, J. (1989) *Strahlen-und Gassterilisation von Arzneidrogen und Gewürzen*. Doctoral thesis, University of Würzburg (German).
Burdock, G.A. (1994) *Fenaroli's Handbook of Flavor Ingredients*, CRC Press, Boca Raton, Florida.
Calame, J.P. and Steiner, R. (1987) In M. Hirata and T. Ishikawa (eds), *Theory and Practice in Supercritical Fluid Technology*, Tokyo Metropolitan Univ., pp. 227–318.
Cardoso, L.A., Moldão-Martins, M., Bernardo-Gil, G. and Beirão da Costa, M.L. (1993) Supercritical fluid extraction of aroma compounds from aromatic herbs (*Thymus zygis* and *Coriandrum sativum*), In *Developments in Food Engineering*, 6th Int. Congress on Engineering and Food, Chiba, Japan, pp. 829–831.
Chipault, J.R., Mizuno, G.R., Hawkins, J.M. and Lundberg, W.O. (1952) *Food Res.*, 17, 16–55.

Chialva, F., Gabri, G., Liddle, P.A.P. and Ulian, F. (1982) Qualitative evaluation of aromatic herbs by direct head-space GC analysis. Application of the method and comparison with the traditional analysis of essential oils. *J. HRC & CC*, 5, 182–188.

Dapkevičius, A., Venskutonis, R., Van Beek, T.A. and Linssen, J.P.H. (1998) Antioxidant activity of extracts obtained by different isolation procedures from some aromatic herbs grown in Lithuania. *J. Sci. Food Agric.*, 77, 140–146.

Deans, S.G., Svoboda, K.P. and Barlett, M.C. (1991) Effect of microwave oven and warm-air drying on the microflora and volatile oil profile of culinary herbs. *J. Essent. Oil Res.*, 3, 341–347.

Delincée, H. (1998) Detection of food treated with ionizing radiation. *Trends in Food Science & Technology*, 9, 73–82.

Eiss, M.I. (1984) Irradiation of spices and herbs, *Food Technol. in Australia*, 36, 362–363, 366.

Farkas, J. (1992) Radiation treatment of spices. *Prehrambeno Tehnol. Biotehnol. Rev.*, 30, 159–163.

Farkas, J. (1988) *Irradiation of Dry Food Ingredients*. CRC Press, Inc., Boca Raton, Florida.

Göksu, H.Y., Regulla, D.F., Hietel, B. and Popp, G. (1990) Thermoluminescent dust for identification of irradiated spices. *Radiation Protection Dosimetry*, 34, 319–322.

Göksu, H.Y. and Regulla, D.F. (1989) Detection of irradiated food. *Nature*, 340, (6228) 23.

Heath, H.B. (1981) *Source Book of Flavors*. The AVI Publishing Company, Inc. Westport, Connecticut, USA.

Heath, H.B. and Reineccius, G. (1986) *Flavor Chemistry and Technology*, Macmillan Publishers Ltd.

Heath, H.B. (1982) Spices and aromatic extracts, influence of technological parameters on quality. In J. Adda and H. Richard (Coord. Scient.), *Int. Symp. on Food Flavors*, Tec. Doc.-Lavoisier, A.P.R.I.A., Paris, pp. 139–175.

Honerlagen, H.J. and Steiner, R. (1990) Verfahren zur Herstellung eines die wasserdampf-flüchtigen und andere lipophile Inhaltsstoffe enthaltenden Teilextraktes aus Heil- und/oder Gewürzpflanzen. *Swiss-Patent* CH 675 685 A5 (German).

Huopalahti, R., Kesälahti, E. and Linko, R.R. (1985) Effect of hot air and freeze drying on the volatile compounds of dill (*Anethum graveolens* L.) herb. *J. Agric. Sci. Finl.*, 57, 133–138.

Huopalahti, R., Lahtinen, R., Hiltunen, R. and Laakso, I. (1988) Studies on the essential oils of dill herb, *Anethum graveolens* L. *Flavour Fragr. J.*, 3, 121–125.

IAEA News Features (1988) December 5.

Jennings, W.G. and Filsoof, M. (1977) Comparison of sample preparation techniques for gas chromatographic analysis. *J. Agric. Food Chem.*, 25, 440–445.

Josimović, L. and Jovanović, M. (1982) The possibility of using ionizing radiation for sterilization of spices. *Hrana I Ishrana*, 23, 55–60.

Kaminski, E., Wàsowicz, E., Zamirska, R. and Wower, M. (1986). The effect of drying and storage of dried carrots on sensory characteristics and volatile constituents. *Nahrung*, 30, 819–828.

Kirsi, M., Julkunen-Tiitto, R. and Rimilainen, T. (1989) The effects of drying methods on the aroma of the herbal tea plant (*Rubus idaeus*). In G. Charalambous (ed.), *Flavors and Off-Flavors*, Elsevier Science Publ. B.V., Amsterdam, pp. 205–211.

Koller, W.D. (1988) Problems with the flavour of herbs and spices. In G. Charalambous (ed.), *Frontiers of Flavor*, Elsevier Science Publ. B.V., Amsterdam, pp. 123–132.

Koller, W.D. and Raghavan, B. (1995). Quality of dried herbs. *Poster, 9th World Congress of Food Science and Technology*, Budapest, Hungary, August.

Lawrence, B.M. (1995) The isolation of aromatic materials from natural plant products. In K. Tuley de Silva (ed.), *A Manual on the Essential Oil Industry*, UNIDO, Vienna, Austria, pp. 57–154.

Leathy, M.M. and Reineccius, G.A. (1984) Comparison of methods for the isolation of volatile compounds from aqueous model systems. In P. Schreier (ed.), *Analysis of Volatiles. Methods and Applications*, Walter de Gruyter & Co., Berlin, New York, pp. 19–47.

Moyler, D.A. (1987) In M. Hirata and T. Ishikawa (eds), *Theory and Practice in Supercritical Fluid Technology*, Tokyo Metropolitan Univ., pp. 319–341.

Moyler, D.A. (1991) Oleoresins, tinctures and extracts. In P.R. Ashurst (ed.), *Food Flavourings*, Blackie, Glasgow, London, pp. 54–86.

Moyler, D.A. (1993) Extraction of flavours and fragrances with compressed CO_2. In M. B. King and T.R. Bott (eds), *Extraction of Natural products Using Near-Critical Solvents*, Blackie Academic & Professional, Glasgow, pp. 140–183.

Müller, J., Reisinger, G., Mühlbauer, W., Martinov, M., Tešić, M. and Kisgeci J. (1989) Trocknung von Heil- und Gewürzpflanzen mit Solarenergie in einem Foliengewächshaus. *Landtechnik*, 44, 58–65 (German).

Munasiri, M.A., Parte, M.N., Ghanekar, A.S., Arun Sharma, Padwal-Desai, S.R. and Nadkarni, G.B. (1987) Sterilisation of ground prepacked Indian spices by gamma irradiation. *J. Food Sci.*, 52, 823–824.

Nguyen, U., Frakman, G. and Evans, D.A. (1991) Process for extracting antioxidants from Labiatae herbs. *USA Patent* 5,0176397.

Nykänen, L. and Nykänen, I. (1987) The effect of drying on the composition of the essential oil of some Labiatae herbs cultivated in Finland. In M. Martens, G.A. Dalen and H. Russwurm Jr. (eds), *Flavour Sci. Tech.*, John Wiley & Sons Ltd., pp. 83–88.

Oszagyan, M., Simandi, B., Sawinsky, J., Kery, A., Lemberkovics, E. and Fekete, J. (1996) Supercritical fluid extraction of volatile compounds from lavandin and thyme. *Flavour Fragr. J.*, 11, 157–165.

Paré, J.R.J., Sigouin, M. and Lapointe, J. (1991) Microwave-assisted natural product extraction. *U.S. Patent* 5,002,784, March 26.

Peyron, L. and Richard H. (1992) L'extraction des épices et herbes aromatiques et les différents types d'extraits. In H. Richard (coordinateur), *Epices et Aromates*, Tec. Doc.-Lavoisier, A.P.R.I.A., Paris, pp. 114–139.

Raghavan, B., Abraham, K.O. and Koller, W.D. (1995) Flavour quality of fresh and dried Indian thyme (*Thymus vulgaris* L.). *Pafai Journal*, 17, 9–14.

Reineccius, G. (1994) *Source Book of Flavors*, 2nd (ed.), Chapman and Hall, New York.

Richard, H. and Loo, A. (1992) La fabrication des extraits, l'extraction par le dioxyde de carbone. In H. Richard (coordinateur), *Epices et Aromates*, Tec. Doc.-Lavoisier, A.P.R.I.A., Paris, pp. 140–155.

Rocha, T., Lebert, A. and Marty-Audouin, C. (1993) Effect of pretreatments and drying conditions on drying rate and colour retention of basil (*Ocimum basilicum*), *Lebensm. Wiss. Technol.*, 26, 456–463.

Sáez, F. (1995) Essential oil variability of *Thymus zygis* growing wild in southeastern Spain. *Phytochemistry*, 40, 819–825.

Schüttler, C., Helle, N. and Bögl, K.W. (1991) Chemische, sensorische und toxikologische Untersuchungen an bestrahlten Gewürzen. 2. Ätherisches Öl, ESR, Piperingehalt bei Pfeffer. *Fleischwirtschaft*, 71, 588, 591–595 (German).

Sjöberg, A.-M., Manninen, M., Härmälä, P. and Pinnioja, S. (1990) Methods for detection of irradiation of spices. *Z. Lebensm. Unters. Forsch.*, 190, 99–103.

Spiro, M. and Chen, S.S. (1995) Kinetics of isothermal and microwave extraction of essential oil constituents of peppermint leaves into several solvent systems. *Flavour Fragr. J.*, 10, 259–272.

Spook, W.J.A. (1993) Stoomsterilisatie van kruiden en specerijen. *Voedingsmiddelen-technologie*, 26, pp. 75–76 (Dutch).

Takruri, H.R. and Daqqaq, R.F. (1986) Effect of drying and storage on the carotenoid content in some leafy vegetables and local herbs. *Dirasat* (Jordan), 13, 87–93 (Arab).

Takruri, H.R. and Daqqaq, R.F. (1984) Effect of drying and storage on the carotenoid content (vitamin A equivalent) in some local leafy Jordanian vegetables and herbs (mint, Jew's mallow and thyme's shadow, sun and oven-drying under 100 °C on preservation of carotene; Jordan). In *Jordan Univ., Amman; Yarmouk Univ., Irbid (Jordan). Abstracts: The Third Arab*

Scientific Conference of Biological Sciences, Amman 3–6 Nov 1984. Amman (Jordan). Nov 1984. Arab (ed.), pp. 27; Engl. (ed.), pp. 31.

Vajdi, M. and Pereira, R. R. (1973) Comparative effects of ethylene oxide, gamma irradiation and microwave treatments on selected spices. *J. Food Sci.*, 38, 893–895.

Van Dijk, L.G.M. (1970) Studie over de toepassingmogelijkheden van bestraling bij het ontsmetten van specerijen. *Proefbedrijf voedselbestraling*, Wageningen, Netherlands, Report nr. 3 (Dutch).

Venskutonis, P.R. (1994) Flavour of irradiated herbs. In *Proceed. of 8th Forum for Applied Biotechnol.*, Medelingen, 1, 1183–1775.

Venskutonis, P.R. (1995) Effect of drying on the aroma constituents of thyme (*Thymus vulgaris*) and sage (*Salvia officinalis*). In M. Rothe and H.-P. Kruse (eds), *Aroma: Perception, Formation, Evaluation*, Eigenverlag Deutsches Institut für Ernährungsforschung, Potsdam-Rehbrücke, pp. 665–670.

Venskutonis, P.R. (1997) Effect of drying on the volatile constituents of thyme (*Thymus vulgaris* L.) and sage (*Salvia officinalis* L.). *Food Chem.*, 59, 219–227.

Venskutonis, P.R. and Dapkevičius, A. (1995) Some aspects of herb aroma research. *Food Chem. and Technol.*, Vilnius "Academia", 28, 68–72.

Venskutonis, P.R., Poll, L. and Larsen, M. (1996) Influence of drying and irradiation on the composition of the volatile compounds of thyme (*Thymus vulgaris* L.). *Flavour Fragr. J.*, 11, 123–128.

9 The genus *Thymus* as a source of commercial products

Brian M. Lawrence and Arthur O. Tucker

INTRODUCTION

The commercial products that are obtained from the genus *Thymus* include essential oils, oleoresins, fresh and dried herbs, and landscape plants. The genus *Thymus* has an estimated 350 species, but only five have achieved any real economic importance (although not all for the same reasons): *Thymus capitatus* (L.) Hoffmanns. et Link (classified most recently as *Thymbra capitata* (L.) Cav., Spanish oregano or conehead thyme), *T. mastichina* L. (Spanish marjoram or mastic thyme), *T. serpyllum* L. (wild thyme, mother-of-thyme), *T. vulgaris* L. (common thyme) and *T. zygis* L. (Spanish thyme). Although essential oils of each of these species are items of commerce, thyme oil is mainly obtained from *T. zygis*, whereas both *T. zygis* and *T. vulgaris* are the main sources of the dried and fresh herb.

PRODUCTION STATISTICS OF *THYMUS* OILS

Thymus oils have been used since the 16th century (Gildemeister and Hoffmann, 1990); however, the data on their production amounts prior to the 1930s could not be found, although it was probably in the 5–10 ton level for many years. These oils were valued because of their aroma character and their richness in a specific constituent. For example, oils of *T. serpyllum*, *T. vulgaris* and *T. zygis* are typically thymol-rich, *T. capitatus* oil is typically carvacrol-rich, and *T. mastichina* oil is typically 1,8-cineole/linalool-rich.

In the early part of the twentieth century, thyme oil (ex *T. vulgaris*) was available from cultivated plants in Germany and wild plants collected from the mountainous regions of southern France. As it became less economically viable to cultivate and distil thyme, harvesting of wild plants became the norm initially in France and then for *T. zygis* in Spain.

Spain is the main country of production for thyme oil from *T. zygis*. The main Spanish-producing areas for thyme oil are Almería, Murcia and Albacete. The crop is harvested from wild plants from July to mid-September. Spanish oregano oil from *T. capitatus* is produced in Huelva and northern Murcia from wild plants harvested between mid-May and August. Spanish marjoram oil (*T. mastichina*), which is also harvested from wild plants between mid-May and August is produced primarily in Murcia and Albacete. In contrast, wild thyme (*T. serpyllum*) is produced almost exclusively in Cuenca (Gaviña Múgica and Torner Ochoa, 1966). A summary of Spanish thyme oil production since 1930 can be seen in Table 9.1 (Lawrence, 1985; Miralles, 1998). Between 1990 and 1998, the amount of oil produced annually has fluctuated between 35 and 45 tons.

Table 9.1 Thyme oil production in Spain (tons)

Year	1930	1935	1940	1945	1947	1958	1955
Amount	12	15	20	25	5	25	14
Year	1960	1965	1970	1975	1980	1985	1990
Amount	14	19	17	22	30	23	25

Table 9.2 Production of oils from other *Thymus* species in Spain (tons)

Year	1936	1946	1947	1950	1970	1980	1993	1994	1995	1996
Spanish marjoram	13		6		15	25	40	30	20	25
Spanish oregano	20	9	3	5	10	10	24	24	15	2

A limited quantity of thymol-rich thyme oil is produced in France (0.6 tons) annually, while smaller quantities are sometimes available from Albania, Algeria, Hungary, Israel, Morocco, Portugal and Yugoslavia.

Wild thyme oil is available only from Spain with its annual production in the 1–3 tons level (Lawrence, 1985). The other two oils that are exclusively produced in Spain are Spanish marjoram and Spanish oregano oils. A summary of their production statistics (Miralles, 1998) can be seen in Table 9.2. Because of an interest in uncommon oils in the aromatherapy trade, a very limited quantity of lemon thyme oil (<100 kg) which is obtained from *T. x citriodorus* (Pers.) Schreb., has become an item of commerce.

Both red and white thyme oils are available commercially. Authentic thyme oil distilled in Spain is usually red in color. This color is caused by the reaction between thymol and the iron in the field stills. White thyme oil is produced from red thyme oil by re-distillation of the red oil in stainless steel equipment. In this re-distillation or rectification process there is generally a small loss of the more volatile materials with a corresponding increase in the thymol content of the oil.

It is estimated that the North American demand for thyme oil is between 18–24 tons. Because of the increasing availability of synthetic thymol, the natural thymol oil demand has remained fairly stable for the past decade. It is postulated that the rising labor costs associated with harvesting and distilling the oil in Spain suggests that production volumes greater than the current levels are unlikely to increase. Nevertheless, assuming usage levels remain constant, current producers should be able to meet the annual oil demands for oils of the commercially important *Thymus* species.

MISCELLANEOUS USES OF THYME OIL

Natural cosmetics or phyto-cosmetics are one of the fastest growing niche markets in Europe and North America (Purohit, 1994). Although most of these products were originally sold in health food stores, they have now found their way into wider distribution channels such as department stores, boutiques, discount stores, salons, etc. and direct sale through the Internet. Within this category of products, materials can be found in which the natural essential oils are purported to be the efficacious components found within them. As a result, thyme oil is used for its antiseptic and aromatherapeutic properties; however, this use has little impact on the production volumes of the oils.

OIL SPECIFICATIONS

To determine whether an oil is a pure product, internationally accepted specifications for the commercially important *Thymus* oils have been developed. The main organizations that have well-recognized specifications are International Organization for Standardization (ISO, 1996a,b) TC-54 (Essential Oils Section), Association Française de Normalisation, French Essential Oils Standards (AFNOR, 1996a,b), the now-defunct Essential Oils Association of the USA (EOA)/the Fragrance Materials Association of the United States Standards (FMA, 1998) and the USA Food Chemical Codex (FCC), National Academy of Sciences (1996). A summary of these specifications can be seen in Tables 9.3–9.6.

Table 9.3 Specification for thyme oil ex *T. vulgaris*

Appearance	A colorless, pale yellow or red mobile liquid possessing a characteristic pleasant odour.
Specific gravity (25 °C)	0.9150–0.9350 (FCC)
Refractive index (20 °C)	1.4950–1.5050 (FCC)
Optical rotation (20 °C)	levorotatory, but not more than -3 ° (FCC)
Solubility in 80% v/v aqueous ethanol (20 °C)	1:2 volumes (FCC)
Phenol content	$\geq$40% (FCC)
Heavy metals (as Pb)	$\leq$0.02% (FCC)
Water soluble phenols	Shake 1 ml of oil with 20 ml of hot water and after cooling pass water layer through a moistened filter. On addition of 1 drop of ferric chloride solution (9 g $FeCl_3.6H_2O$), no transient blue or violet color should be produced (FCC)

Table 9.4 Specification for thyme oil ex *T. zygis*

Appearance	Red to very intense brown-red, almost black mobile liquid with a characteristic phenolic, spicy aroma
Density (20 °C)	0.9120–0.9350 (ISO/AFNOR)
Refractive index (20 °C)	1.4950–1.5050 (ISO/AFNOR)
Optical rotation	Because of colour, it could not be measured, generally levorotatory
Solubility in 80% v/v aqueous ethanol (20 °C)	1:2 vols. (ISO/AFNOR)
Flash point (c/c)	+62 °C (ISO)
Phenol content	38–56% v/v (ISO/AFNOR)
GC analysis (ISO)	α-thujene (0.5–1.6%), α-pinene (0.6–2.1%), myrcene (1.0–2.8%), α-terpinene (0.9–2.6%), γ-terpinene (5.0–10.3%), *p*-cymene (15.0–28.0%), *tr*-sabinene hydrate (trace-0.5%), linalool (4.0–7.0%), methyl carvacrol (0.1–1.5%), thymol (36–55%), carvacrol (1.2–4.0%), β-caryophyllene (0.6–1.8%)

Table 9.5 Specification for Spanish oregano oil ex *T. capitatus* (today *Thymbra capitata*)

Appearance	A yellowish to dark brown almost black mobile liquid with a characteristic phenolic, spicy odor
Specific gravity (20 °C)	0.9380–0.9630 (FMA)
Density (20 °C)	0.9300–0.9550 (ISO/AFNOR)
	0.9350–0.9660 (FCC)
Refractive index (20 °C)	1.5000–1.5130 (ISO)
	1.5020–1.5080 (FMA)

Optical rotation (20 °C)	−2 ° to +3 ° (FMA)
	−10 ° to +2 ° (ISO)
Phenol content	60–75% (ISO)
Solubility in 70% v/v	1:4 vols. (ISO)
aqueous ethanol	1:2 vols. (FMA)
Flash point (closed cup)	+65 °C (ISO)
GC analysis (ISO)	α-thujene (0.5–2.0%), α-pinene (0.5–1.5%), mycrene (1.0–3.0%), α-terpinene (0.5–2.5%), γ-terpinene (3.5–8.5%), *p*-cymene (5.5–9.0%), linalool (0.5–3.0%), terpinen-4-ol (0.5–2.0%), thymol (0–5.0%), carvacrol (60–75%), β-caryophyllene (2.0–5.0%)

Table 9.6 Specification for Spanish marjoram oil ex. *T mastichina*

Appearance	A colourless to pale yellow liquid with a characteristic, agreeable, spicy, eucalyptus-like odour
Density (20 °C)	0.9000–0.9200 (ISO/AFNOR)
Refractive index (20 °C)	1.4620–1.4680 (ISO/AFNOR)
Optical rotation (20 °C)	−6 ° to +10 ° (ISO/AFNOR)
	−5 ° to +10 ° (FCC)
Solubility in 70% v/v aqueous ethanol	1:3 vols. (ISO/AFNOR)
Phenol content	≤4.0% (ISO/AFNOR)
1,8-Cineole content	40–65% (ISO/AFNOR)
	49–65% (FCC)
Heavy metals (as Pb)	≤0.002% (FCC)
GC analysis (AFNOR)	α-pinene, camphene, β-pinene, sabinene, myrcene, limonene, 1,8-cineole, γ-terpinene, *p*-cymene, linalool and α-terpineol

As the oil of wild thyme (*T. serpyllum*) is neither produced nor used in large quantities, no international standard exists for this oil at the present time.

OIL ADULTERATION

In the 1920s, adulteration of red thyme oil with turpentine to produce white thyme oil that had a phenol content of 1–2 per cent was a common practice (Parry, 1925).

According to Guenther (1945), in the mid-1940s thyme oil was frequently adulterated by the addition of terpenes or 'thymene' and synthetic thymol and carvacrol. Thymene is the by-product mixture obtained from ajowan oil (ex. *Trachyspermum copticum* (L.) Link) after removal of thymol. Prior to the advent of modern instrumental analytical techniques and the use of column chromatography or thin-layer chromatography (TLC), the oil was evaporated to yield crystalline thymol, which was free from a creosote-like off-odour associated with synthetic carvacrol. Also, if the oil did not crystalize on evaporation the use of synthetic carvacrol as an adulterant was concluded. Thyme oil adulteration is practiced even today. Such evidence is especially true when white thyme oil can be found on the market at prices lower than red thyme oil. This is an impossible situation because red thyme oil is the crude product used to make white thyme oil by treatment with tartaric acid and re-distillation.

In the past, Spanish oregano oil was also subjected to adulteration generally by the addition of synthetic *p*-cymene and/or synthetic carvacrol. Again, the detection of a creosote-like off-odor associated with synthetic carvacrol on evaporation of the oil was used to determine adulteration (Gildemeister and Hoffmann, 1990).

Since the early 1960s, the use of gas chromatography (GC) combined with other techniques has been used to determine the composition of an oil. More recently the use of GC with flame ionization detection, electronic integration, automatic injection, capillary columns of a polar and non-polar nature for retention index determination and GC/mass spectrometry (MS) has led to a more accurate detailed analysis of an oil composition. As a result, the addition of synthetic thymol, carvacrol or 1,8-cineole to thyme oil, Spanish oregano oil or Spanish marjoram oil, respectively, is readily detected because of the corresponding decrease in the minor oil constituents.

To further assist the analyst in determining the genuineness of an oil, the introduction and subsequent use of chiral GC columns for enantiomer separation has become a more common technique. Considering thyme oil, although thymol is not optically active, some of the other constituents are, for example, examination of enantiomers of α-pinene, β-pinene and limonene in thyme oil by Hener *et al.* (1990) revealed the following distribution:

(1S)-(−)α -pinene (89 per cent) : (1R)-(+)-α-pinene (11 per cent)
(1S)-(−)-β-pinene (96 per cent) : (1R)-(+)-β-pinene (4 per cent)
(4S)-(−)-limonene (70 per cent) : (4R)-(+)-limonene (30 per cent)

Analysis of a thyme oil whose enantiomeric distribution of particularly α- and β-pinene falls outside the levels shown above is indicative of oil adulteration. In addition to the monoterpene hydrocarbons of thyme oil, the monoterpene alcohols are optically active. Using chiral GC analysis, Casabianca *et al.* (1998) revealed that adulteration of the oil with a coupage (a mixture of components) containing synthetic linalool was detectable from examination of the distribution of linalool enantiomers. The results of this study are shown in Table 9.7. As can be seen from these results, two of the French commercial samples of thyme oil appear to be adulterated.

Linalyl acetate is also present in *T. vulgaris* oil at a very low level (<0.2 per cent); nevertheless, Casabianca *et al.* (1998) determined that its enantiomeric distribution was as follows:

(3R)-(−)-linalyl acetate (93.8–99.2 per cent) : (3S)-(−)-linalyl acetate (0.8–6.2 per cent)

Table 9.7 Enantiomeric distribution of linalool in thyme oils

Botanical origin	(3R)-(−) Linalool	(3S)-(+)-Linalool
T. vulgaris (France) (7 lab distilled samples)	92.0–99.4	0.6–8.0
Commercial sample 1	95.2	4.8
Commercial sample 2	50.0	50.0
Commercial sample 3	83.0	17.0
T. zygis (Spain)	99.6	0.4
T. zygis (Portugal)	98.3	1.7
T. serpyllum (France) 1	89.4	10.6
T. serpyllum (France) 2	96.4	3.6

Again, the enantiomeric distribution of linalyl acetate along with linalool can be used as one of the indicators to determine oil adulteration.

The adulteration of Spanish marjoram oil with a 1,8-cineole-rich eucalyptus oil is not uncommon. Detection of this adulteration is not easy especially if the level of adulteration is less than 10 per cent. Nevertheless, if the hydrocarbons are separated from the oxygenated compounds of an oil suspected to be adulterated and aromadendrene is found in the hydrocarbon fraction, and *tr*-pinocarveol and globulol are found in the oxygenated fractions, then the oil is probably adulterated with a 1,8-cineole-rich eucalyptus oil.

In 1995, Ravid *et al.* determined that, although α-terpineol was only present in *T. capitatus* (today *Thymbra capitata*) oil as a minor component (ca. 0.2 per cent), its enantiomeric distribution was as follows:

(4R)-(+)-α-terpineol (61.5 per cent) : (4S)-(−)-α-terpineol (38.5 per cent)

If an oil of Spanish oregano oil possessed distribution of α-terpineol enantiomers different to that shown above, this is probably indicative of adulteration.

OTHER EXTRACTIVES

In addition to the oils a small amount of thyme (ex *T. vulgaris*) oleoresin is also produced. It is impossible to obtain volumes on the production of this minor commodity as production is done by spice oleoresin and seasonings manufacturers in the US and Europe. No oleoresins appear to be produced from *T. zygis*, *T. capitatus*, *T. mastichina*, or *T. serpyllum*.

WHOLE LEAF AND GROUND THYME

Although there is a concern about pesticide/fungicide residues, aflatoxins and microbiological contamination found in imported spices and herbs, the major concern by various government agencies around the world is cleanliness. As a result, contaminant levels for animal hairs and excreta, insect fragments and foreign material, standards have been established in many countries (Tainter and Grenis, 1993).

In the US, the American Spice Trade Association (ASTA) is the advisory organization that helps the spice and seasoning industry develop acceptability standards for whole and ground spices and herbs. Examples of the recommended physical and chemical specifications of whole thyme leaves and ground thyme can be seen in Tables 9.8 and 9.9 (Tainter and Grenis, 1993).

DRIED HERB PRODUCTION STATISTICS

Although thyme (*T. vulgaris*) is native to South Europe, it is both collected from the wild in France, Albania, Spain, Morocco, Lebanon, Syria, Turkey, Tunisia, Greece, Yugoslavia, etc. and widely cultivated in France, Germany, Morocco, India, Spain, Bulgaria, Hungary, Russia, Canada, US, etc. while the so-called Spanish thyme (ex *T. zygis*) is collected

Table 9.8 Whole thyme: physical and chemical specifications

Cleanliness specifications	
Whole dead insects	8/kg
Mammalian excreta	2/kg
Other excreta	10/kg
Mold (w/w)	1.0%
Insect infested/contaminated (w/w)	0.5%
Insect fragments	ca. 325/25 g
Rodent hairs	ca. 2/25 g
Chemical specifications	
Volatile oil	≥0.8%
Moisture	≤10.0%
Ash	≤10.0%
Acid insoluble ash	≤3.0%
Bulk index	ca. 400 mg/100 g

Table 9.9 Ground thyme: physical and chemical specifications

Cleanliness specifications	
Insect fragments	ca. 925/10 g
Rodent hairs	ca. 2/10 g
Chemical specifications	
Volatile oil	≥0.5%
Moisture	≤10.0%
Ash	≤10.0%
Acid insoluble ash	≤3.0%
Sieve test	95% through a 200 mesh
Bulk density	250 ml/100 g

from the wild in Spain and Portugal. Accurate up-to-date export and import statistics for whole thyme leaves are not accessible because most countries group their minor spices and herbs together into one statistic. For example, thyme is often grouped with laurel (bay) leaves, marjoram, oregano, etc. in published government statistics.

In Europe, France is the largest producer of cultivated herbs destined for the culinary and seasonings trade as dried herbs. Maffei (1992) reported that in France 25 tons of dried whole thyme leaves were produced from wild collection in Drôme, Var and Languedoc-Roussilon, whereas 250–280 tons of dried leaves were produced from plants mainly cultivated in the Provence-Alpes-Côte d'Azur areas. Over the same time period 700–770 tons of dried whole leaf thyme were imported into France.

It was further reported (Maffei, 1992) that in 1990 Germany imported 500 tons of dried thyme while an additional 50 tons were produced internally. Most of the thyme was imported from Spain, although smaller quantities were imported from Poland and Morocco. It should be noted here that the wild thyme of Moroccan origin is obtained from *T. satureioides* Cas. and Bal. and not *T. vulgaris* or *T. zygis*.

In 1990, the Netherlands imported 90 tons of dried thyme mainly from Spain, while the UK imported 220 tons during the same time period (Maffei, 1992). Like the Netherlands, Spain was the major source of UK thyme, although ca. 32 tons were produced domestically.

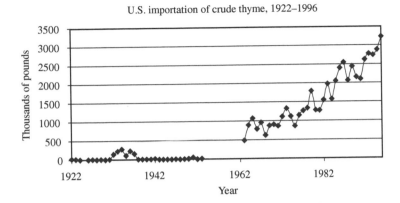

U.S. importation of crude thyme, 1922–1996

Figure 9.1 US importation of crude thyme, 1922–1996; data are missing for 1925 and 1954–1962. In 1989–1996, statistics for crude thyme were reported together with bay laurel, so an estimate of 2/3 of this total was used for crude thyme alone.

Importation of crude thyme into the US, 1922–1996, is illustrated in Figure 9.1. Although large quantities of thyme are imported into the US, there were also three commercial herb producers in California in 1974 (Tyner, 1974) that grow a variety of herbs on approximately 125 ha of which ca. 10–15 per cent is devoted to thyme. This production is then used in bottle or jar trade of grocery store herbs in the US for a more up-market product.

FRESH HERB PRODUCTION

Thyme has been grown since medieval times as a garden herb which was used at that time to flavor a potage which was served with meat for its dual purpose of adding flavour and its curative properties (Freeman, 1943). In modern times, the production of fresh and deep frozen thyme has become a moderate-sized business in the fresh herb trade. In France, fresh thyme is grown on ca. 220 ha in the Rhone-Alpes region in the departments of Drôme, Ardeche, Loire, Vaucluse, Alpes de haute Provence, Essone, Yvelines and Seine et Marne (Garnon, 1992). It is an integral component of 'bouquet garni' along with laurel leaves and rosemary. In fact, it is sold in fresh bunches all over France where it is the largest item in the fresh herb marketplace (Verlet, 1989). Further cultivation of thyme in France is being promoted by ITEIPMAI (Bouverat-Bernier 1992; Institut Technique Interprofessionel des Plantes Medicinales, Aromatiques et Industrielles, 1983) and the Office National Interprofessionnel des Plantes Aromatiques et Medicinales colloquially known as ONIPPAM (Garnon, 1992).

Thyme is also grown in other European countries for fresh herb production. For example, thyme is grown on ca. 2.5 ha in the Canton du Valais in Switzerland (Rey, 1992). It is also grown in Austria on 40 ha (Dachler, 1992), and in the former East Germany on 85 ha (Pank, 1992). Unfortunately, there are no statistics on the amounts of fresh thyme produced.

Since the mid-1980s fresh herb production in Israel including thyme has become a more important minor industry (Putievsky, 1988). For example in 1989, 25 tons of fresh thyme was exported to France (30 per cent), UK (30 per cent), Germany (20 per cent), Switzerland (10 per cent) and other European countries. Although no new figures on this production are available, it can safely be assumed that this market has not only been maintained but has grown.

There is also a cottage-scale trade in fresh herbs including thyme for the restaurant trade. The herbs are generally raised by small landowner-herb growers who reside in small towns and villages close to the metropolitan areas in the US and Europe so that there is a rapid delivery of fresh thyme to the restaurants.

Finally, organically grown herbs have started to appear in speciality markets in the US; however, the amount of fresh thyme sold is minuscule compared to the amount of dry thyme sold.

THYMUS SPECIES AS GARDEN PLANTS

In addition to the industrial and commercial uses of *Thymus* species and their extractives, various species have been and are still grown as decorative, aromatic border, pathway or rock garden plants. As a result, a flourishing herb trade exists in North America and western Europe in which numerous *Thymus* species and hybrids are traded for their decorative foliage and diverse aromatic properties. The most popular *Thymus* species and hybrids cultivated in the US (Flannery, 1982) are: *T. broussonetii* Boiss., *T. caespititius* Brot., *T. camphoratus* Hoffm. et Link, *T. capitatus* Hoffm. et Link, *T. carnosus* Boiss., *T. cherlerioides* Vis., *T. cilicicus* Boiss. et Bal., *T. comptus* Friv., *T. doerfleri* Ronn., *T. herba-barona* Loisel., *T. leucotrichus* Halacsy, *T. mastichina* L., *T. membranaceus* Boiss., *T. nummularius* Bieb., *T. praecox* Opiz, *T. pulegioides* L., *T. quinquecostatus* Celak., *T. richardii* Pers., *T. vulgaris* L., *T. zygis* (Loefl.) L., *T.* x *citriodorus* (Pers.) Schreb., *T.* 'Argenteus', *T.* 'Broad-leaf English', *T.* 'Doone Valley', *T.* 'Longwood', *T.* 'Pinewood', *T.* 'Porlock', *T.* 'Variegated English', *T.* 'Wedgewood English', *T.* 'Woolly-Stemmed Sharp' and *T.* 'Woolly-Stemmed Sweet'.

REFERENCES

American Spice Trade Association (1960) *Official Analytical Methods*, American Spice Trade Association, Englewood Cliffs, New Jersey.

Association Française de Normalisation (1996a) *Huiles Essentielles. Tome 2, Specifications 5e Edn. Huile essentielle de thym sauvage d'Espagne* (Thymus mastichina Linnaeus). NFT 75–343, Avril 1987, AFNOR, Paris, pp. 483–441.

Association Française de Normalisation (1996b), *Huiles Essentielles. Tome 2, Specifications 5e Edn. Huile essentielle de thym à thymol* (Thymus zygis (Loefl.) L., type Espagne, NFT 75–3459, Mars 1993, AFNOR, Paris, pp. 473–481.

Bouverat-Bernier, J.P. (1992) Les travaux d'amelioration genetique a l'ITEIPMAI. In N. Verlet (ed.), *3èmes Rencontres Techniques et Economiques. Plantes Aromatiques et Medicinales, Nyons, 2–3–4 Decembre 1991.* Centre Formation Professionnelle Promotion Agricole (C.F.P.P.A.) de Nyons, pp. 140–155.

Casabianca, H., Graff, J.B., Faugier, V., Fleig F. and Grenier, C. (1998) Enantiomeric distribution studies of linalool and linalyl acetate. *J. High Resol. Chromatogr.*, 21, 107–112.

Dachler, M. (1992) La production des plantes medicinales et aromatiques en Autriche, en particulier celle du pavot et du carvi. In N. Verlet (ed.), *3èmes Rencontres Techniques et Economiques. Plantes Aromatiques et Medicinales, Nyons, 2–3–4 Decembre 1991.* Centre Formation Professionnelle Promotion Agricole (C.F.P.P.A.) de Nyons, pp. 122–128.

Flannery, H.B. (1982) *A Study of the Taxa of* Thymus L. *(Labiatae) Cultivated in the United States.* Ph.D. thesis, Cornell Univ.

Fragrance Materials Association (1998), *FMA Monographs. Vol. 2, Origanum Oil Spanish Type. (replaces EOA #142),* FMA, Washington, DC (02–06–98).

Freeman, M.B. (1943) *Herbs for the Medieval Household for Cooking, Healing and Divers Uses,* Metropolitan Museum of Art, New York.

Garnon, P. (1992) 10 ans de production en France: Bilan et perspectives. In N. Verlet (ed.), *3èmes Rencontres Techniques et Economiques. Plantes Aromatiques et Medicinales, Nyons, 2–3–4 Decembre 1991.* Centre Formation Professionnelle Promotion Agricole (C.F.P.P.A.) de Nyons, pp. 216–231.

Gaviña Múgica, M. and Torner Ochoa, J. (1966) *Contribución al estudio de los aceites esenciales españoles I. Aceites de la Provincia de Cuenca.* Ministerio de Agricultura, Instituto Forestal de Investigaciones y Experiencias, Madrid, pp. 257–265.

Gildemeister, E. and Hoffmann, F. (1990) *The Volatile Oils.* Transl. E. Kremers. Pharmaceutical Rev. Publ. Co., Milwaukee, Wisconsin, p. 32.

Guenther, E.A. (1945) Survey of Spanish essential oils. *Amer. Perfum.,* 60, 43–44.

Hener, U., Kreis, P. and Mosandl, A. (1990) Enantiomeric distribution of α-pinene, β-pinene and limonene in essential oils and extracts. Part 2. Oils, perfumes and cosmetics. *Flavour Fragr. J.,* 5, 201–204.

Institut Technique Interprofessionel des Plantes Medicinales, Aromatiques et Industrielles (1983) *Domestication de la production conditionnement et définition du thym* (Thymus vulgaris L.), I.T.E.P.M.A.I., Milley-la-Fôret, France.

International Organization for Standardization (1992), *ISO/TC 54 47298 Oil of Wild Thyme* (Thymus mastichina L.) (15–02–1992).

International Organization for Standardization (1996a), *ISO/TC54/SCN1684, Essential Oils. Oil of* Thymbra capitata *(L.) Cav., Spanish type.* (2–26–1996).

International Organization for Standardization (1996b), *ISO/TC54/SCN1680, Essential Oils. Thyme oil containing thymol* (Thymus zygis (Loefl.) L.), *Spanish type.* (2–26–1996).

Lawrence, B.M. (1985) A review of the world production of essential oils 1984. *Perfum. Flavor.,* 10, 1–16.

Maffei, M. (1992) Dry culinary herbs: an overview of selected western Europe markets. In N. Verlet (ed.), *3èmes Rencontres Techniques et Economiques. Plantes Aromatiques et Medicinales, Nyons, 2–3–4 Decembre 1991.* Centre Formation Professionnelle Promotion Agricole (C.F.P.P.A.) de Nyons, pp. 249–292.

Miralles, J. (1998) Spanish distilled essential oils. Today's production. In *Sevilla 1997. IFEAT Proceedings.* IFEAT, London, pp. 171–183.

National Academy of Sciences (1996) *Food Chemical Codex (FCC).* 4th (ed.) Natl. Acad. Press, Washington, DC.

Pank, F. (1992) Medicinal and Spice Plant Production and Research in the Eastern part of Germany. In N. Verlet (ed.), *3èmes Rencontres Techniques et Economiques. Plantes Aromatiques et Medicinales, Nyons, 2–3–4 Decembre 1991.* Centre Formation Professionnelle Promotion Agricole (C.F.P.P.A.) de Nyons, pp. 129–139.

Parry, E.J. (1925) *Parry's Cyclopedia of Perfumery M-Z.* P. Blakiston's & Son, Philadelphia, pp. 746–754.

Purohit, P. (1994) Phyto-cosmetics and the expanding market. In *4èmes Rencontres Techniques et Economiques. Plantes Aromatiques et Medicinales, Nyons, 5–6–7 Decembre 1993.* Centre Formation Professionnelle Promotion Agricole (C.F.P.P.A.) de Nyons, pp. 80–82.

Putievsky, E. (1988) Production of aromatic plants in Israel. In J.E. Simon and L.Z. Clavio (eds), *Proceedings of the Third National Herb Growing and Marketing Conference, Baton Rouge, Louisiana*. International Herb Growers and Marketers Association, pp. 130–133.

Ravid, U., Putievsky, E. and Katzir, I. (1995) Determination of the enantiomeric composition of α-terpineol in essential oils. *Flavour Fragr. J.*, 10, 281–284.

Rey, Ch. (1992) Recherche et production des plantes medicinales et aromatiques en Suisse: Situation actuelle. In N. Verlet (ed.), *3èmes Rencontres Techniques et Economiques. Plantes Aromatiques et Medicinales, Nyons, 2–3–4 Decembre 1991*. Centre Formation Professionnelle Promotion Agricole (C.F.P.P.A.) de Nyons, pp. 116–121.

Tainter, D.R. and Grenis, A.T. (1993) *Spices and Seasonings. A Food Technology Handbook*. VCH Publ. Inc., New York.

Tyner, G.E. (1974) *The Dispersal of Culinary Herbs in Relation to Contemporary California Herb Industry*. Ph.D. Thesis, Univ. California, Los Angeles.

Verlet, N. (1989) New markets for herbs in France and Europe. In J. E Simon, A. Kestner, M.A. Buehrle (eds), *Herbs 1989. Proceedings of the Fourth National Herb Growing and Marketing Conference, San Jose, California*. International Herb Growers and Marketers Association, pp. 80–84.

10 The medicinal and non-medicinal uses of thyme

Antonio Zarzuelo and Esperanza Crespo

INTRODUCTION

The uses of thyme, *Thymus vulgaris* and other *Thymus* species are well known, and extensive parts of the world get benefit from this plant group in medicinal and non-medicinal respects. Following the development of the medicinal uses of thyme we can see that thyme has changed from a traditional herb to a serious drug in rational phytotherapy. This is due to many pharmacological *in vitro* experiments carried out during the last decades, and even a few clinical tests. The studies have revealed well defined pharmacological activities of both, the essential oils and the plant extracts, the antibacterial and spasmolytical properties being the most important ones. The use of thyme in modern phytotherapy is based on this knowledge, whereas the traditional use of thyme describes only empirical results and often debatable observations. Therefore it seems necessary to present here the data available on the pharmacodynamics of thyme and thyme preparations in order to substantiate the use of thyme in modern medicine.

The non-medicinal use of thyme is no less important, because thyme (mainly *T. vulgaris*) is used in the food and aroma industries. It serves as a preservative for foods and is a culinary ingredient widely used as seasoning in many parts of the world. This use is due to the typical aroma of the plant which the essential oil is responsible for, and most people are very familiar with its typical smell. The special aroma also causes the role of the essential oil as a raw material in perfumery and in everyday cosmetics. The use of thyme as a preservative of food can be put down to the antioxidative effects of the plant extracts, which were of increasing interest in the last years. The results of these studies may be of further importance assuming that free-radical generation causes oxidative stress when it exceeds the capacity of antioxidant defenses in the human body. This way it may significantly contribute to pathogenesis.

Irrespective of the pharmacological and non-pharmacological effects when we examine and discuss the use of thyme we must take into account the chemistry of the plant. Therefore most of the studies described here refer to the chemical components of thyme. The structural details of the compounds mentioned in this chapter are given in Chapter 3 of this book.

THE THERAPEUTICAL USES OF THYME

The Traditional Use of Thyme in Folk Medicine

The widespread use of thyme dates back to ancient Egypt where various species of thyme were grown to perfume unguents and for embalming and, one can suppose, for medicinal purposes as well. The Greeks and the Romans used it in the same way, how and for what, we know thanks to Plinius (First century), Dioscorides (First century) and Theophrastus Bombastus von Hohenheim (Paracelsus 1493/94–1541). However the use of this plant did not extend further than the Alps until the eleventh century. The first northern chronicles, relating to this period, can be read in the "Physika" by the abbess Hildegard von Bingen (1098–1179) and in the works of Albertus Magnus (1193–1280). All the posterior works come from the Herbal written by the herbalist P. Mathiolus (1505–1577). The knowledge of traditional medicine is based on his works, in which the strength and efficacy of thyme were mentioned first. From then on numerous therapeutical properties have been attributed to thyme on a more or less empirical and debatable basis.

So many beneficial effects have been attributed to thyme that Bardeau (1973) stated thyme to be "an indispensable plant which should be consumed to conserve health. Furthermore, if one could replace one's morning cup of coffee with an infusion of thyme, you would quickly appreciate its positive effects: animation of spirit, sensation of lightness in the stomach, absence of morning cough, and its euphoriant and tonic effect". The readership will understand that it is impossible and not useful to enumerate all therapeutical benefits attributed to thyme which can be found in numerous popular works in folk medicine. Therefore only a selection of external and internal applications and indications shall be described, especially those which seem to be the most plausible ones.

Externally, infusions and the essential oil of thyme is traditionally used in treating injuries, bruises, infected ulcers, abscesses, cutaneous ulcers, various types of dermatitis, and in certain cases pruritis (Schauenberg and Paris, 1977). Emulsions are useful when used in massage on rheumatic types of pain (sciatica, arthritis, lumbago), gout and neuritic pains (Furlenmeier, 1984). As regards the capillaries, thyme improves blood flow and the oxygenation of the scalp, reduces seborrhoea, regenerates the capillary glands, improves the state of the hair, prevents baldness, and is therefore useful in cases of alopecia (Poletti, 1979). Baths of thyme possess invigorative and sedative properties, also behaving as coadjuvants in slimming cures (Pahlow, 1979; Furlenmeier, 1984; Volák and Stodola, 1989).

The internal use of thyme is highlighted in the treatment of a variety of illnesses of the respiratory tract because of its expectorant, spasmolytic and antiseptic properties, such as flu, colds, sinusitis, chronic and acute bronchitis, tuberculosis, calming convulsive coughs (pertussis), and irritable and spasmodic coughs (asthma) (Bardeau, 1973; Pahlow, 1979; Poletti, 1979; González and Muñoz, 1980; Fernández and Nieto, 1982; Furlenmeier, 1984; Volák and Stodola, 1989). It is attributed as having general stimulant properties, acting as a nervous tonic and being used in asthenic states. It is also useful in combatting insomnia, anxiety and depression (Valnet, 1964; Perrot and Paris, 1971; Bardeau, 1973).

Moreover it is used in a wide variety of gastro-intestinal troubles like dyspepsia (slow digestion), colic, fermentation, flatulence, diarrhea, gastritis and gastric ulcers. It is also useful in treating bacterial and parasitical infectious processes (*Ascaris, Oxyuris, Taenia*) (Perrot and Paris, 1971; Schauenberg and Paris, 1977; González and Muñoz, 1980; William and Thomson, 1980; Fernández and Nieto, 1982; Furlenmeier, 1984). In the genitourinary system the use of thyme for its diuretic, antiseptic and emenagogic properties is appreciated (Benigni *et al.*, 1964; Bardeau, 1973; Pahlow, 1979; González and Muñoz, 1980). In the cardiovascular system thyme helps the circulation, behaving like a hypocholesterolemic agent (Valnet, 1964; Brasseur, 1983).

Thyme in phytotherapy

Although for thousands of years plants have been used as remedies, a controversial discussion of their usefulness has arisen in the last two decades. This is due to the knowledge that chemical substances are responsible for the pharmacological effects in the human body and additionally due to the detailed knowledge on their pharmacological mechanisms in human cells, tissues, and organs. Therefore in modern medicine the tendency has developed that only chemically defined and pure substances should be applied, exclusively such substances whose positive (curing) effects have been proved in clinical tests. Moreover, an optimal "package" for the chemicals is required which guarantees the liberation of the drug after application meaning highly developed drug technologies such as tablets, capsules, suppositoria, etc. Within this concept medicinal plants and herbal drugs cannot easily find consideration, because their chemical compositions represent very heterogeneous systems and, additionally, these complex mixtures are inadequately packed in vegetable cells.

However the increasing consciousness for health and environment results in the fact that people have not forgotten the herbal drugs. Quite the opposite has happened, globally an increasing demand for herbal medicines can be registered. This trend has also been accepted by the orthodox medicine remembering the benefit of traditional medicine lasting for centuries; but in modern medicine one cannot stick only to traditional uses. In order to concede the herbal remedies a "real-drugs" status, many scientific efforts have been and have to be made, and phytotherapy has developed into "rational phytotherapy", a part of a scientifically-oriented medicine.

Legal regulations in Germany and Europe for drug registration ask for a proof of quality, safety and effectiveness. The latter has been evaluated for many herbal drugs, also for the herbs of *T. vulgaris* (common thyme) and *T. serpyllum* (wild thyme). A critical investigation of all the bibliography dealing with the chemistry and pharmacology of these two herbal drugs resulted in two "positive" (approved) monographs elaborated by a German expert group. The so-called Commission E Monographs represent a valuable therapeutic guide to herbal medicines in phytotherapy (Blumenthal, 1998). Within the paragraphs "Uses" of these monographs the application of thyme and thyme preparations is only recommended for the treatment of some clearly-defined diseases, reading as follows:

(Common) Thyme – Thymi herba. External use: bath additive as a supporting cure of acute and chronical diseases of the upper respiratory tract; in addition against pruritus of dermatosis. Internal use: can be applied against symptoms of bronchitis and whooping cough. Catarrhs of the upper respiratory tract.

Wild thyme – Serpylli herba. External use: bath additive as supporting cure of acute and chronical diseases of the upper respiratory tract. Internal use: catarrhal infections of the upper respiratory tract.

Compared with the traditional uses of thyme the Commission E monographs obviously allow only restricted applications of thyme. These regulations must be followed when drug manufacturers apply for a registration of their phytotherapeutics as drugs. However, in serious publications on phytotherapy (Wagner and Wiesenauer, 1995; Schulz and Hänsel, 1996; Reuter, 1997; Loew *et al.*, 1999) the area of application of thyme preparations is described similarly limited with special reference to the essential oil which can be used as bath additive (in mixtures) or for inhalation. It is recommended for a treatment of cough and sinusitis and its effects are described as secretomotoric, bronchospasmolytic and antibacterial. Thymol is proved to be responsible for these effects.

The essential oil of thyme can be administered in diverse galenic forms when used in therapy. Administered externally it can be applied directly by means of pomades, emulsions, poultices, and liniments. Alternatively it can be ingested in liquid form (drops, syrups and elixirs) or administered in solid forms (capsules). Recently the pharmaceutical industry has developed new ways of administering essential oils to facilitate their dosage and handling. These new ways come in the form of buffered microcapsules and consist of powder impregnated with essential oil (40 per cent). The fact that they are buffered gives them greater gastric tolerance. The essential oil of thyme can be administrated rectally in the form of suppositories and microenemas; vaginally, in the form of ovules; in sublingual administration in the form of solutions and nasally, in the form of drops and pomades. Finally we must not forget the inhalers and aerosols which are commonly used for the treatment of respiratory conditions (Güenechea, 1992).

Thyme oils (from e.g. *T. vulgaris*, *T. serpyllum*), especially thymol and carvacrol, provide an antiseptic action when eliminated via the lungs (Didry *et al.*, 1993), but also has a mild irritant effect which stimulates the secretory cells of the mucosa and increases the movement of the ciliated epithelium in the bronchial tree (bronchi). This produces an increase of secretions which causes the decongestion of the entire respiratory system. The spasmolytic properties which these essential oils possess, capable of relaxing smooth bronchial muscle, determine their usefulness in the treatment of respiratory tract obstructive processes, and along with its expectorant properties make them effective against different types of coughs: cough caused by thick and viscous secretions, irritable cough, spasmodic cough (Errera, 1978; Forster *et al.*, 1980; Schäfer and Schäfer, 1981; van den Broucke and Lemli, 1981; Furlenmeier, 1984).

Thyme in aromatherapy

A pleasant odour has always been, and still is, an important factor for people to feel good, and feeling well is synonymous with good health. The most ancient way to treat a patient in the sense of aromatherapy was the fumigation which was practised in all ancient civilisations. Although this was done by religious worship, it was nevertheless useful in the treatment of patients because the air became disinfected and the good aroma induced calmness. Up to the nineteenth century, the disinfectant effect of fumes was often used and people tried to banish the bad air in sick rooms by igniting good smelling candles and the doctors tried to protect themselves against infections by sniffing essential oils.

The term aromatherapy was coined in the late 1920s by the French cosmetic chemist R.M. Gattefossé, who noticed the excellent antiseptic properties and skin permeability of essential oils. Anticipating the trends of the 1980s, "New Age" and esoterics, Tisserand revived this term by including it in a general natural healing method with elements of wholistic, cosmic, anthroposophic and other phenomena (Tisserand, 1980). Confusingly, aromatherapy is used whenever good smelling plants or drugs are used to cure diseases not asking if the "aroma" is really responsible for the effect and additionally it is often combined with mystic elements.

A scientific clarification was necessary and it is due to Buchbauer (Buchbauer, 1990, 1996) that nowadays aromatherapy has become a scientific discipline. It is strictly based on his definition given as follows: "Aromatherapy: therapeutic uses of fragrances or at least mere volatiles to cure and to mitigate or prevent diseases, infections and indispositions only by means of inhalation". To avoid misunderstandings here "aromatherapy" is used according to Buchbauer's definition, although we are aware of the fact that in popular works "aromatherapy" is more often used in the traditional definition.

Concerning the application of thyme, a big overlap of aromatherapy with phytotherapy is inevitable, because the essential oil (the "aroma") is the most important and effective principle of the herbal drug in both cures. Essential oils of thyme hold a privileged position because they have been demonstrated to have various pharmacological activities, the antimicrobial and spasmolytic ones being the most utilised in therapy. The essential oil of thyme, whichever administration route (orally, rectally or cutaneously) is chosen, is eliminated via the lungs (Penso, 1980; Arteche *et al.*, 1994) and there it develops its capacity to act on the respiratory system.

Thyme in homoeopathy

Homoeopathy is based on an independent principle of therapy which was founded by the German doctor Samuel Hahnemann (1755–1843) who was teaching in Leipzig. This principle can be understood as a therapy which targets the internal regulation (stimulation) of the human body itself by a (special) drug whose reactivity corresponds to each patient individually. The methodical concepts base on the so-called "Ähnlichkeitsregel", the *Simile* principle – Similia similibus curentur, meaning that "similar can be cured by something similar", not as a law of nature but as an instruction for behaving. The feature of the remedy must be similar to the feature of the illness concerning location, form and character. Typically the drugs are administered in diluted forms which in Hahnemann's idea corresponds to an increase in "potency".

In homoeopathy numerous plants and parts of plants have a use for a multitude of indications. In order to preserve the homoeopathic remedies it became necessary, like in phytotherapy, to review the knowledge on their effects. This was performed by a group of experts in Germany (Commission D). The fresh aerial parts of Thyme of two species, *T. vulgaris* and *T. serpyllum*, have traditionally been used in homoeopathy and both plants are described in the German homoeopathic pharmacopoeia (Homoeopathisches Arzneibuch, HAB 2000). The critical evaluation of published data in 1989 resulted in the decision that both plants lack any substantial effectiveness in homoeopathy (negative monographs). That means that the application of thyme in homoeopathy can no longer be recommended.

PHARMACOLOGICAL EFFECTS OF THYME

Antimicrobial effects of thyme essential oils and thyme preparations

Antibacterial effects

The first researcher who attributed antibacterial properties to thyme (without specifying the species) was Chamberlain in 1887, after observing the antibacterial effect of its "vapours" on *Bacillus anthracis*. Since then, numerous studies with essential oils of different species of *Thymus* have been carried out. They were shown to inhibit a broad spectrum of bacteria, generally Gram-positive bacteria being more sensitive than Gram-negative bacteria. This became obvious in some screening studies administering *Thymus* oils to a variety of bacteria (Blakeway, 1986; Farag *et al.*, 1986; Deans and Ritchie, 1987; Knobloch *et al.*, 1988).

Recently the antibacterial activity of thyme *(T. vulgaris)* oil against some important food-borne pathogens, namely *Salmonella enteritidis*, *Escherichia coli*, *Staphylococcus aureus*, *Listeria monocytogenes*, and *Campylobacter jejuni*, was tested. The latter was found to be the most resistant of the bacteria investigated (Smithpalmer *et al.*, 1998). In another study it was shown that the essential oil of thyme and especially its phenols, thymol and carvacrol, have antibacterial acivity against periodontopathic bacteria including *Actinobacillus*, *Capnocytophaga*, *Fusobacterium*, *Eikenella*, and *Bacteroides* species (Osawa *et al.*, 1990), and may therefore be suitable for plaque control, although few essential oils have been found to possess clinical efficacy (Marsh, 1992). Furthermore, the essential oil of thyme showed a wide antibacterial activity against microorganisms that had developed resistance to antibiotics such as methicillin-resisting *Staphylococcus aureus* and vancomycin-resisting *Enterococcus faecium* (Nelson, 1997).

Several studies have focused on the antimicrobial activity of the essential oils of thyme in order to identify the responsible compounds. Thymol and carvacrol seem to play an outstanding role. These terpene phenols join to the amine and hydroxylamine groups of the proteins of the bacterial membrane altering their permeability and resulting in the death of the bacteria (Juven *et al.*, 1994). In addition, thymol and carvacrol were shown to induce a decrease of the intracellular adenosine triphosphate (ATP) pool of *Escherichia coli* and an increase of the extracellular ATP (Helander *et al.*, 1998). Antibacterial activity was also observed for the aliphatic alcohols, especially geraniol, and ester components. A variety of activities was presented by the esters, in some cases they were more active than their corresponding free alcohols, but sometimes less active (Megalla *et al.*, 1980).

Crespo *et al.* (1990) have evaluated the antimicrobial activity exhibited by the main chemical groups found in the essential oil of *Thymus serpylloides* ssp. *gadorensis* including hydrocarbons, alcohols, acetates, and phenols (Table 10.1). Again the phenols turned out to be the most effective against all microorganisms tested, the activity of the alcohols was on lower levels. Hydrocarbons proved to be effective only against *Bacillus megaterium* and *Mycobacterium phlei*, against the latter also the acetates showed weak activity. The higher sensitivity of *Bacillus megaterium* and *Mycobacterium phlei* to the essential oil of *T. serpylloides* ssp. *gadorensis* may be interpreted as the joint effectiveness of three and four active fractions respectively.

Studies on the structure–activity relationships of 32 terpenoids resulted in the following observations (Table 10.2, Hinou *et al.*, 1989): (a) α-isomers were inactive as

Table 10.1 Antibacterial activity of the main chemical groups in the essential oil of *T. serpylloides* ssp. *gadorensis*

Test microorganism	Essential oil	Hydrocarbon mixture	Alcohol mixture	Acetate mixture	Phenol mixture
Pseudomonas fluorescens	31	–	12	–	11
Escherichia coli	32	–	13.6	–	28.5
Bacillus megaterium	48	7	17	–	40.3
Staphylococcus aureus	32	–	8	–	17.3
Micrococcus luteus	60	–	–	–	35
Mycobacterium phlei	75	13	10.5	14	67

Note
Data expressed in mm of growth inhibition in an agar overlay technique assay.
(–) means an inhibition area minor than 7 mm.
Source: Crespo *et al.*, 1990.

opposed to the β-isomers which showed a pronounced activity, e.g. α-pinene; (b) *cis*-isomers proved to be inactive contrary to the active *tr*-isomers, e.g. geraniol versus its *cis*-isomer nerol; (c) compounds with a methyl-isopropyl cyclohexane ring like some alcohols and ketones were the most active, e.g. pulegone; (d) unsaturation of the cyclohexane ring further increased the antimicrobial activity, e.g. terpinolene and α-terpineol which proved to be the most active of the compounds examined against all the bacteria of the test. Negative results were found in case of α- or *cis*-isomers or when the compounds lack the common terpene C10-structure, e.g. citronellol or nerolidol.

With respect to the botanical species one can classify the essential oils of thyme, in general terms, into two main groups (Crespo *et al.*, 1991): (i) The first group contains those species in which phenols (thymol and carvacrol) are the predominant components. These oils show remarkable antimicrobial activities. (ii) In the oils of the second group phenols are scarce or lacking, whereas other components, such as monoterpene hydrocarbons, non-phenolic oxygenated monoterpenes or even sesquiterpene hydrocarbons, predominate. Such oils usually demonstrate lower antimicrobial activities than those in the first group.

The results obtained by the evaluation of the antimicrobial activity of a non-phenolic essential oil of thyme from *Thymus granatensis* may serve as an example of the above statement (Cabo *et al.*, 1986b). Although this essential oil presented activity against all the germs tested, with the exception of *Escherichia coli*, it proved to be only weakly active, in some cases practically inactive as was the case of *Candida albicans* and *Pseudomonas fluorescens* (Table 10.3). Similar results were obtained when other non-phenolic essential oils of thyme were tested, e.g. *T. hyemalis* (Cabo *et al.*, 1982), *T. longiflorus* (Cruz *et al.*, 1989a) and *T. baeticus* (Cruz *et al.*, 1993). A remarkably stronger antimicrobial activity was observed when typically phenolic oils, such as e.g. from *T. serpylloides* ssp. *gadorensis* were administered (Crespo *et al.*, 1990) (Table 10.3), a fact which was confirmed by studies with further phenolic essential oils such as that from *T. zygis* (Cabo *et al.*, 1978) and from *T. orospedanus* (Cabo *et al.*, 1987).

Two research groups evaluated the different antimicrobial (antibacterial and antifungal) effects of the essential oils of the seven chemotypes of *T. vulgaris* containing 1,8-cineole, geraniol/geranyl acetate, linalool, α-terpineol/α-terpinyl acetate, thymol,

Table 10.2 Antibacterial activity of individual components of thyme oil

Components	Pseudomonas aeruginosa	Escherichia coli	Staphylococcus aureus	Bacillus cereus	References
Hydrocarbons					
Myrcene	–	–	–		Megalla *et al.*, 1980
Ocimene	+	++	+++	+	Megalla *et al.*, 1980
Limonene	–	–	–		Megalla *et al.*, 1980
Dipentene	–	–	–		Megalla *et al.*, 1980
Phellandrene	–	–	–		Megalla *et al.*, 1980
Δ3-Carene	+	++	+	+	Megalla *et al.*, 1980
β-Pinene	–	–	+++		Hinou *et al.*, 1989
α-Pinene	–	–	–		Hinou *et al.*, 1989
Camphene	–	–	+++		Hinou *et al.*, 1989
Sabinene	–	–	+++		Hinou *et al.*, 1989
Terpinolene	+	+++	+++		Hinou *et al.*, 1989
Caryophyllene	–	–	–		Megalla *et al.*, 1980
Alcohols					
Octyl alcohol	+++	–	+++		Hinou *et al.*, 1989
Linalool	–	+++	+++		Hinou *et al.*, 1989
Geraniol	+++	+++	+++	+++	Hinou *et al.*, 1989 Megalla *et al.*, 1980
Nerol	–	–	–		Hinou *et al.*, 1989
Citronellol	–	–	–		Hinou *et al.*, 1989
Terpineol	–	+++	+++		Hinou *et al.*, 1989
Borneol	+	++	++	++	Megalla *et al.*, 1980
Nerolidol	–	–	–		Hinou *et al.*, 1989
Farnesol	+	+	+++	++	Megalla *et al.*, 1980
Aldehydes					
Citral	–	+++	+++		Hinou *et al.*, 1989
Citronellal	+	+++	+	++	Megalla *et al.*, 1980
Myrtenal	–	–	+++		Hinou *et al.*, 1989
Ketones					
Carvone	–	–	+++		Hinou *et al.*, 1989
α-Thujone	–	–	–		Hinou *et al.*, 1989
Pulegone	–	+++	+++		Hinou *et al.*, 1989
Camphor	–	–	–		Megalla *et al.*, 1980
Phenols					
Thymol	+	+++	++	+++	Megalla *et al.*, 1980
Carvacrol	+	+++	+++	+++	Megalla *et al.*, 1980 Hinou *et al.*, 1989
Eugenol	+	+++	+	+++	Megalla *et al.*, 1980
Esters					
Neryl acetate	+++	–	+++		Hinou *et al.*, 1989
Geranyl acetate	–	–	+++		Hinou *et al.*, 1989
Linalyl acetate	+++	+++	+++		Hinou *et al.*, 1989
Bornyl acetate	–	–	–		Hinou *et al.*, 1989
Isobornyl acetate	–	–	+++		Hinou *et al.*, 1989

Terpenyl acetate	–	+++	+++		Hinou *et al.*, 1989
Methyl benzoate	+	++	+	++	Megalla *et al.*, 1980
Oxides					
Cineole	–	–	–		Megalla *et al.*, 1980

Notes

Results are expressed as growth inhibition in an agar overlay technique assay.

(–) area of inhibition minor than 7 mm.

(+) area of inhibition between 7–10 mm.

(++) area of inhibition between 11–16 mm.

(+++) area of inhibition greater than 16 mm.

Table 10.3 Antimicrobial activity of essential oils from *Thymus granatensis* and *Thymus serpylloides* ssp. *gadorensis*

Test micro-organisms	Thymus granatensis	Thymus serpylloides ssp. gadorensis
Escherichia coli	(–)	(+++)
Pseudomonas fluorescens	(–)	(+++)
Citrobacter freundii	(+)	(+++)
Micrococcus luteus	(+)	(+++)
Staphylococcus aureus	(+)	(+++)
Bacillus cereus	(+)	(+++)
Bacillus macerans	(+)	(+++)
Bacillus megaterium	(+)	(+++)
Mycobacterium phlei	(+++)	(+++)
Candida albicans	(–)	(+++)

Notes

Results are expressed as growth inhibition in an agar overlay technique assay.

(–) area of inhibition minor than 7 mm.

(+) area of inhibition between 7–10 mm.

(++) area of inhibition between 11–16 mm.

(+++) area of inhibition greater than 16 mm.

Source: Cabo *et al.*, 1986a; Crespo *et al.*, 1990.

carvacrol, and *tr*-sabinene hydrate/*cis*-myrcenol-8 respectively as main compounds. Logically the oils of these chemotypes vary in their minimal inhibitory concentration (MIC) values obtained. The most active oil was proven to be that of the thymol chemotype followed by the carvacrol and geraniol type; the linalool type showed similar activity to that of the geraniol type. The oils of the other chemotypes were much less active (Simeon de Bouchberg *et al.*, 1976; Lens-Lisbone *et al.*, 1987).

When accepting that the chemistry of the essential oils is responsible for their antimicrobial activities, as was explained above, it becomes obvious that all factors influencing the chemical composition of essential oils within the plant indirectly influence their antimicrobial activity. Climatic conditions are known to modify the chemical composition of the essential oils and therefore, climatic conditions indirectly influence the antibacterial activities. For example, Kowal and Kuprinska (1979) found large differences in the MIC values for the essential oils of *T. pulegioides* from different regions in Croatia. Also Cabo *et al.* (1986b) have found considerable differences in the antimicrobial activity of the essential oils obtained from different populations of *T. granatensis* collected in southeastern Spain. Another factor determining the variability of the oil composition of thyme is the stage of the plant growth during harvest. Arras and Grella (1992) studied the influence of the harvest time of *T. capitatus* (*Thymbra capitata*) on the fungistatic effects of its essential oil.

Although the antimicrobial action is to a large extent attributed to the essential oils, non-volatile constituents also have been described to possess antimicrobial activity, such as saponins and resins. These two constituents from *T. capitatus* (*Thymbra capitata*) inhibited the growth of several bacteria and fungi (Kandil *et al.*, 1994). It has recently been demonstrated that the watery extract of thyme (*T. vulgaris*) showed a strong inhibition of *Helicobacter pylory* reducing both, its growth and potent urease activity (Tabak *et al.*, 1996). Clinical trials confirming this property would be of significant interest, since this microorganism plays a role in the etiology of gastroduodenal ulcers.

Finally we will refer to the techniques commonly used to evaluate antibacterial activity. They can be classified depending on whether they require a homogeneous dispersion in water or not (Janssen *et al.*, 1987). (a) The agar overlay technique does not require a homogeneous dispersion in water. Discs, holes or cylinders are used as reservoirs containing the essential oils to be tested and are brought into contact with an inoculated medium and, after incubation, the diameter of the transparent zone around the reservoir (inhibition diameter), where the microorganisms have been destroyed by the action of the essential oil, is measured. The method can be modified when the reservoir is placed in the lid of the Petri dish, thus excluding transport by diffusion (Pellecuer *et al.*, 1980). (b) Dilution techniques require a homogeneous dispersion of the oil. Since essential oils are insoluble in the watery liquid culture medium, non-ionic emulsifying agents such as Tween or Spans are needed (Allegrini and Simeon de Bouchberg, 1972). Although the addition of an emulsifying agent introduces an extra component with respect to activity and possible interactions, this method has proved easily reproducible (Janssen *et al.*, 1987). However, other dilution methods that avoid the use of tensio-actives have been proposed, such as the solution of the essential oil in DMSO (Melegari *et al.*, 1985), or in the formation of a stable suspension for 24 h in a sterile watery solution of agar (Lens-Lisbonne *et al.*, 1987).

The techniques to determine the qualitative bacteriostatic activity are based on the agar overlay technique. Quantification of the antimicrobial activity is generally established by determining the MIC values. The MIC values can be determined either by using the agar overlay technique or the dilution techniques.

Antifungal effects

Several *in vitro* and *in vivo* screenings have shown that volatile oils, especially those of the genus *Thymus*, may be used against fungal diseases (Roussel *et al.*, 1973; Blakeway, 1986; Farag *et al.*, 1986; Deans and Ritchie, 1987). Different screenings focused on the essential oil of *T. vulgaris* and its effect on food spoiling yeasts (Conner and Beuchat, 1984; Ismaiel and Pierson, 1990), especially *Aspergillus* (Farag *et al.*, 1989), on various dermatophytes (Janssen *et al.*, 1988), and on some phytopathogenic fungi, e.g. *Rhizoctonia solani, Pythium ultimum, Fusarium solani*, and *Calletotrichum lindemthianum* (Zambonelli *et al.*, 1996). Not only the oil of *T. vulgaris* but also the oils of other *Thymus* species showed antifungal activity, e.g. that of *T. zygis* against *Botrytis cinerea*

(Wilson *et al.*, 1997). The oil of *T. serpyllum* was found to be highly active against various species of *Penicillium*, *Fusarium* and *Aspergillus* (Agarwal and Mathela, 1979; Agarwal *et al.*, 1979). Various oils, namely the oils of *T. zygis* (Cabo *et al.*, 1978), *T. hyemalis* (Cabo *et al.*, 1982), *T. vulgaris* (Menghini *et al.*, 1987), *T. serpylloides* (Crespo *et al.*, 1990), and *T. baeticus* (Cruz *et al.*, 1993), inhibited the growth of *Candida albicans*.

The essential oil of *T. vulgaris* inhibits both mycelial growth and aflatoxin synthesis by *Aspergillus parasiticus* (Tantaoui-Elaraki and Beraoud, 1994) at only 0.1 per cent in the medium. Therefore it is used as a preservative in agriculture, completely inhibiting aflatoxin production on lentil seeds up to eight weeks of incubation (El-Maraghy, 1995). In addition, the essential oil of *T. vulgaris* exerts a protective effect in corn against *Aspergillus flavus*, without producing phytotoxic effects on germination or corn growth (Montes and Carvajal, 1998).

According to Agarwal and Mathela (1979) and Agarwal *et al.* (1979) the antifungal activity of the essential oil of *T. serpyllum* is attributable to thymol and carvacrol. They cause the degeneration of the fungal hyphae which seems to empty their cytoplasmic content (Zambonelli *et al.*, 1996). The terpenic alcohols as well as the aldehydes, ketones and some esters, also presented considerable activities, whereas the hydro-carbons showed only low activities (Table 10.4). Terpenic alcohols which display monoterpenic structure and hydroxyl group at terminal carbon (i.e. geraniol, nerol and citronellol) have shown the highest activity. No difference was observed in the antifungal activity between *cis* or *trans* isomer forms of these molecules. The components which display carbonyl groups were also active in inhibiting fungal growth, showing the alde-hydes (i.e. citral and citronellal) a higher activity than ketones. Similarly in terpenic alcohols, this effect could be attributed to the presence of the functional group at a terminal carbon.

Antiviral effects

Only few studies demonstrate the antiviral effects of thyme extracts. In 1967, Herrmann and Kucera reported on the antiviral effects of *Thymus serpyllum* and Spanish and French thymes (*Thymus* sp.) against Newcastle disease virus (NDV). The antiviral activity was concentrated in the tannin fraction although non-tannin extracts also showed effects attributed to the polyphenol precursor compounds of tannins. However the activity was smaller compared to that observed with *Melissa officinalis* extracts. More recently other studies failed to demonstrate antiviral effects of Thyme extracts against Rubella virus (Zeina *et al.*, 1996). The antiviral activity recently observed in other members of the Labiatae family has been attributed to new di- and tri-terpenoid compounds that appear to be specific inhibitors of HIV-1 protease (Min *et al.*, 1998, 1999). Those com-pounds have not yet been detected in the genus *Thymus*.

Spasmolytic effects

The spasmolytic properties are commonly considered as the major action of thyme preparations. In this regard *T. vulgaris* is the most representative species. Therefore many publications have focused on the effects of thyme preparations on smooth muscles, especially rat and guinea pig intestines, such as duodenum and ileum, guinea pig trachea, seminal vesicles and rabbit jejunum. Two different protocols are typically followed: (i) The isolated smooth muscle is first contracted using several agonists (acetylcholine,

Table 10.4 Antifungal activity of individual components of the essential oils of thyme

Components	Penicillum cyclopium	Fusarium moniliforme	Aspergillus aegyptiacus	Trichoderma viride	References
Hydrocarbons					
Myrcene	−		−	−	Megalla et al., 1980
Ocimene	++		++	+	Megalla et al., 1980
Limonene	+		+++	+	Megalla et al., 1980
Dipentene	−		−	−	Megalla et al., 1980
Phellandrene	−		−	−	Megalla et al., 1980
Δ3-Carene	−	−	−	−	Agarwal and Mathela, 1979; Megalla et al., 1980
β-Pinene		−			Agarwal and Mathela, 1979
α-Pinene		−			Agarwal and Mathela, 1979
Camphene	−	−	−	−	Agarwal and Mathela, 1979; Megalla et al., 1980
Caryophyllene	−		−	−	Megalla et al., 1980
Alcohols					
Linalool	−		−	−	Megalla et al., 1980
Geraniol	+++	+	+++	+++	Agarwal and Mathela, 1979; Megalla et al., 1980
Nerol	+++		+++	+++	Megalla et al., 1980
Citronellol	+++		+++	+++	Megalla et al., 1980
Terpineol	−	+	−	−	Agarwal and Mathela, 1979; Megalla et al., 1980
Borneol	++		+	+	Megalla et al., 1980
Farnesol	++		++	++	Megalla et al., 1980
Aldehydes					
Citral	+++		+++	++	Megalla et al., 1980
Citronellal	+++		+++	+++	Megalla et al., 1980
Ketones					
Carvone	+		++	++	Megalla et al., 1980
Camphor	++		+	+	Megalla et al., 1980
Phenols					
Thymol	+++	+++	+++	+++	Agarwal and Mathela, 1979; Megalla et al., 1980
Carvacrol	+++	+++	+++	+++	Agarwal and Mathela, 1979; Megalla et al., 1980
Eugenol	+++		+++	+++	Megalla et al., 1980
Esters					
Linalyl acetate	++		+++	+	Megalla et al., 1980
Geranyl acetate	−		−	−	Megalla et al., 1980
Oxides					
Cineole	+	−	++	+	Agarwal and Mathela, 1979; Megalla et al., 1980

Notes
Results are expressed as growth inhibition in an agar overlay technique assay.
(−) halo of inhibition minor than 7 mm.
(+) halo of inhibition between 7–10 mm.
(++) halo of inhibition between 11–16 mm.
(+++) halo of inhibition greater than 16 mm.

histamine, adrenaline, nicotine and $BaCl_2$) and the thyme preparations are subsequently added until maximum relaxation is achieved. The spasmolytical effect is evaluated by measuring the maximum relaxant effect and the ED50 (contraction that produces 50 per cent of the maximum spasmolytic response). (ii) The isolated smooth muscle is first incubated with the thyme preparations; the modification of the dose-response curves produced by the contracting agents are calculated. In this protocol, the relaxant agent remains in the bath throughout the experiment. The use of various spasmogens with different mechanisms of action causing muscle contraction can provide information on the pharmacological basis of the spasmolytic properties. As reference substances atropine, papaverine, and isoprenaline are used.

Spasmolytic activity of the essential oil of thyme (T. vulgaris)

Debelmas and Rochat (1964) have shown that thyme oils in which phenols are the predominant components have antispasmodic activity on intestinal smooth muscle contracted by several agents. Later the same authors (Debelmas and Rochat, 1967) studied the spasmolytic activity of the oils of different plants using various isolated smooth muscles and contractor agents (Table 10.5). They found that thyme oil was the most active, presenting an antispasmodic action of an unspecific type inhibiting the contractions induced by all agents tested. These initial studies were carried out with water saturated with the essential oil, trying to overcome the problems to bring the hydrophobic essential oil into contact with the isolated organ.

Reiter and Brandt (1985) studied the effects of the volatile oils of 22 plants from 11 different families on tracheal and ileal smooth muscles of the guinea pig (Table 10.6). All the oils showed relaxant effects on the tracheal smooth muscle developing shortly after addition. The most potent oils were (in the order of potency): angelica root, clove, elecampane, basil and balm. Sixteen oils inhibited the phasic contraction of ileal preparations with ED50 values between 4.5 and 76 mg/l. With regard to the relaxant effects, the majority of oils were 1.4–8.4 times more potent on the ileal than on the tracheal

Table 10.5 Relaxant effect of volatile-oil-saturated water from different plants, versus the contractions induced by several contractor agents

Animal	Rat			Guinea pig	Rabbit	
Isolated muscle Contractor agent	Seminal vesicles Adrenaline	Duodenum Acetylcholine	Duodenum BaCl2	Ileum Histamine	Aorta Adrenaline	Jejunum Nicotine
Plants						
Chenopodium	0	±	++	+++	0	+
Clove	±	+	++	+++	±	+++
Caraway	0	+	±	+	0	+++
Sage	0	+	0	+++	0	+
Thyme	++	++	++	+++	++	+++
Balm	0	++	±	++	+	++

Notes
±, <20% of inhibition.
+, 20–40% of inhibition.
++, 40–60% of inhibition.
+++, >60% of inhibition.
Source: Debelmas and Rochat, 1967.

Table 10.6 Relaxant effect (ED50) of different volatile oils on tracheal and ileal longitudinal smooth muscle of the guinea pig

Species; pharmaceutical preparation	Trachea ED50 (mg/l)	Ileum ED50 (mg/l)	ED50 tracheal/ ED50 ileum
Thymus vulgaris L. thyme oil	56±1.6	6.9±1.1	8.1
Melissa officinalis L. balm leaf oil	22±1.5	7.8±1.4	2.8
Mentha piperita L. peppermint oil	87±1.6	26±1.5	3.4
Ocimum basilicum L. basil oil	19±1.7	32±2.2	0.6
Salvia officinalis L. sage oil	371±1.8	44±1.4	8.4
Reference drug	trachea ED50 (nmol/l)	ileum ED50 (nmol/l)	ED50 trachea/ ED50 ileum
Isoprenaline	3.9±0.7	21±1.0	0.19
Papaverine	240±0.9	3700±1.6	0.06

Note
Volatile oils were solubilised in water by means of polyoxyethylene fatty ester, Arlatone 285.
Source: Reiter and Brandt, 1985.

muscle. This ratio was almost eight-fold for the essential oil of *T. vulgaris*. The authors divided the 22 volatile oils investigated into three groups according to their different effects on the mechanical behaviour of the stimulated ileal mysenteric plexus-longitudinal muscle preparations. Together with 15 others thyme oil belongs to the first group which had predominantly relaxing effects.

Spasmolytic activity of the essential oils of different Thymus species

Between 1985 and 1990 our research group carried out several studies aimed at studying spasmolytic activities of a variety of essential thyme oils from plants collected in eastern Andalusia, relating them with their components. In Table 10.7, the ED50 values of these essential oils and the chemical compositions (only functional groups) are given. All the oils produced a relaxant effect against acetylcholine-induced contractions in isolated rat duodenum. However marked differences existed which might partly be explained by the differences of the chemical composition. One may highlight the following:

- Oils containing higher portions of phenolic components have shown higher spasmolytic potency, namely the oils of *T. zygis* (Cabo *et al.*, 1986a) and *T. orospedanus* (Cabo *et al.*, 1987).
- The oil of *T. granatensis* although lacking phenolic compounds showed a powerful relaxant effect (Cabo *et al.*, 1986a). This could be due to the high content of terpene hydrocarbons, which also may explain the high effect of *T. orospedanus* containing both, phenols and hydrocarbons.

Table 10.7 Relaxant effects (ED50) versus contractions induced by acetyl-choline in isolated rat duodenum and quantitative composition of different essential oils of thyme when collected in full bloom

Sample	ED50 (μg/ml)	Components	%
T. zygis	6.5	Hydrocarbons	29.1
		Alcohols /esters	15.4
		Ketones	3.0
		Aldehydes	
		Ethers	10.3
		Phenols	41.4
T. orospedanus	8.3	Hydrocarbons	61.8
		Alcohols/esters	12.1
		Ketones	4.1
		Aldehydes	
		Ethers	
		Phenols	22.9
T. granatensis	12.5	Hydrocarbons	43.2
		Alcohols/esters	27.6
		Ketones	3.4
		Aldehydes	
		Ethers	7.5
		Phenols	
T. baeticus	26.7	Hydrocarbons	28.1
		Alcohols/esters	44.5
		Ketones	
		Aldehydes	6.9
		Ethers	27.0
		Phenols	
T. longiflorus	94.1	Hydrocarbons	25.6
		Alcohols/esters	10.1
		Ketones	10.2
		Aldehydes	1.3
		Ethers	40.6
		Phenols	

Note
Volatile oils were emulsified in water with Tween 20 at a proportion of 9/1 p/p.

- Essential oils lacking phenolic components and with low hydrocarbon levels are less potent, documented by the results obtained with oils of *T. baeticus* (Cruz *et al.*, 1989b) and *T. longiflorus* (Zarzuelo *et al.*, 1989).
- The lower spasmolytic effect of *T. longiflorus* (Zarzuelo *et al.*, 1989) in comparison with that of *T. baeticus* (Cruz *et al.*, 1989b) may be due to a higher level of ethers (1,8-cineole).

The above results prompted our group to investigate the spasmolytic potency in more detail (Table 10.8, Cabo *et al.*, 1986c). Taking all the data obtained into consideration one can say that indeed the phenolic components (thymol and carvacrol) as well as the terpene hydrocarbons (myrcene and caryophyllene) presented higher spasmolytic

Table 10.8 Relaxant effects (ED50) of different components of essential oils versus contractions induced by acetylcholine and BaCl$_2$ in isolated rat duodenum. The relative potency was calculated in comparison to papaverine

Agonist	Antagonist	ED50 (µM)	Relative potency
Acetylcholine	Carvacrol	6.67 ± 0.86	1.28
	Thymol	4.88 ± 0.74	1.75
	Myrcene	5.48 ± 1.08	1.56
	Caryophyllene	9.00 ± 2.14	0.95
	Camphor	inactive	–
	Papaverine	8.55 ± 1.09	1
BaCl$_2$	Carvacrol	7.73 ± 1.22	0.41
	Thymol	7.25 ± 0.81	0.44
	Myrcene	inactive	–
	Caryophyllene	1.34 ± 0.25	2.37
	Camphor	inactive	–
	Papaverine	3.18 ± 0.68	1

Note
The components of volatile oils were emulsificied in water by means of tween 20 at a proportion of 9/1 p/p.
Source: Cabo *et al.*, 1986c.

potency. Camphor was shown to be inactive whereas 1,8-cineole acted as a spasmogenic and showed to increase rat duodenum contractions up to a maximum of 75 per cent of acetylcholine induced contractions. This effect was competitively antagonised by atropine. Consequently these results favour the hypothesis that 1,8-cineole acts as a partial agonist at the level of the acetylcholine receptors (Zarzuelo *et al.*, 1987; Gámez *et al.*, 1990) and that the presence of 1,8-cineole in the oils diminishes their spasmolytic capacity.

The mechanism of the spasmolytic effect was studied by the modification of the dose-response curves induced by acetylcholine and CaCl$_2$. All the essential oils included in Table 10.7 modified the dose-response curves induced by acetylcholine, showing a dosage-dependent decrease in the maximum effect suggesting a non-competitive mechanism (Cabo *et al.*, 1986a; Cruz *et al.*, 1989b; Zarzuelo *et al.*, 1989). Such oils are capable of inhibiting the contractions induced by CaCl$_2$ in high-KCl Ca^{2+}-free solution, also diminishing their maximum effect in a dose-dependent way (Cruz *et al.*, 1989b; Zarzuelo *et al.*, 1989). Godfraind *et al.* (1986) demonstrated that Ca^{2+}-induced contractions of a KCl-depolarized smooth muscle are due to an increased Ca^{2+} influx through voltage-stimulated type-L Ca^{2+} channels. Therefore, the inhibitory effects of essential oils in these concentrations may also be explained by a) an inhibition of Ca^{2+} entry through voltage-stimulated channels into the smooth muscle and/or b) blocking release of intracellular bound Ca^{2+}.

Several authors have investigated the spasmolytic mechanism of some phenolic compounds of essential oils. Van den Broucke and Lemli (1982) and Cabo *et al.* (1986c) studied the antagonistic effect of thymol and carvacrol on guinea pig ileum and rat duodenum against the contractions induced by carbachol, histamine and BaCl$_2$ and concluded that contractions induced by these spasmogenic agents were inhibited by the phenols in a non-competitive antagonistic way.

There are very few studies which demonstrate the relaxant effect of the essential oil of thyme on vascular smooth muscle. In experiments on mice, guinea pigs and rabbits

Guseinov *et al.* (1987) found that the essential oil of Kochi Thyme (*T. kotchyanus*) was non-toxic and produced hypotensive effects in rabbits at concentrations of 1 mg/kg.

Spasmolytic activity of T. vulgaris *extracts*

The therapeutic value of the thyme herb, indeed, depends on the quantity of phenols in the essential oil (see above). Therefore, for all the *Thymus* species, one has to prefer those that contain a high concentration of thymol and/or carvacrol. However the presence of a non-volatile principle in the *Thymus* species has always been supposed. This was established by Van den Broucke and Lemli (1981) who examined the correlation between the phenol content of *T. vulgaris* liquid extracts and the spasmolytic activity on the muscles of the guinea pig ileum and trachea. The extracts showed a high spasmolytic action, but no correlation between phenol content and activity could be observed. The thymol and carvacrol concentration of extracts was much too low (<0.001 per cent) and could not be responsible for the antispasmodic activity.

The authors continued their experiments testing flavonoids isolated from *T. vulgaris* (Van den Broucke and Lemli, 1983) *in vitro* for their spasmolytic activity on the smooth muscles of the guinea pig ileum and of the rat vas deferens (Table 10.9). Both the flavones and the thyme extracts inhibited responses to agonists which stimulate specific receptors (acetylcholine, histamine, noradrenaline) as well as to agents whose actions are not mediated via specific receptors (BaCl$_2$). Flavonoids appeared to act as musculotropic agents. Musculotropic spasmolysis is complex and the results of this study could not clarify the events in the muscle completely. However, inhibition of Ca^{2+} induced contractions in K$^+$ depolarized muscles pointed to a possible decrease in the avaibility of Ca^{2+}. Flavones induced relaxation of the carbachol-contracted tracheal strip without stimulation of the β_2-receptors, which were blocked by propranolol. This relaxation

Table 10.9 Relaxant effects (pD'2) and potency relative to papaverine of thymonin, 8-methoxycirsilineol, cirsilineol, luteolin, apigenin, papaverine and phentolamine on the guinea pig ileum and the rat vas deferens versus different smooth muscle agonist

Antagonist	*Guinea pig ileum*			*Rat vas deferens*
	Carbachol	*Histamine*	*BaCl$_2$*	*Noradrenaline*
Thymonin	4.55±0.12	4.73±0.04	5.02±0.22	4.38±0.07
	30	56	250	239
8-MeOH-Cirsilineol	4.46±0.23	4.60±0.12	4.14±0.19	4.15±0.16
	27	41	33	140
Cirsilineol	4.69±0.10	4.54±0.15	4.25±0.07	3.98±0.15
	46	36	43	96
Luteolin	4.76±0.10	4.74±0.14	4.52±0.14	3.80±0.18
	54	57	79	63
Apigenin	4.75±0.23	5.00±0.10	4.46±0.17	3.72±0.09
	52	104	69	52
Papaverine	5.03±0.16	4.98±0.12	4.62±0.24	4.01±0.10
	100	100	100	100
Phentolamine				6.76±0.13

Source: Van den Broucke *et al.* 1982, Van den Broucke and Lemli, 1983.

could probably be due to an inhibition of the phosphodiesterase, followed by an increase of the intracellular c-AMP level (Beretz *et al.*, 1980).

The flavonoid pattern of *T. satureioides* differs from that of the other *Thymus* species in the high portion of polymethoxylated flavones. The *in vitro* tracheal relaxant activity of thyme extracts prepared from *T. satureioides* compared with that of the pure flavones (same quantity of flavones), supports the action of a thyme extract to be almost completely explained by the content on flavones independently from their degree of permethoxylation (Van den Brouke *et al.*, 1982). The affinity of methoxylated flavones in the guinea pig ileum does not differ significantly from that of luteolin and apigenin. However, methylation of the hydroxy groups of the flavone skeleton increases the relaxant activity.

Further evidence of the spasmolytic effects of flavonoids was provided by Capasso *et al.* (1991a). They screened 13 flavonoids for their effects on contractions in guinea pig ileum induced by prostaglandin $E_2(PGE_2)$, Leukotriene D_4 (LTD_4), acetycholine and $BaCl_2$. The flavonoids showed spasmolytic effects that may be due to a non-specific action since they were found to be active against contractions induced by several agents. Flavonoids have also been shown to inhibit electrically-induced contractions (transmural electrical stimulations) (Capasso *et al.*, 1991b). Flavonoids inhibit the contractile responses probably through a reduction of calcium influx by way of calcium channels and through inhibition of calcium release from intracellular stores, decreasing the calcium concentration available for contractile machinery (Gálvez *et al.*, 1996).

Spasmolytic activity of extracts of different Thymus *species*

Blázquez *et al.* (1989) examined the spasmolytic activity of *T. webbianus* and *T. leptophyllus* extracts. Air-dried leaves and stems of these species were extracted with 70 per cent aqueous methanol. The methanol was removed and the remaining aqueous fraction was successively treated with diethyl ether, ethyl acetate and butanol. The results of this study showed that the extracts of both plants have significant spasmolytic effects on isolated rat duodenum. In general, a difference in the responses produced between differently polar extracts was observed. In fact, the diethyl ether extracts (low polarity) produced a dose-dependent reduction in acetylcholine-induced contractions at 1, 10, 100 µg/ml concentrations, whereas the ethyl acetate extracts (middle polarity) and the butanolic (high polarity) were inactive at the same concentrations, but active only at higher concentrations. *T. webbianus* consistently proved to be more active than *T. leptophyllus*. Zafra-Polo *et al.* (1990) isolated two steroidal compounds with relaxant properties from the diethyl ether extracts of *T. webbianus*. The compounds could be considered as derivatives of stigmastenone and its isomer β-sitostenone.

Blázquez *et al.* (1995) investigated the effects of a diethyl ether extract of the leaves of *T. leptophyllus* on rat uterine and aorta strip muscle. It inhibited the contraction of uterine smooth muscle at lower concentrations than those observed for aorta strips. Rat uterus experiments with and without extracellular calcium yielded similar ED50 values suggesting a non-specific mechanism for the relaxant activity. In the presence of extracellular calcium, the extract inhibited the contractile response of rat aorta induced by K^+ depolarising solution, and had a lower inhibitory effect on noradrenaline-induced contraction.

An ethanolic extract of the leaves of *T. orospedanus* has been shown to significantly decrease arterial blood pressure in Wistar rats, at a dose of 150 mg/kg during the first hour following the administration, whilst the 300 mg/kg dose continued to show hypotensive effects for up to 5 h (Jiménez *et al.*, 1988). Flavonoids have also been shown to have vasodilator effects in isolated rat and rabbit vascular smooth muscle (Duarte *et al.*, 1993a; Herrera *et al.*, 1996). This vascular smooth muscle relaxation has been attributed to several mechanisms including decreased transmembrane $^{45}Ca^{2+}$ uptake (Ko *et al.*, 1991) and inhibitory effects on cAMP and cGMP-phosphodiesterases (Beretz *et al.*, 1980) or on protein kinase C (Duarte *et al.*, 1993b).

Antioxidant effects

It was only in the 1970s that scientists realised that the human body constantly creates free radicals and eliminates them by a series of antioxidant defense mechanisms. When free-radical generation exceeds the capacity of antioxidant defenses, the result is "oxidative stress". It occurs in many human diseases and sometimes makes a significant contribution to their pathogenesis. In literature several publications are dedicated to the antioxidative effects of plant extracts with phenolic compounds (Alscher and Hess, 1993). Recently Chung *et al.* (1997) studied the effects of methanol extracts from 51 plant species on OH-radical scavenging. Mustard varieties, thyme, oregano, and clove all exhibited strong scavenging activity.

There are several papers showing that both the essential oil and the flavonoids of *T. vulgaris* are potent antioxidant agents. Dorman *et al.* (1995) studied the antioxidant properties of the essential oil of *T. vulgaris* (0.75–100 ppm) among others. The antioxidant properties were evaluated in three avian thiobarbituric acid reactive substances (TBARS) assays using egg yolk, one-day-old chicken liver or muscle from mature chicken. *T. vulgaris* was one of the most effective antioxidants in the egg yolk assay besides *Monarda citriodora* var. *citriodora* and *Myristica fragrans*.

Deans *et al.* (1993) investigated the protection of polyunsaturated fatty acids within the liver of old mice by ingestion of culinary and medicinal plant volatile oils obtained by hydrodistillation. This protection effectively reverses the normal trend in polyunsaturated fatty acid metabolism during aging where a decrease in level is concomitant with a reduction in tissue function and integrity. The essential oil of thyme was overall the most effective agent in this protective effect.

Ternes *et al.* (1995) showed that carvacrol, thymol and *p*-cymene-2,3-diol, all the three components of thyme oil, exhibit antioxidant activities. The authors determined the concentration of each substance in different foodstuffs containing thyme extracts as well as their stabilities at different temperatures. Schwarz *et al.* (1996) assessed the antioxidant activity of thyme phenols (Rancimat method 110 °C, Schal test 60 °C) and showed that *p*-cymene-2,3-diol was the most active one being more active than α-tocopherol and butylated hydroxyanisole. Five thyme species (*T. vulgaris*, *T. pseudolanuginosus*, *T. citriodorus*, *T. serpyllum* and *T. doerfleri*) were analysed by means of High-Performance Liquid Chromatography (HPLC) for all 3 compounds. The highest amounts were found in *T. vulgaris*.

Pearson *et al.* (1997) investigated the potential antioxidant activity of various plant phenolics, namely carnosic acid, carnosol, and rosmarinic acid (in rosemary extracts), thymol (in thyme extracts), carvacrol (in origanum extracts), and zingerone (in ginger extracts), using aortic endothelial cells to mediate the oxidation of low-density lipoprotein (LDL). The extent of oxidation was determined spectrometrically by measuring the absorbance at 234 nm of the conjugated dienes. Their relative antioxidant activities decreased in the order carnosol > carnosic acid approximate to rosmarinic acid > thymol > carvacrol > zingerone.

Flavonoids are abundant in diverse species of thyme. They are potent antioxidants which are responsible for many of the beneficial effects of these plants. The electron-donating properties of flavonoids have been repeatedly emphasised as the basis of their antioxidant action. Common antioxidative flavonoids, like luteolin and quercetin, isolated from various species of thyme, have shown potent antioxidant activity *in vivo* as much as *in vitro* (Gálvez *et al.*, 1995a,b).

From leaves of *T. vulgaris* Haraguchi *et al.* (1996) isolated two antioxidative components by a bioassay-directed fractionation: a biphenyl compound, 3,4,3',4'-tetra-hydroxy-5,5'-diisopropyl-2,2'-dimethylbiphenyl, and a flavonoid, eriodictyol. Their antioxidant effects on biological systems were studied in three different biological systems: inhibition of superoxide anion production in the xanthine/xanthine oxydase system, inhibition of microsomal peroxydation, inhibition of hemolysis of human erythrocytes. Both the new biphenyls as well as eriodictyol showed outstanding antioxidant effects.

Further effects

Antiparasitic effects

An extract of *T. vulgaris* shows antiparasitic properties against *Leishmania mexicana* (Schnitzler *et al.*, 1995), because it inhibits the mitochondrial DNA polymerase (IC50 value was 0.82 mg/ml). Thymol was mainly responsible for this effect (Khan and Nolan, 1995). The essential oil of this species was also active against diverse phytonematodes, the oxygenated compounds being partially responsible for the nematicidal effects (Abd-Elgawad and Omer, 1995).

Perrucci *et al.* (1995) evaluated the *in vitro* ascaricidal properties of some natural monoterpenoid constituents of several essential oils against rabbit mange mite (*Psoroptes cuniculi*) by direct, external contact and by inhalation. The natural terpenoids assayed were: hydrocarbons (limonene, myrcene, γ-terpinene), alcohols (linalool, geraniol, nerol, terpinen-4-ol, α-terpineol) and phenols (thymol, eugenol), an ester (linalyl acetate) and an ether (estragole). Because the test components represent different chemical classes, it was also possible to discern in a preliminary fashion a correlation between chemical structure and ascaricidal activity. All the monoterpene hydrocarbons, either acyclic (i.e. myrcene) or cyclic (i.e. limonene and γ-terpinene) did not show any miticidal activity at the doses tested (1.0 per cent, 0.25 per cent and 0.125 per cent). The double-bond position and/or number seems to be unimportant for this kind of biological activity.

In contrast, the terpene alcohols, such as linalool, geraniol, nerol, menthol, terpinen-4-ol, and α-terpineol, were able to kill nearly 100 per cent of the mites at the dose tested. Therefore, the oxygenated functional groups potentiate the ascaricidal properties among these compounds. Neither the acyclic (i.e. linalool, geraniol, nerol) nor cyclic (i.e. menthol, terpinen-4-ol, α-terpineol) nature of the compound appeared to influence the miticidal activity. Similarly, the site of linkage to the ring or to a side chain, as well as the nature of the hydroxyl group (primary, secondary, or tertiary), does not influence the activity. The *cis/trans* isomerism represented by nerol and geraniol seems to be important.

Thymol and eugenol killed nearly 100 per cent of the parasites at all dosages assayed in the direct contact test, indicating that a phenolic function can enhance the miticidal characteristics of terpenes. The low susceptibility of mites to linalyl acetate, particularly at the lowest doses, could be related to the esterification of the oxygenated function. Estragole, structurally close to eugenol, but with a methylated phenolic group, exhibited, at a concentration of one per cent, an activity comparable to that of the same dose of eugenol. However, this action decreased (63 per cent) at 0.25 per cent and disappeared completely at 0.125 per cent. These results indicate that the best miticidal activity of the monoterpenes examined in the direct contact test can be related to compounds with free alcoholic or phenolic functional groups.

Insecticidal effects

Recently it was demonstrated that aromatic plants present a double insecticidal effect: by direct toxicity on adult insects and by inhibiting reproduction. The most efficient plant in this regard belongs to the Labiatae family (Regnaultroger and Hamraoui, 1997). Therefore one can profit using the essential oils of *T. vulgaris* and *T. serpyllum* in addition to a fumigant against *Anathoscelides obtectus* Say (Coleoptera, Bruchidae), a frequent pest that damages its host plant, the kidney bean (*Phaseolus vulgaris* L.) in the field and during storage (Regnaultroger *et al.*, 1993). The oils have a toxic effect on adult insects and also inhibit the reproduction through ovicidal and larvicidal effects (Regnaultroger and Hamraoui, 1994). This insecticidal action is also produced by other components of the species such as non-volatile phenols, non-proteinic amino acids, and flavonoids (Regnaultroger and Hamraoui, 1995).

The essential oil of *T. vulgaris* and thymol shows activity against *Tetranychus urticae*. Thymol was shown to be more potent than thyme oil as a deterrent factor for reducing egg laying by the mite. Mortality percentage reached 100 per cent with both materials used; however, at low concentrations the effect again was more pronounced applying thymol than applying thyme oil (El-Gengaihi *et al.*, 1996).

Karpouhtsis *et al.* (1998) have demonstrated the genotoxic effect of thymol on the somatic mutation and recombination test on *Drosophila* and that this effect could contribute to the insecticidal action of essential oils such as *T. vulgaris*. Other components of the oils such as carvacrol, γ-terpinene and *p*-cymene have been found to be ineffective. Another pest species sensitive to the essential oil of *T. vulgaris* is *Spodoptera littoralis*. Feeding larvae with leaves treated with the essential oil reduced the successful development and egg production (Farag *et al.*, 1994).

Lee *et al.* (1997) evaluated the acute toxicity of 34 naturally occurring monoterpenoids against three important arthropod pest species: the larvae of the Western corn rootworm, *Diabrotica virgifera* LeConte, the adult two-spotted spider mite, *Tetranychus urticae* Koch, and the adult house fly, *Musca domestica* L. Thymol was the most topically toxic against the house fly, and citronellol and thujone were the most effective on the Western corn rootworm. Most of the monoterpenoids were lethal to the two-spotted spider mite at high concentrations; terpinen-4-ol was especially effective.

PHARMACOKINETICS OF THYMOL AND CARVACROL

Data about the pharmacokinetics of essential oils are scarcely available, but there can be found a few on the phenolic terpenes, thymol and carvacrol. In an early publication Schröder and Vollmer (1932) described thymol and carvacrol to redistribute rapidly to the blood and kidneys following oral administration. These observations were made in experiments with animals.

The metabolism of carvacrol and thymol has been studied in rats. It was found that the urinary excretion of metabolites was rapid and only very small amounts were excreted after 24 h. Although large quantities of carvacrol and, especially, thymol were excreted unchanged (or as their glucuronide and sulphate conjugates), extensive oxidation of the methyl and isopropyl groups also occurred. This resulted in the formation of derivatives of benzyl alcohol and 2-phenylpropanol and their corresponding carboxylic acids. In contrast, ring hydroxylation of the two phenols was a minor reaction (Austgulen *et al.*, 1987).

Takada *et al.* (1979) investigated the metabolism of thymol in rabbits and humans. Thymol glucuronide featuring an intact aglycone was isolated from the urine of thymol-medicated rabbits and identified as an acetyl derivative of methyl glucuronate. The hydroxylated product of thymol, thymohydroquinone, was detected in small amounts in the urines of thymol-medicated humans. It was presumed that thymolhydroquinone is excreted as an ethereal sulfuric acid conjugate.

TOXICOLOGY OF THYME OIL

Toxic effects of the vegetable parts of thyme, *T. vulgaris*, have not been published, but it is important to mention that a certain level of toxicity can be found in the essential oil of thyme (acute oral LD50 = 4.7 g/Kg rat). This toxicity has been attributed by some authors to thymol and carvacrol (Dilaser, 1979), their acute oral LD50 being 0.88–1.8 g/Kg and 0.1–0.18 g/Kg, respectively. Furthermore, these phenols cause skin irritations and especially irritations of the mucosa, which precludes patients with gastroduodenal ulcers from the use of the essential oil. Thus undiluted thyme oil was found to be severely irritating to both mouse and rabbit skin, however it produced no irritation on human subjects when tested at 12 per cent (Tisserand and Balacs, 1995). Hypersensitivity reactions have also been reported for this essential oil (Tisserand and Balacs, 1995; Benito *et al.*, 1996; Lemier *et al.*, 1996); therefore it is strongly recommended to perform a tolerance test prior to attempting internal administration.

Various essential oils from *Thymus* contain considerable quantities of camphor and other terpenic ketones, which are known to produce convulsions and epileptic/neurotoxic crises (Dupeyron *et al.*, 1976; Steinmetz *et al.*, 1980). Limonene, which is a common component of thyme oil, is capable of diminishing the incidence of tumors in experimental animals treated with tumor-inducing agents (Tisserand and Balacs, 1995). However upon oxidation of the molecule the risk of carcinogenesis is increased and it behaves as a catalysing agent. Therefore it is important to use fresh, non-oxidized essential oil of thyme in phytotherapy and aromatherapy.

Due to the toxic effects the use of the essential oil during pregnancy and lactation is contraindicated (Peris *et al.*, 1995).

THE NON-MEDICINAL USE OF THYME

Thyme as a food preservative

Due to their antimicrobial and antioxidant qualities numerous aromatic plants, such as thyme, have been used and are still being used as food preservatives (Shelef, 1983;

Nakatani, 1992; Amr, 1995). As was described before, the essential oils of thyme present a marked antimicrobial activity. This activity has been demonstrated to include bacteria responsible for alterations in food (Essen and Karapinar, 1986; Akgül and Kivanç, 1988a). Aureli *et al.* (1992) carried out a study on the antimicrobial activity of diverse essential oils of plants widely used in the food industry against *Listeria monocytogenes* (bacteria implicated in alterations in food). Only the essential oils of cinnamon, clove, marjoram, pepper and thyme presented antimicrobial activity.

Researchers have also demonstrated that a number of aromatic plants, including thyme, have a marked antifungal activity against food spoiling fungi (Akgül and Kivanç, 1988b; Salmerón *et al.*, 1990). The high antimycotic activity of clove and thyme was tested for their possible use as preservatives for agricultural commodities by El-Maraghy (1995). Both species completely inhibited aflatoxin production in lentil seeds for an eight week incubation period.

Antioxidant activity can also be responsible for a preservative activity, especially in preventing oxidation of lipids in food. This was studied by Budincevic *et al.* (1995) who tested ethanol extracts of *T. marschallianus* using tallow and lard as the substrates, at 60 °C in the Rancimat apparatus. The extracts showed antioxidant effects with the substrates processed at 60 °C but not at 100 °C. Adding citric and malic acid a synergistic effect could be observed. Dorman et al. (1995) demonstrated the antioxidant activity of the essential oil of *T. vulgaris* in TBARS using egg yolk, one-day old chicken liver or muscles from mature chickens.

Botsoglou *et al.* (1997) evaluated the effect of dietary thyme on the oxidative stability of egg shells over a 60-day refrigerated storage period. In addition, the influence of dietary thyme and of the storage time on the oxidative stability of liquid yolks adjusted to various pH values and agitated in the presence of light was investigated. Results show that malonaldehyde was not produced during the storage of egg shells. It was also evident that thyme treatment reduced the oxidation of liquid yolk, which was significantly increased by light and acidity. The authors proposed that thymol is the most important antioxidant component of thyme, but that there must be other components in thyme which act synergisticly with thymol.

The cosmetic uses of thyme

Thyme oil, in general, is used in many cosmetic preparations, such as deodorants, because of its capacity to suppress smells (González and Muñoz, 1980; Brasseur, 1983) and for its antimicrobial properties. The oil finds some use in soap perfumes where its power and freshness can introduce a hint of medicinal notes, often desirable in certain types of soap or detergent. The oil exerts an excellent masking effect over tarry odors (Arctander, 1960). Added to lotions perfumes or colognes in trace amounts, thyme oil may lend body and sweet freshness. Therefore it is used in the composition of cosmetic creams and milks, eau-de-cologne (frequently accompanied by lemon and bergamot), and soapy solutions to disinfect surgeons' hands (Valnet, 1964). These cosmetic products are useful to fight acne and skin complaints.

The essential oil of thyme and thymol are also used in the production of toothpaste and mouthwashes (Marsh, 1992; Banoczy *et al.*, 1995). This essential oil peroxidised to 10 per cent in a soapy solution destroys the microbial flora in the oral cavity in 3 min.

The culinary uses of thyme

Thyme is widely used as a seasoning especially in the Mediterranean kitchen. It has a strong but agreeable aroma and is pleasant in greasy or fatty food, such as sausages, bacon and other fatty meats and even strong cheeses. Along with rosemary it constitutes a highly recommended seasoning for pizzas and similar products (Pahlow, 1979). It can be added (with care) to sauces, soups, meat and fish dishes. The fresh leaves give flavour to salads. In the liquor industry thyme is used to give flavour and aroma, *T. moroderi* being the aromatic plant used to produce the 'Licor de Cantueso' and *T. vulgaris* participating in the formulation of several liquorices produced in the Spanish Eastern and Balearic regions.

ACKNOWLEDGEMENTS

The authors would like to thank Dr Elisabeth Stahl-Biskup for her contribution to the present chapter with the sections entitled "Thyme in Phytotherapy", "Thyme in Aromatherapy" and "Thyme in Homoeopathy", whose authorship is this way recognised.

REFERENCES

Abd-Elgawad, M.M. and Omer, E.A. (1995) Effect of essential oils of some medicinal plants on phytonematodes. *Anz. Schädlingskd. Pfl.*, 68, 82–84.
Agarwal, I. and Mathela, C.S. (1979) Study of antifungical activity of some terpenoids. *Indian Drugs Pharm. Ind.*, 14, 19–21.
Agarwal, I., Mathela, C.S. and Sinha, S. (1979) Studies on the antifungical activity of some terpenoids against Aspergilli. *Indian Phytopathol.*, 32, 104–105.
Akgül, A. and Kivanç, M. (1988a) Inhibitory effects of the six Turkish thyme-like spices on some common food-borne bacteria. *Nahrung*, 32, 201–203.
Akgül, I.A. and Kivanç, M. (1988b) Inhibitory effects of selected Turkish spices and oregano components on some food-borne fungi. *Int. J. Food Microbiol.*, 6, 263–268.
Allegrini, J. and Simeon de Bouchberg, M. (1972) Une technique d'etude du pouvoir antibacterien des huiles essentielles. *Prod. Probl. Pharm.*, 27, 891–897.
Alscher, R.G. and Hess, J.L. (1993) *Antioxidants in Higher Plants*, CRC Press, Boca Raton, pp. 135–169.
Amr, A. (1995) Antioxidative role of some aromatic herbs in refrigerated ground beef patties. *DIRASAT (Pure and Applied Sciences)*, 22 B, 1475–1487.
Arctander, S. (1960) *Perfum and Flavor Materials of Natural Origin*, Elizabeth, New Jersey.
Arras, G. and Grella, G.C. (1992) Wild thyme, *Thymus capitatus*, essential oil seasonal changes and antimycotic activity. *J. Hortic. Sci.*, 67, 197–202.
Arteche, A., Güenechea, J.I., Uriarte, C. and Vanaclotxa, B. (1994) *Fitoterapia. Vademecum de prescripción*, Publicaciones y Documentación, Bilbao.
Aureli, P., Costantini, A. and Zolea, S. (1992) Antimicrobial activity of some plant essential oils against *Listeria monocytogenes. J. Food Prot.*, 55, 344–348.
Austgulen, L.T., Solheim, E. and Scheline R.R. (1987) Metabolism in rats of *p*-cymene derivates: carvacrol and thymol. *Pharmacol. Toxicol.*, 61, 98–102.
Banoczy, J., Gombik, A., Szoke, J. and Nasz, I. (1995) Effect of an antibacterial varnish and amine-fluoride/stannous fluoride (AmF/SnF2) toothpaste on *Streptococcus* mutans counts in saliva and dental plaque of children. *J. Clin. Dent.*, 6, 131–134.
Bardeau, F. (1973) *La pharmacie de bon Dieu*, Stock, Paris, pp. 279–281.

Benigni, R., Capra, C. and Cattorini, P.E. (1964) *Piante Medicinali. Clinica Farmacologia e Terapia*, Inverni & Della Beffa, Miláno, pp. 1618–1628.

Benito, M., Jorro, C., Morales, C., Peláez, A. and Fernández, A. (1996) Labiatae allergy: systemic reactions due to ingestion of oregano and thyme. *Ann. Allerg. Asthma Im.*, 6, 416–418.

Beretz, A., Stoclet, J. and Anton, R. (1980) Inhibition of isolated rat aorta contraction by flavonoids. Possible correlation with cyclic AMP phosphodiesterase inhibition. *Planta Med.*, 39, 236–237.

Blakeway, J. (1986) The antimicrobial properties of essential oils. *Soap Perfum. Cosmet.*, 59, 201–203.

Blázquez, M.A., Zafra-Polo, M.C. and Villar, A. (1989) Effects of *Thymus* species extracts on rat duodenum isolated smooth muscle contraction. *Phytother. Res.*, 3, 41–42.

Blázquez, M.A., Catret, M. and Zafra-Polo, M.C. (1995) Effects on rat uterine and aorta strip smooth muscle of *Thymus leptophyllus* extract. *J. Ethnopharmacol.*, 45, 59–66.

Blumenthal, M. (1998) *The Complete German Commission E Monographs – Therapeutic Guide to Herbal Medicines*, American Botanical Council, Austin, Texas and Integrative Medicine Communications, Boston, Massachusetts.

Botsoglou, N.A., Yannakopoulos, A.L., Fletouris, D.J., Tservenigoussi, A.S. and Fortomaris, P.D. (1997) Effect of dietary thyme on the oxidative stability of egg yolk. *J. Agric. Food Chem.*, 45, 3711–3716.

Brasseur, T. (1983) Etudes botaniques, phytochimiques et pharmacologiques consacrées au thym. *J. Pharm. Belg.*, 38, 261–271.

Buchbauer, G. (1990) Aromatherapy: do essential oils have therapeutic properties? *Perf. &Flav.*, 15 (May/June), 47–50.

Buchbauer, G. (1996) Methods in aromatherapy research. *Perf. &Flav.*, 21 (May/June), pp. 31–36.

Budincevic, M., Vrbaski, Z., Turkulov, J. and Dimic, E. (1995) Antioxidative effect of plant extracts on feed fats. *Fett Wiss. Technol.*, 97, 461–466.

Cabo, J., Jiménez, J., Miro, M. and Toro, M.V. (1978) Determinación de la actividad antimicrobiana de los componentes de la esencia de *Thymus zygis* L. *Pharm. Med.*, 12, 393–399.

Cabo, J., Cabo, M.M., Jiménez, J. and Navarro, C. (1982) *Thymus hyemalis* Lange. III. – Determinación cuali y cuantitativa de la actividad antimicrobiana de su aceite esencial. *Pharm. Med.*, 13, 446–449.

Cabo, J., Cabo, M.M., Crespo, M.E., Jiménez, J. and Zarzuelo, A. (1986a) *Thymus granatensis*. IV. – Pharmacodynamic study of its essential oil. *Fitoterapia*, 57, 173–178.

Cabo, J., Cabo, M.M., Crespo, M.E., Jiménez, J., Navarro, C. and Zarzuelo, A. (1986b) *Thymus granatensis* Boiss. III. – Étude comparative de différents échantillons d'origine géographique divers. *Plant. Méd. Phytothér.*, 20, 135–147.

Cabo, J., Crespo, M.E., Jiménez, J. and Zarzuelo, A. (1986c) The spasmolytic activity of various aromatic plants from the province of Granada. The activity of the major components of their essential oils. *Plant. Méd. Phytothér.*, 20, 213–218.

Cabo, J., Crespo, M.E., Jiménez, J., Navarro, C. and Zarzuelo, A. (1987) A pharmacodynamic study of the *Thymus orospedanus*. *Fitoterapia*, 58, 39–44.

Capasso, A., Pinto, A., Mascolo, N., Autore, G. and Capasso, F. (1991a) Reduction of agonist-induced contractions of guinea-pig isolated by flavonoids. *Phytother. Res.*, 5, 85–87.

Capasso, A., Pinto, A., Sorentino, R. and Capasso, F. (1991b) Inhibitory effects of quercetin and other flavonoids on electrically-induced contractions of guinea-pig ileum. *J. Ethnopharmacol.*, 34, 279–281.

Chung, S.K., Osawa, T. and Kawakishi, S. (1997) Hydroxyl radical-scavenging effects of spices and scavengers from brown mustard (*Brassica nigra*). *Biosci. Biotech. Bioch.*, 61, 118–123.

Conner, D.E. and Beuchat, R.L. (1984) Effects of essential oils from plants on growth of food spoilage yeasts. *J. Food Sci.*, 49, 429–434.

Crespo, M.E., Jiménez, J., Gomis, E. and Navarro, C. (1990) Antibacterial activity of the essential oil of *Thymus serpylloides* subspecies *gadorensis*. *Microbios*, 61, 181–184.

Crespo, M.E., Jiménez, J. and Navarro, C. (1991) Special methods for the essential oils of the genus *Thymus*. In H.F. Linskens and J.F. Jackson (eds), *Modern Methods of Plant Analysis. New Series*, Vol. 12: *Essential Oils and Waxes*, Springer-Verlag, Berlin, pp. 41–61.

Cruz, T., Cabo, M.P., Cabo, M.M., Jiménez, J., Cabo, J. and Ruiz, C. (1989a) *In vitro* antibacterial effect of the essential oil of *Thymus longiflorus* Boiss. *Microbios*, 60, 59–61.

Cruz, T., Jiménez, J., Zarzuelo, A. and Cabo, M.M. (1989b) The spasmolytic activity of the essential oil of *Thymus baeticus* Boiss. in rats. *Phytother. Res.*, 3, 106–109.

Cruz, T., Cabo, M.M., Castillo, M.J., Jiménez, J., Ruiz, C. and Ramos-Cormezana, A. (1993) Chemical composition and antimicrobial activity of the essential oils of different samples of *Thymus baeticus* Boiss. *Phytother. Res.*, 7, 92–94.

Deans, G.G. and Ritchie, G. (1987) Antibacterial properties of plant essential oils. *Int. J. Food Microbiol.*, 5, 165–180.

Deans, G.G., Noble, R.C., Penzes, L. and Imre, G.G. (1993) Promotional effects of plant volatile oils on the polyunsaturated fatty-acid. *Age (Chester, Pa.)*, 16, 71–74.

Debelmas, A.M. and Rochat, J. (1964) Étude comparée sur la fibre lisse de solutions aqueuses saturées d'essence de thym, de thymol et de carvacrol. *Bull. Trav. Soc. Pharm. Lyon*, 8, 163–172.

Debelmas, A.M. and Rochat, J. (1967) Étude pharmacologique des huiles essentielles. Activité antispasmodique étudiée sur une cinquantaine d'echantillons différents. *Plant. Méd. Phytothér.*, 1, 23–27.

Didry, N., Dubreuil, L. and Pinkas, M. (1993) Antibacterial activity of thymol, carvacrol and cinnamaldehyde alone or in combination. *Pharmazie*, 48, 301–304.

Dilaser, M. (1979) Intoxication par le camphre et le menthol par voie trans-cutanée, d'un nourrisson de six semaines. *Bull. Sign.*, 40, 194.

Dorman, H.J., Deans, S.G., Noble, R.C. and Sera, H. (1995) Evaluation *in vitro* of plant essential oils as natural antioxidants. *J. Essent. Oil Res.*, 7, 645–650.

Duarte, J., Pérez-Vizcaino, F., Jiménez, J., Tamargo, J. and Zarzuelo, A. (1993a) Vasodilatory effects of flavonoids in rat aortic smooth muscle. Structure-activity relationships. *Gen. Pharmacol.*, 24, 857–862.

Duarte, J., Pérez-Vizcaino, F., Zarzuelo, A., Jiménez, J. and Tamargo, J. (1993b) Vasodilatadory effects of quercetin in isolated rat vascular smooth muscle. *Eur. J. Pharmacol.*, 239, 1–7.

Dupeyron, J.P., Quattrocchi, F., Castaing, H. and Fabiani, P. (1976) Intoxication aiguë du nourrisson par application cutanée d'une pommade révulsive locale et antiseptique pulmonaire. *Eur. J. Toxicol. Environ. Hyg.*, 9, 313–320.

El-Gengaihi, S.E., Amer, S.A.A. and Mohamed, S.M. (1996) Biological activity of thyme oil and thymol against *Tetranychus urticae* Koch. *Anz. Schädlingskd. Pfl.*, 69, 157–159.

El-Maraghy, S.S.M. (1995) Effect of some species as preservatives for storage of lentil (*Lens esculenta* L.) seeds. *Folia Microbiol.*, 40, 490–495.

Errera, H. (1978) *Como curarse con las plantas*, Argos Bergara, Barcelona, pp. 196–197.

Essen, S. and Karapinar, M. (1986) Sensitivity of some common food poisoning bacteria to thyme, mint and bay leaves. *Int. J. Food Microbiol.*, 3, 349–354.

Farag, R.S., Daw, Z.Y. and Abo-Raya, S.H. (1989) Inluence of some spice essential oils on *Aspergillus parasiticus* growth and production of aflatoxins in a synthetic medium. *J. Food Prot.*, 54, 74–76.

Farag, R.S., Salem, H., Badei, A.Z.M.A. and Hassanein, D.E. (1986) Biochemical studies on the essential oils of some medicinal plants. *Fette Seifen Anstrichm.*, 88, 69–72.

Farag, R.S., Abd-El-Aziz, O., Abd-El-Moein, N.M. and Mohamed, S.M. (1994) Insecticidal activity of thyme and clove essential oils and their basic compounds on cotton leaf worm (*Spodoptera littoralis*). *Bull. Fac. Agric., Cairo*, 45, 207–230.

Fernández, M. and Nieto, A. (1982) *Plantas Medicinales*, Eunsa, Pamplona, pp. 168–170.

Forster, H.B., Niklas, H. and Lutz, S. (1980) Antispasmodic effects of some medicinal plants. *Planta Med.*, 40, 309–319.

Furlenmeier, M. (1984) *Plantas curativas y sus propiedades medicinales*, Schwittez, Zug, Switzerland, pp.168ff.

Gálvez, J., de la Cruz, J.P., Zarzuelo, A. and Sánchez de la Cuesta, F. (1995a) The flavonoid inhibition of enzymic and non-enzymic lipid peroxidation differs from its influence on the glutathione related enzymes. *Pharmacology*, 51, 127–133.

Gálvez, J., de la Cruz, J., Zarzuelo, A., Sánchez de Medina, F., Jiménez, J. and Sánchez de la Cuesta, F. (1995b) Oral administration of quercetin modifies intestinal oxidative status in rat. *Gen. Pharmacol.*, 25, 1237–1243.

Gálvez, J., Duarte, J., Sánchez de Medina, S., Jiménez, J. and Zarzuelo, A. (1996) Inhibitory effects of quercetin on guinea-pig ileum contractions. *Phytother. Res.*, 10, 66–69.

Gámez M.J., Jiménez, J., Navarro, C. and Zarzuelo, A. (1990) Study of the essential oil of *Lavandula dentata* L. *Pharmazie*, 45, 69–70.

Godfraind, T., Miller, R. and Wibbo, M. (1986) Calcium antagonism and calcium entry blockade. *Pharmacol. Rev.*, 38, 324.

González, A. and Muñoz, F. (1980) *Secretos y virtudes de las plantas medicinales*, Selecciones del Reader's Digest, Madrid, p. 277.

Güenechea, J.I. (1992) Fitoterapia y Farmacia Galénica. In A. Arteche, J.I. Güenechea and B. Vanaclotxa (eds), *Fitoterapia. Vademecum de prescripción*, Publicaciones y documentación, Bilbao, pp. 33–48.

Guseinov, D.I., Kagramanova, K.M., Kasumov, F.Yu. and Akhundov, R.A. (1987) Research on the chemical composition and aspects of the pharmacological action on the essential oil of Kochi thyme. *Farmakol. Toksicol.*, 50, 73–74.

Haraguchi, H., Saito, T., Ishikawa, H., Kataoka, S., Tamura, Y. and Mizutani, K. (1996) Antiperoxidative components in *Thymus vulgaris*. *Planta Med.*, 62, 217–221.

Helander, I.M., Alakomi, H.-L., Latva-Kala, K., Mattila-Sandholm, T., Pol, I., Smid, E.J., Gorris, L.G.M. and Von Wright, A. (1998) *J. Agric. Food Chem.*, 46, 3590–3595.

Herrera, M.D., Zarzuelo, A., Jiménez, J., Marhuenda, E. and Duarte, J. (1996) Effects of flavonoids on rat aortic smooth muscle contractility: structure–activity relationship. *Gen. Pharmacol.*, 27, 273–277.

Herrmann, E.C. and Kucera, L.S. (1967) Antiviral substances in plants of the Mint Family (*Labiatae*) III. Peppermint (*Mentha piperita*) and other Mint plants. *Proc. Soc. Exp. Biol. Med.*, 124, 874–878.

Hinou, J.B., Harvala, C.E. and Hinou, E.B. (1989) Antimicrobial activity screening of 32 common constituents of essential oils. *Pharmazie*, 44, 302–303.

Homöopathisches Arzneibuch, Ausgabe 2000, Deutscher Apotheker Verlag, Stuttgart.

Ismaiel, A. and Pierson, M.D. (1990) Inhibition of growth and germination of *C. botulinum* 33A, 40B, and 1623E by essential oil of spices. *J. Food Sci.*, 55, 1676–1678.

Janssen, A.M., Scheffer, J.J.C. and Baerheim Svendsen, A. (1987) Antimicrobial activity of essential oils: A 1976–1986 literature review. Aspects of the test methods. *Planta Med.*, 53, 395–398.

Janssen, A.M., Scheffer, J.J.C., Parhan-van Atten, A.W. and Baerheim Svendsen, A. (1988) Screening of some essential oils for their activities on dermatophytes. *Pharm. Weekbl. Sci.*, 10, 277–280.

Jiménez, J., Zarzuelo, A. and Crespo, M.E. (1988) Hypotensive activity of *Thymus orospedanus* alcoholic extract. *Phytother. Res.*, 2, 152–153.

Juven, B.J., Kanner, J., Schued, F. and Weisslowicz, H. (1994) Factors that interact with the antibacterial action of thyme essential oil and its active constituents. *J. Appl. Bacteriol.*, 76, 626–631.

Kandil, O., Radwan, N.M., Hassan, A.B., Amer, A.M.M., El-Banna, H.A. and Amer, W.M.M. (1994) Extracts and fractions of *Thymus capitatus* exhibit antimicrobial activities. *J. Ethnopharmacol.*, 44, 19–24.

Karpouhtsis, I., Pardali, E., Feggu, E., Kokkini, S., Scouras, Z.G. and Mavraganitsipidou, P. (1998) Insecticidal and genotoxic activities of oregano essential oils. *J. Agric. Food Chem.*, 46, 1111–1115.

Khan, N.N. and Nolan, L.L. (1995) Screening of natural products for antileishmanial chemo-therapeutic potential. *Acta Hortic.*, 426, 47–56.

Knobloch, K., Pauli, A., Iberl, B., Weis, N. and Weigand, H. (1988) Mode of action of essential oil components on whole cell of bacteria and fungi in plate tests. In P. Schreier (ed.), *Bioflavour 87*, de Gruyter, Berlin, New York, pp. 287–299.

Ko, F., Huang, T. and Teng, C. (1991) Vasodilatory action mechanisms of apigenin isolated from *Apium graveolens* in rat thoracic aorta. *Biochim. Biophys. Acta*, 1115, 69–74.

Kowal, T. and Kuprinska, A. (1979) Antibacterial activity of the essential oil from *Thymus pulegioides*. *Herba Pol.*, 25, 303–310.

Lee, S., Tsao, R., Peterson, C. and Coats, J.R. (1997) Insecticidal activity of monoterpenoids to western corn rootworm (*Coleoptera*: Chrysomelidae), twospotted spider mite (*Acari*: Tetranychidae), and house fly (*Diptera*: Muscidae). *J. Econ. Entomol.*, 90, 883–892.

Lemiere, C., Cartier, A., Lehrer, S.B. and Malo, J.L. (1996) Occupational asthma caused by aromatic herbs. *Allergy*, 51, 647–649.

Lens-Lisbonne, C., Cremieux, A., Maillard, C. and Balansard, G. (1987) Methodes d'evaluation de l'activite antibacterienne des huiles essentielles: application aux essences de Thym et de cannelle. *J. Pharm. Belg.*, 42, 297–302.

Loew, D., Habs, M., Klimm, H.-D. and Trunzler, G. (1999) *Phytopharmaka-Report*, 2nd ed., Steinkopf, Darmstadt.

Marsh, P.D. (1992) Microbiological aspects of the chemical control of plaque and gingivitis. *J. Dent. Res.*, 71, 1431–1438.

Megalla, S.E., El-Keltawi, N.E.M. and Ross, S.A. (1980) A study of antimicrobial action of some essential oil constituents. *Herba Pol.*, 26, 181–186.

Melegari, M., Albasini, A., Provisionato, A., Bianchmi, A., Vampa, G., Pecorari, P. and Rinaldi, M. (1985) Richerche su caratteristiche chimie e propietá antibatteriche di olii essenziali di *Satureja montana*. *Fitoterapia*, 56, 85–91.

Menghini, A., Savino, A., Lollini, M.N. and Caprio, A. (1987) Activité antimicrobienne en contact direct et en microatmosphere de certains huiles essentielles. *Plant. Méd. Phytothér.*, 42, 21–36.

Min, B.S., Nakamura, N., Miyashiro, H., Bae, K.W. and Hattori, M. (1998) Triterpenes from the spores of *Ganoderma lucidum* and their inhibitory activity against HIV-1 protease. *Chem. Pharm. Bull. (Tokyo)*, 46, 1607–1612.

Min, B.S., Hattori, M., Lee, H.K. and Kim, Y.H. (1999) Inhibitory constituents against HIV-1 protease from *Agastache rugosa*. *Arch. Pharm. Res.*, 22, 75–77.

Montes, R. and Carvajal, M. (1998) Control of *Aspergillus flavus* in maize with plant essential oils and their components. *J. Food Prot.*, 61, 616–619.

Nakatani, N. (1992) Natural antioxidants from spices. *ACS Symposium Series*, 507, 72–86.

Nelson, R.R. (1997) *In vitro* activities of five plant essential oils against methicillin-resistant *Staphylococcus aureus* and vancomycin-resistant *Enterococcus faecium*. *J. Antimicrob. Chemother.*, 40, 305–306.

Osawa, K., Matsumoto, T., Maruyama, T., Takiguchi, T., Okuda, K. and Takazoe, I. (1990) Studies on the antibacterial activity of plant extracts and their constituents against periodontophatic bacteria. *Bull. Tokyo Dent. Coll.*, 31, 17–21.

Pahlow, M. (1979) *Das große Buch der Heilpflanzen – gesund durch Heilkräfte der Natur*, Gräfe und Unzer, München, pp. 333–334.

Pearson, D.A., Frankel, E.N., Aeschbach, R. and German, J.B. (1997) Inhibition of endothelial cell-mediated oxidation of low-density lipoprotein by rosemary and plant phenolics. *J. Agric. Food Chem.*, 45, 578–582.

Pellecuer, J., Jacob, M., Simeon de Bouchberg, M., Dusart, G., Attisso, M., Barthez, M., Gourgas, L., Pascal, B. and Tomei, B. (1980) Essais d'utilisation d'huiles essentielles de plantes aromatiques méditerranéennes en odontologie conservatrice. *Plant. Méd. Phytothér.*, 14, 83–98.

Penso, G. (1980) *Piante medicinali nella terapia medica*, Organizzazione editoriale medico-farma-ceutica, Milan, pp. 89–213.

Peris, J.B., Stübing, G. and Vanaclocha, B. (1995) *Fitoterapia Aplicada*, M.I.C.O.F., Valencia.

Perrot, E. and Paris, R. (1971) *Les plantes medicinales*, Presses Universitaires de France, Paris, p. 233.

Perrucci, S., Macchioini, G., Cioni, P.L., Flamini, G. and Morelli, I.(1995) Structure–activity relationship of some natural monoterpenes as acaricides against *Psoroptes cuniculi. J. Nat. Prod.*, 58, 1261–1264.

Poletti, A. (1979) *Plantas y Flores Medicinales*, Instituto Parramon, Barcelona, pp. 103–104.

Regnaultroger, C. and Hamraoui, A. (1997) Defense against phytophagic insects by aromatic plants allelochemicals. *Acta Bot. Gall.*, 144, 401–412.

Regnaultroger, C., Hamraoui, A., Holeman, M., Theron, E. and Pinel, R. (1993) Insecticidal effect of essential oils from Mediterranean plants upon *Acanthoscelides obtectus* Say (Coleoptera, Bruchidae), a pest of kydney bean (*Phaseolus vulgaris* L.) *J. Chem. Ecol.*, 19, 1233–1244.

Regnaultroger, C. and Hamraoui, A. (1994) Inhibition of reproduction of *Acanthoscelides obtectus* Say (Coleoptera), a kidney bean (*Phaseolus vulgaris*) bruchid, by aromatic essential oils. *Crop. Prot.*, 13, 624–628.

Regnaultroger, C. and Hamraoui, A. (1995) Comparison of the insecticidal effects of water extracted and intact aromatic plants on *Acanthoscelides obtectus*, a bruchid beetle pest of kidney beans. *Chemoecology*, 5/6, 1–5.

Reiter, M. and Brandt, W. (1985) Relaxant effects on tracheal and ileal smooth muscles of the guinea pig. *Arzneim.-Forsch.*, 35, 408–414.

Reuter, H.D. (1997) *Therapie mit Phytopharmaka: Pharmakologie, Indikationen, Dosierungen*, G. Fischer Verlag, Ulm Stuttgart Jena Lübeck.

Roussel, J.L., Pellecuer, J. and Andary, C. (1973) Propriétés antifongiques comparées des essences de trois labiées méditerranéennes: romarin, sarriette et thym. *Trav. Soc. Pharm. Montp.*, 33, 587–592.

Salmerón, J., Jordano, R. and Pozo, R. (1990) Antimycotic and antiaflatoxigenic activity of oregano (*Origanum vulgare* L.) and thyme (*Thymus vulgaris* L.) *J. Food Prot.*, 53, 697–700.

Schäfer, D. and Schäfer, W. (1981) Pharmacological studies with an ointment containing menthol, camphene and essential oils from broncholytical and secretolytical effects. *Arzneim. – Forsch.*, 31, 82–86.

Schauenberg, P. and Paris, F. (1977) *Guia de las plantas medicinales*, Omega, Barcelona, pp. 316–317.

Schnitzler, A.C., Nolan, L.L. and Labre, R. (1995) Screening of medicinal plants for anti-leishmanial and antimicrobial activity. *Acta Hortic.*, 426, 235–241.

Schröder, V. and Vollmer, H. (1932) The excretion of thymol, carvacrol, eugenol and guaiacol and the distribution of these substances in the organism. *Naunyn Schmiedebergs Arch. Exp. Path. Parmak.*, 168, 331–353.

Schwarz, K., Ernst, H. and Ternes, W. (1996) Evaluation of antioxidative constituents from thyme. *J. Sci. Food Agric.*, 70, 217–233.

Schulz, V. and Hänsel, R. (1996) *Rationale Phytotherapie*, Springer Verlag, Berlin, Heidelbeg, New York.

Shelef, L.A. (1983) Antimicrobial effects of spices. *J. Food Safety*, 6, 29–44.

Simeon de Bouchberg, M., Allegrini, J., Bessiere, C., Attisto, M., Passet, J. and Granger, R. (1976) Propietés microbiologiques des huiles essentielles de chimiotypes de *Thymus vulgaris* L. *Riv. Ital. E.P.P.O.S.*, 58, 527–536.

Smithpalmer, A., Stewart, J. and Fyfe, L. (1998) Antimicrobial properties of plant essential oils and essences against five important food-borne pathogens. *Lett. Appl. Microbiol.*, 26, 118–122.

Steinmetz, M.D., Tognetti, D., Mourgue, M., Jouglard, J. and Millet, Y. (1980) Toxicity of certain commercial essential oils: oil of hyssop and oil of sage. *Plant. Méd. Phytothér.*, 14, 34–35.

Tabak, M., Armon, R., Potasman, I. and Neeman, I. (1996) In vitro inhibition of *Helicobacter pylori* by extracts of thyme. *J. Appl. Bacteriol.*, 80, 667–672.

Takada, M., Agata, I., Sakamoto, M., Yagi, N. and Hayashi, N. (1979) On the metabolic detoxication of thymol in rabbit and man. *J. Toxicol. Sci.*, 4, 341–350.

Tantaoui-Elaraki, A. and Beraoud, L. (1994) Inhibition of growth and aflatoxin production in *Aspergillus parasiticus* by essential oils of selected plant materials. *J. Food Prot.*, 61, 616–619.

Ternes, W., Gronemeyer, M. and Schwarz, K. (1995) Determination of p-cymene-2,3-diol, thymol and carvacrol in different foodstuffs. *Z. Lebensm.-Unters.Forsch.*, 201, 544–577.

Tisserand, R.B. (1980) *Aromatherapie, Heilung durch Duftstoffe*, Verlag H. Bauer, Freiburg/Breisgau.

Tisserand, R. and Balacs, T. (1995) Essential Oil Safety. A guide for health care professionals, Churchill Livingstone, Edinburg.

Valnet, J. (1964) *Aromathérapie. Traitement des maladies par les essences des plantes*, Maloine, Paris, pp. 270–275.

Van den Broucke, C.O. and Lemli, J.A. (1981) Pharmacological and chemical investigation of thyme liquid extract. *Planta Med.*, 41, 129–135.

Van den Broucke, C.O. and Lemli, J.A. (1982) Antispasmodic activity of *Origanum compactum*. Part 2. Antagonist effect of thymol and carvacrol. *Planta Med.*, 45, 188–190.

Van den Broucke, C.O., Lemli, J.A. and Lamy, J. (1982) Action spasmolytique des flavones de différents spèces de *Thymus*. *Plant. Méd. Phytothér.*, 16, 310–317.

Van den Broucke, C.O. and Lemli, J.A. (1983) Spasmolytic activity of the flavonoids from *Thymus vulgaris*. *Pham. Weekbl.*, 5, 9–14.

Volák, J. and Stodola, J. (1989) *El gran libro de las plantas medicinales*, Susaeta, España, pp. 288–289.

Wagner, H. and Wiesenauer, M. (1995) Phytotherapie – Phytopharmaka und pflanzliche Homöopathika. Gustav Fischer Verlag, Stuttgart Jena, New York, p. 95.

William, A.R. and Thomson, D.M. (1980) *Guia práctica ilustrada de las plantas medicinales*, Blume, Barcelona, pp. 104.

Wilson, C.L., Solar, J.M., Elghaouth, A. and Wisniewski, M.E. (1997) Rapid evaluation of plant extracts and essential oils for antifungal activity against *Botrytis cinerea*. *Plant Dis.*, 81, 204–210.

Zafra-Polo, M.C., Blázquez, M.A. and Villar, A. (1990) Relaxant properties of two steroid compounds isolated from *Thymus webbianus*. *Planta Med.*, 56, 685–686.

Zambonelli, A., Daulerio, A.Z., Bianchi, A. and Albasini, A. (1996) Effects of essential oils on phytopathogenic fungi *in vitro*. *J. Phytopathol.*, 144, 491–494.

Zarzuelo, A., Navarro, C., Crespo, M.E., Ocete, M.A., Jiménez, J. and Cabo, J. (1987) Spasmolytic activity of *Thymus membranaceus* essential oil. *Phytother. Res.*, 1, 114–116.

Zarzuelo, A., Cabo, M.M., Cruz, T. and Jiménez, J. (1989) Spasmolytic action of the essential oil of *Thymus longiflorus* Boiss. in rats. *Phytother. Res.*, 3, 36–38.

Zeina, B., Othman, O. and Al-Assad, S. (1996) Effect of honey versus thyme on Rubella virus survival *in vitro*. *J. Altern. Complement. Med.*, 2, 345–348.

11 Thyme as a herbal drug – pharmacopoeias and other product characteristics

Elisabeth Stahl-Biskup

INTRODUCTION

Traditional medicine has been using thyme for many centuries. In the past the plants were collected in the countryside and only the herb collectors were responsible for the quality of the herbs, which differed considerably. Nowadays, in modern phytotherapy, increasing requirements concerning the safety of drugs must be fulfilled. The current status of drugs, including herbal drugs, has to take into consideration various legal regulations so that the products can obtain the "drug status" and be sold in the pharmaceutical market according to the Medicines Acts of European and other countries. An increased use of herbal medicines requires a thorough evaluation of the quality, overall safety and effectiveness of phytomedicines.

The status of herbal drugs is not the same all over the world. In Europe, particularly in Germany, herbs and phytomedicines are accepted and integrated into medicine and pharmacy. In 1978, the German Ministry of Health established the Commission E, a panel of experts charged with evaluating the safety and efficacy of the herbs available in pharmacies for general use. The Commission reviewed over 300 herbal drugs and published its results in the form of monographs in the *Bundesanzeiger*, the German *Federal Gazette*. These monographs provide guidelines for the general public, health practitioners, and companies applying for the registration of herbal drugs.

In the US herbs and phytomedicines are also experiencing explosive growth in pharmacies and other mass-market retail outlets. It is said that the herb sector of the dietary supplement market represents one of the biggest financial investment opportunities since the advent of the high-technology industry. However, such a product cannot make a statement that is deemed 'therapeutic' or imply that is useful to diagnose, treat, cure, or prevent any disease. A petition by European and American phytomedicines manufacturers has requested that the Food and Drug Administration (FDA) grants well-researched European phytomedicines the status of old drugs so that they would not have to be evaluated by the prohibitively costly new drug application process, but the FDA has not responded to this petition.

The edition of the German Commission E Monographs (English version: Blumenthal, 1998) is a result of intensive efforts to preserve and maintain the herbal remedies in the status of drugs in accordance with the German Drug Law. Further intensive efforts on herbal remedies were undertaken by the European Scientific Cooperative on Phytotherapy (ESCOP) as well as by the World Health Organization (WHO). Both organizations evaluated herbal remedies for their therapeutic benefit and safety in order to promote harmonization in the use of herbal medicines with respect to levels of safety, efficacy, and quality control. The herbal drug monographs of all three organizations consider a clear definition of the herbal drug, the effectivity, the side-effects, interactions, toxicological data and dosage. Neither the German Commission E nor the ESCOP monographs contain standards for assaying the quality and purity of herbal drugs. This is left to the Pharmacopoeias, and quality standards can be found in the European Pharmacopoeia (3rd ed. 1997, Supplement 2001) or in the national Pharmacopoeias of Germany (Deutsches Arzneibuch, DAB 2000, and Deutscher Arzneimittel Codex, DAC 2000), the British Pharmacopoeia (BP 2000), the British Herbal Pharmacopoeia (BHP 1979), the Pharmacopée Française (PF X 2000), and the Swiss Pharmacopoeia (Pharmacopoea Helvetica 8, Suppl. 2000). Pharmacopoeial summaries for quality assurance can also be found in the WHO monographs.

Such a detailed introduction is necessary to understand the role of the monographs quoted in the following paragraphs. The fact that thyme has found consideration in the monographs of the Commission E, the ESCOP and the WHO as well as in the European Pharmacopoeia reflects the importance of thyme among the herbal remedies. This is also documented in the various sample product formulations which can be found on the drug market.

MONOGRAPHS FOR THYME IN PHYTOPHARMACY

Commission E monographs

The work of the expert group of the Commission E (1984–1994) is closely connected with the laws regulating the registration of drugs in Germany (German Drug Law). The task of the experts was to evaluate scientific knowledge published on herbal remedies resulting in a decision whether an herbal drug is approved ("positive monographs", 186 monographs, monopreparations only) or not approved ("negative monographs", 110 monographs). The latter category concerns herbals with no plausible evidence of efficacy, or with potential benefits outweighed by safety concerns. Applying for the drug registration of an herbal product the manufacturers can refer to the positive monographs as a proof of effectiveness and safety of their products, but they have to complete their documents by further publications or their own experiments, because the Commission E monographs are no longer up to date.

Among the Commission E Monographs thyme is considered with two approved (positive) monographs: *Thymi herba* (*Thymus vulgaris*) and Serpylli herba (*T. serpyllum*). Both will be quoted in the following paragraphs. The category "uses" shows common thyme to be therapeutically more beneficial than wild thyme.

Commission E: Thyme – Thymi herba *(Thymiankraut)*

Bundesanzeiger, published December 5, 1984; revised March 13, 1990, and December 2, 1992.

> *Composition of drug*: Thyme is constituted of the stripped and dried leaves and flowers of *Thymus vulgaris* L., *Thymus zygis* L. (family Lamiaceae), or both species as well

as their preparations in effective dosage. The herb contains at least 0.5 per cent phenols, calculated as thymol (C10H14O, MW = 150.2) based on dried herb.

Uses: Symptoms of bronchitis and whooping cough. Catarrhs of the upper respiratory tract.

Contraindications, side effects, interaction with other drugs: None known.

Dosage: Unless otherwise prescribed: 1 to 2 g of herb for 1 cup of tea, several times a day as needed; 1 to 2 g fluid-extract, 1 to 3 times daily; 5 per cent infusion for compresses.

Mode of administration: Cut herb, powder, liquid extract or dry extract for infusions and other galenical preparations. Liquid and solid medicinal forms for internal and external application. Note: Combinations with other herbs that have expectorant action could be appropriate.

Actions: Bronchoantispasmotic, expectorant, antibacterial.

Commission E: Wild thyme herb – Serpylli herba (Quendelkraut)

Bundesanzeiger, published October 15, 1987; revised March 13, 1990.

Composition of drug: Wild thyme consists of the dried, flowering, above-ground parts of *Thymus serpyllum* L. (family Lamiaceae), as well as its preparations in effective dosage. The drug contains essential oil, principally carvacrol and/or thymol.

Uses: Catarrhs of the upper respiratory tract.

Contraindications, side effects, interaction with other drugs: None known.

Dosage: Unless otherwise prescribed: Average daily dose: 6 g of herb; equivalent preparations.

Mode of administration: Cut herb for infusions and other preparations for internal use.

Action: Antimicrobial

(Ed. Note: Commercially, *Thymus pulegioides* L. and *T. praecox* Opiz subsp. *arcticus* (Dur.) Jalas are also offered as mixed with *T. serpyllum* L.)

ESCOP monographs

The ESCOP was founded in 1989 in Cologne in order to harmonise the evaluation criteria for phytomedicines in Europe. This scientific committee, which includes experts on phytotherapy of all members of the European Union (EU), had the assignment to develop monograph drafts on herbal drugs as guidelines for the European market of herbal drugs. Up to date 60 monograph drafts in the form of Summaries of Product Characteristics have been published in the time from 1994 to 1999, including one monograph for thyme – Thymi herba. It is the result of 38 substantial publications on thyme which are quoted at the end of the monograph.

ESCOP-proposal: Thyme – Thymi herba

Published by ESCOP, 1996.

Name of the medicinal product: To be specified for the individual finished product.

Qualitative and quantitative composition

Active ingredient: Thyme in the crude or processed state in appropriate dosage units.

Definition: Thyme consists of the whole leaves and flowers separated from the previously dried stems of *Thymus vulgaris* L. or *Thymus zygis* Loefl. ex L. or a mixture of both species. It contains not less than 1.2 per cent (V/m) of essential oil and not less than 0.5 per cent of volatile phenols, expressed as thymol ($C10H14O$; M_r 150.2), both calculated with reference to the anhydrous drug.

The material complies with the European Pharmacopoeia. Fresh material may also be used, provided that when dried it complies with the European Pharmacopoeia.

Constituents: Essential oil containing phenols, predominantly thymol and/or carvacrol, and terpenoids (Weiss and Flück, 1970; Adzet *et al.*, 1977; Stahl-Biskup, 1991); glycosides of phenolic monoterpenoids, eugenol and aliphatic alcohols (Skopp and Hörster, 1976; Van den Dries and Baerheim Svendsen, 1989); flavonoids, among which thymonin, cirsilineol and 8-methoxy-cirsilineol are characteristic (Van den Broucke *et al.*, 1982; Adzet *et al.*, 1988) biphenyl compounds of monoterpenoid origin (Nakatani *et al.*, 1989); caffeic and rosmarinic acid (Hegnauer, 1966; Litvinenko *et al.*, 1975; Lamaison *et al.*, 1990); saponins (García Marquina and Gallardo Villa, 1949; Hegnauer, 1966).

Pharmaceutical form

Crude or processed drug in appropriate dosage forms (to be specified for the individual finished product).

Clinical particulars

Therapeutic indications: Catarrh of the upper respiratory tract, bronchial catarrh and pertussis (whooping cough). Stomatitis and halitosis (Czygan, 1989).

Posology and method of administration: Dosage (internal use). Herb: Adults and children from 1 year: 1 to 2 g of the dried herb or the equivalent amount of fresh herb as an infusion several times a day (Van Hellemont, 1988; Czygan, 1989; Dorsch *et al.*, 1993); children up to 1 year: 0.5 to 1 g (Dorsch *et al.*, 1993). Fluid extract: Adults and children: Dependant on the herb-extract ratio dosage to be calculated according to the dosage to the herb (Hochsinger, 1931). Tincture (1:10, 70 per cent ethanol): 40 drops up to three times daily (Van Hellemont, 1988). Other preparations accordingly.

Dosage (topical use). A 5 per cent infusion as a gargle or mouth-wash (Van Hellemont, 1988; Czygan, 1989).

Method of administration: For oral or topical administration.

Duration of administration: No restriction.

Contraindications: None known.

Special warnings and special precautions for use: None required.

Interactions with other medicaments and other forms of interaction: None reported.

Pregnancy and lactation: No data available. In accordance with general medical practice, the product should not be used during pregnancy and lactation without medical advice.

Effects on ability to drive and use machines: None known.

Undesirable effects: None reported.

Overdose: No toxic effects reported.

Pharmacological properties

Pharmacodynamic properties: *In vitro* experiments: Bronchospasmolysis is attributed to the flavonoids, thymonin, cirsilineol and 8-methoxycirsilineol, shown to be potent spasmolytics by *in vitro* experiments in guinea-pig trachea (Van den Broucke *et al.*, 1983; Van den Broucke and Lemli, 1983).

The essential oil is highly antibacterial and antifungal, when tested in Gram-positive and Gram-negative bacteria, fungi, and yeasts, e.g. *Candida albicans*. The activity is mainly attributed to thymol and carvacrol (Allegrini and Simeon de Bouchberg, 1972; Patákova and Chládek, 1974; Simeon de Bouchberg *et al.*, 1976; Farag *et al.*, 1986; Janssen *et al.*, 1986; Lens-Lisbonne *et al.*, 1987; Menghini *et al.*, 1987; Deans and Ritchie, 1987; Vampa *et al.*, 1988; Janssen, 1989; Chalchat and Garry, 1991). Thyme oil inhibits prostaglandin biosynthesis (Wagner *et al.*, 1986). Rosmarinic acid has anti-inflammatory activity due to inhibition of classical complement pathway in rats and inhibition of some human PMN functions, when tested at several dosage levels and by several application methods (Gracza *et al.*, 1985; Englberger *et al.*, 1988). *In vivo* experiments: Rosmarinic acid exhibited inhibitory activity in three *in vivo* models in which complement activation plays a role: reduction of oedema induced by cobra venom factor in the rat; inhibition of passive cutaneous anaphylaxis; impairment of *in vivo* activation by heat-killed *Corynebacterium parvum* of mouse macrophages. Rosmarinic acid did not inhibit t-butylhydroperoxide-induced paw oedema in rat, indicating selectivity for complement-dependent processes (Englberger *et al.*, 1988).

Pharmacokinetic properties: No data available.

Preclinical safety data: Acute toxicity: A concentrated extract produced decreased locomotor activity and slight slowing down of respiration in mice in an acute toxicity test. Oral doses were 0.5 to 3.0 g extract/kg body weight corresponding to 4.3 to 26.0 g dried plant material and these effects were produced at all dose levels (Qureshi *et al.*, 1991). The LD50 of the essential oil is 2.84 g/kg body weight in rats (Von Skramlik, 1959).

Subchronic toxicity: An increase in liver and testes weight was observed after oral administration of a concentrated 95 per cent ethanol extract of plant material to mice. A dose corresponding to 0.9 g dried plant was administered daily for three months. 30 per cent of the male animals died while in the female and control group only 10 per cent died (Qureshi *et al.*, 1991).

Mutagenicity: Thyme oil had no mutagenic or DNA-damaging activity in either the Ames or *Bacillus subtilis* rec-assay (Zani *et al.*, 1991).

WHO monographs

During the fourth International Conference of Drug Regulatory Authorities (ICDRA) held in Tokyo in 1986, WHO was requested to compile a list of medicinal plants and to establish international specifications for the most widely used medicinal plants and simple preparations. Guidelines for the assessment of herbal medicines were subsequently prepared by WHO. As a result of ICDRA recommendations and in response to requests from WHO member States for assistance in providing safe and effective herbal medicines for use in national health-care systems, WHO has published 28 monographs on selected medicinal plants which are widely used and important in all regions, and for each sufficient scientific information seemed available to substantiate safety and efficacy. One of these monographs deals with thyme (*Thymus vulgaris, T. zygis*). In 1994 an advisory group selected thyme to be an important plant in all WHO regions and recommended an evaluation of 38 substantial publications (quoted within the monograph). The monograph also includes the quality standards of the Pharmacopoeias (Part 1). In the following the paragraphs concerning the clinical applications, pharmacology, contraindications, warnings, precautions, potential adverse reactions, and posology are quoted (Part 2).

WHO Monograph: Herba Thymi (Part 2)

Published in 1999.

Definition: Herba Thymi is the dried leaves and flowering tops of *Thymus vulgaris* L. or *Thymus zygis* Loefl. ex L. (Lamiaceae) (Pharmacopoeia Europaea 1995, Materia medika Indonesia 1980).

Selected vernacular names: Common thyme, tomillo, farigola, garden thyme, herba timi, herba thymi, mother of thyme, red thyme, rubbed thyme, ten, thick leaf thyme, thym, Thymian, thyme, time, timi, za ater (Youngken, 1950; British Herbal Pharmacopoeia 1979; Ghazanfar, 1994; Farnsworth, 1995; Pharmacopoeia Europaea 1995; Deutsches Arzneibuch 1996).

Description: An aromatic perennial sub-shrub, 20 to 30 cm in height, with ascending, quadrangular, greyish brown to purplish brown lignified and twisted stems bearing oblong-lanceolate to ovate-lanceolate greyish green leaves that are pubescent on the lower surface. The flowers have a pubescent calyx and a bilobate, pinkish or whitish, corolla and are borne in verticillasters. The fruit consists of 4 brown ovoid nutlets (Youngken, 1950; Mossa *et al.*, 1987; Bruneton, 1995).

Major chemical constituents: Herba Thymi contains about 2.5 per cent but not less than 1.0 per cent of volatile oil. The composition of the volatile oil fluctuated depending on the chemotype under consideration. The principal components of Herba Thymi are thymol and carvacrol (up to 64 per cent of oil), along with linalool, p-cymol, cymene, thymene, α-pinene, apigenin, luteolin, and 6-hydroxyluteolin glycosides, as well as di-, tri- and tetramethoxylated flavones, all substituted in the 6-position (for example 5,4'-dihydroxy-6,7-dimethoxyflavone, 5,4'-dihydroxy-6,7,3'-trimethoxyflavone and its 8-methoxylated derivative 5,6,4'-trihydroxy-7,8,3'-

trimethoxyflavone) (Youngken, 1950; British Herbal Pharmacopoeia 1979; Mossa *et al.*, 1987; Ghazanfar, 1994; Pharmacopoeia Europaea 1995; Deutsches Arzneibuch 1996).

Dosage forms: Dried herb for infusion, extracts, and tincture (Pharmacopoeia Europaea 1995).

Medicinal uses: Uses supported by clinical data: None.

Uses described in pharmacopoeias and in traditional systems of medicine:
Thyme extract has been used orally to treat dyspepsia and other gastrointestinal disturbances; coughs due to colds, bronchitis and pertussis; and laryngitis and tonsillitis (as a gargle). Topical applications of thyme extract have been used in the treatment of minor wounds, the common cold, disorders of the oral cavity, and as an antibacterial agent in oral hygiene (Youngken, 1950; British Herbal Pharmacopoeia 1979; Petersson *et al.*, 1992; Bruneton, 1995; Twetman *et al.*, 1995). Both the essential oil and thymol are ingredients of a number of proprietary drugs including antiseptic and healing ointments, syrups for the treatment of respiratory disorders, and preparations for inhalation. Another species in the genus, *T. serpyllum* L., is used for the same indications (Bruneton, 1995).

Uses described in folk medicine, not supported by experimental or clinical data: as an emmenagous, sedative, antiseptic, antipyretic, to control menstruation and cramps, and in the treatment of dermatitis (Farnsworth, 1995).

Pharmacology – experimental pharmacology

Spasmolytic and antitussive activities: The spasmolytic and antitussive activity of thyme has been most often attributed to the phenolic constituents thymol and carvacrol, which make up a large percentage of the volatile oil (Reiter and Brand, 1985). Although these compounds have been shown to prevent contractions induced in the ileum and the trachea of the guinea pig, by histamine, acetylcholine and other reagents, the concentration of phenolics in aqueous preparations of the drug is insufficient to account for this activity (Van den Broucke, 1980; Van den Broucke and Lemli, 1981). Experimental evidence suggests that the *in vitro* spasmolytic activity of thyme preparations is due to the presence of polymethoxyflavones (Van den Broucke and Lemli, 1983). *In vitro* studies have shown that flavones and thyme extracts inhibit responses to agonists of specific receptors such as acetylcholine, histamine and L-norepinephrine, as well as agents whose actions do not require specific receptors, such as $BaCl_2$ (Van den Broucke and Lemli, 1983). The flavones of thyme were found to act as non-competitive and non-specific antagonists (Van den Broucke and Lemli, 1983); they were also shown to be Ca^{2+} antagonists and musculotropic agents that act directly on smooth muscle (Van den Broucke and Lemli, 1983).

Expectorant and secretomotor activities: Experimental evidence suggests that thyme oil has secretomotoric activity (Gordonoff and Merz, 1931). This activity has been associated with a saponin extract from *T. vulgaris* (Vollmer, 1932). Stimulation of ciliary movements in the pharynx mucosa of frogs treated with diluted solutions of thyme oil, thymol or carvacrol has also been reported (Freytag, 1933). Furthermore, an increase in mucus secretion of the bronchi after treatment with thyme extracts has been observed (Schilf, 1932).

Antifungal and antibacterial activities: *In vitro* studies have shown that both thyme essential oil and thymol have antifungal activity against a number of fungi, including *Cryptococcus neoformans*, *Aspergillus*, *Saprolegnia*, and *Zygorhynchus* species (Tantaoui-Elaraki and Errifi, 1994; Vollon and Chaumont, 1994; Perrucci *et al.*, 1995; Paster *et al.*, 1995). Both the essential oil and thymol had antibacterial activity against *Salmonella typhimurium*, *Staphylococcus aureus*, *Escherichia coli*, and a number of other bacterial species (Janssen *et al.*, 1987; Juven *et al.*, 1994). As an antibiotic, thymol is 25 times more effective than phenol, but less toxic (Czygan, 1989).

Contraindications: Pregnancy and lactation (See Precautions, below).

Warnings: No information available.

Precautions: General: Patients with a known sensitivity to plants in the Lamiaceae (Labiatae) should contact their physician before using thyme preparations. Patients sensitive to birch pollen or celery may have a cross-sensitivity to thyme (Wüthrich *et al.*, 1992).

Carcinogenesis, mutagenesis, impairment of fertility: Thyme essential oil did not have any mutagenic activity in the *Bacillus subtilis* rec-assay or the *Salmonella* microsome reversion assay (Zani *et al.*, 1991; Azizan and Blevins, 1995). Recent investigations suggest that thyme extracts are antimutagenic (Natake, 1989) and that luteolin, a constituent of thyme, is a strong antimutagen against the dietary carcinogen Trp-P-2 (Samejima *et al.*, 1995).

Pregnancy: non-teratogenic effects.

The safety of Herba Thymi preparations during pregnancy or lactation has not been established. As a precautionary measure, the drug should not be used during pregnancy or lactation except on medical advice. However, widespread use of Herba Thymi has not resulted in any safety concerns.

Nursing mothers: (See pregnancy – non-teratogenic effects, above).

Other precautions: No information available concerning drug interactions, drug and laboratory test interactions, paediatric use, or teratogenic effects on pregnancy.

Adverse reactions: Contact dermatitis has been reported. Patients sensitive to birch pollen or celery may have a cross-sensitivity to thyme (Wüthrich *et al.*, 1992).

Posology: Adults and children from 1 year: 1 to 2 g of the dried herb or the equivalent amount of fresh herb as an oral infusion several times a day (Czygan, 1989; Dorsch *et al.*, 1993); children up to 1 year: 0.5 to 1 g (Dorsch *et al.*, 1993). Fluid-extract: dosage calculated according to the dosage of the herb (Hochsinger, 1931). Tincture: (1:10, 70 per cent ethanol): 40 drops up to 3 times daily (Van Hellemont, 1988).

Topical use: a 5 per cent infusion as a gargle or mouth-wash (Van Hellemont, 1988; Czygan, 1989).

QUALITY CONTROL – THYME IN THE PHARMACOPOEIAS

According to all drug regulations, drugs (including herbal drugs) have to fulfil a high qualitative standard and they are subject to a strict quality control. Quality standards are given in the pharmacopoeias. The pharmacopoeias comprise pharmaceutical rules acknowledged concerning quality, assay, storage, dispensation and labelling of drug and drug raw materials. The fulfilling of these quality requirements is obligatory when drugs are produced or treated respectively.

On the pharmacopoeial level thyme is represented by 4 monographs. The Pharmacopoeia Europaea (3rd edition and Supplement 2001), which is in force in most of the European countries, contains 2 monographs, namely *Thymi herba* (thyme herb) and *Thymi aetheroleum* (thyme oil), the German Pharmacopoeia (DAB 2000) contains another 2 monographs, namely *Serpylli herba* (Quendelkraut) and *Thymi extractum fluidum* (Thymianfluidextrakt). In the Swiss Pharmacopoeia (Pharmacopoea Helvetica 8, Supplement 2000) a monograph on *Thymi extractum liquidum normatum* (Eingestellter Thymianliquidextrakt) as well as one on *Thymi sirupus* (Thymiansirup) can be found. The United States Pharmacopoeia (USP) has no record of thyme. As mentioned above the WHO monograph on thyme also includes the pharmacopoeial quality standards (Part 1).

European Pharmacopoeia (Ph. Eur. 2001)

Thyme (Thymi herba)

Definition: Thyme consists of the whole leaves and flowers separated from the previously dried stems of *Thymus vulgaris* L. or *Thymus zygis* Loefl. ex L. or a mixture of both species. It contains not less than 12 ml/kg of essential oil and not less than 0.5 per cent m/m volatile phenols, expressed as thymol ($C_{10}H_{14}O$; M 150.2), both calculated with reference to the anhydrous drug.

Characters: Thyme has a strong aromatic odour reminiscent of thymol. It has the macroscopic and microscopic characters described under identification tests A and B.

Identification: A. The leaf of *Thymus vulgaris* is usually 4 mm to 12 mm long and up to 3 mm wide: it is sessile or has a very short petiole. The lamina is tough, entire, lanceolate to ovate, covered on both surfaces by a grey to greenish-grey indumentum; the edges are markedly rolled up towards the abaxial surface. The midrib is depressed on the adaxial surface and is very prominent on the abaxial surface. The calyx is green, often with violet spots and is tubular; at the end are two lips of which the upper one is bent back and at the end has three lobes, the lower is longer and has two hairy teeth. After flowering, the calyx tube is closed by a crown of long, stiff hairs. The corolla, about twice as long as the calyx, is usually brownish in the dry state and is slightly bilabiate. The leaf of *Thymus zygis* is usually 1.7 mm to 6.5 mm long and 0.4 mm to 1.2 mm wide; it is acicular to linear-lanceolate and the edges are markedly rolled towards the abaxial surface. Both surfaces of the lamina are green to greenish-grey and the midrib is sometimes violet; the edges, in particular at the base, have long, white hairs. The dried flowers are very similar to those of *Thymus vulgaris*.

B. Reduce to a powder. The powder of the two species is greyish-green to greenish-brown. Examine under a microscope using chloral hydrate solution. The epidermises of the leaves have cells with anticlinal walls which are sinuous and beaded and the stomata are of the diacytic type; numerous secretory trichomes made up of twelve secretory cells, the cuticle of which is generally raised by the secretion to form a globular to ovoid bladder-like covering; the glandular trichomes have a unicellular stalk and a globular to ovoid head; the covering trichomes of the adaxial surface are common to both species; they have warty walls and are shaped as pointed teeth; the warty covering trichomes of the abaxial surface are of many types: unicellular, straight or slightly curved, and bicellular or tricellular, and often elbow-shaped (*Thymus vulgaris*); bicellular or tricellular, more or less straight (*Thymus zygis*). Fragments of calyx are covered by numerous, uniseriate trichomes with five or six cells and with a weakly striated cuticle. Fragments of the corolla have numerous uniseriate covering trichomes, often collapsed, and secretory trichomes with generally twelve cells. Pollen grains are relatively rare, spherical and smooth with six germinal slit-like pores, measuring about 35 µm in diameter. The powder of *Thymus zygis* also contains numerous thick bundles of fibres from the main veins and from fragments of stems.

C. Examine by thin-layer chromatography, using silica gel with a fluorescent indicator having an optimal intensity at 254 nm as the coating substance.

Test solution: To 1.0 g of the powdered drug add 5 ml of methylene chloride and shake for 3 min, filter through about 2 g of anhydrous sodium sulphate. Use the filtrate as the test solution. Reference solution: Dissolve 5 mg of thymol and 10 µl of carvacrol in 10 ml of methylene chloride.

Apply separately to the plate as bands, 20 µl of each solution. Develop twice over a path of 12 cm using methylene chloride. Allow the plate to dry in air and examine in ultraviolet light at 254 nm. Mark the quenching zones. The chromatograms obtained with the reference solution and the test solution show in the central part a quenching zone due to thymol. The chromatogram obtained with the test solution shows slightly above the zone due to thymol a prominent quenching zone and other quenching zones in the lower third of the chromatogram. Spray with anisaldehyde solution using 10 ml for a plate 200 mm square and heat at 100 °C to 105 °C for 10 min. The chromatogram obtained with the reference solution shows in the central part a brownish-pink zone corresponding to thymol and, immediately below it, a pale violet zone corresponding with the test solution to carvacrol. The chromatogram obtained with the test solution shows these two zones in the central part of the plate; they are more or less prominent, depending upon the species examined. Between these two zones and the starting-line are four zones of similar intensity; in order of decreasing Rf value these bands are: pink, violet (1,8-cineole and linalool), greyish-brown (borneol) and violet-blue. Near the solvent front, an intense violet-red to greyish-violet band is visible. Other bands are also present adjacent to the starting-line.

Tests: Foreign matter: Not more than 10 per cent of stem. Stems must not be more than 1 mm in diameter and 15 mm in length. Leaves with long trichomes at their base and with weakly pubescent other parts are not allowed (*Thymus serpyllum* L.).

Water: Not more than 10.0 per cent, determined by distillation of 20.0 g of powdered drug.

Total ash: Not more than 15.0 per cent.

Ash insoluble in hydrochloric acid: Not more than 3.0 per cent.

Assay: Essential oil: Carry out the determination of essential oils in vegetable drugs. Use 30.0 g of the drug, a 1000 ml round-bottomed flask and 400 ml of water as the distillation liquid. Distill at a rate of 2 ml/min to 3 ml/min for 2 h without xylene in the graduated tube.

Phenols: Taking care that as little water as possible is transferred, transfer the essential oil obtained in the assay of essential oil to a 50.0 ml volumetric flask with the aid of small portions of alcohol (90 per cent V/V) rinsing the graduated tube of the apparatus with the same solvent and dilute to 50.0 ml with the same solvent. To 5.0 ml of the solution add 40 ml of alcohol (90 per cent (V/V) and dilute to 100.0 ml with water. Place 5.0 ml of the solution in a separating funnel and add 45 ml of water, 0.5 ml of dilute ammonia and 1 ml of a 20 g/l solution of aminopyrazolone. Mix and add 4 ml of a freshly prepared 20 g/l solution of potassium ferricyanide and mix again. Allow to stand for 5 min, add 25 ml of methylene chloride and shake. Separate the methylene chloride layer and filter through a plug of absorbent cotton moistened with methylene chloride into a 100 ml volumetric flask. Shake the aqueous layer with two quantities, each of 25 ml, and with 10 ml of methylene chloride, filter the methylene chloride layers through the plug of absorbent cotton. Rinse the plug with methylene chloride and dilute to 100.0 ml with the same solvent. Measure the absorbance at 450 nm using methylene chloride as the compensation liquid.

Calculate the percentage content of phenols, expressed as thymol, taking the specific absorbance to be 805.

Storage: Store in a well-closed container, protected from light and moisture.

Thyme oil (Thymi aetheroleum)

Definition: Thyme oil is obtained by steam distillation from the fresh flowering aerial parts of *Thymus vulgaris* L., *T. zygis* Loefl. ex L. or a mixture of the two species.

Characters: A. clear, yellow or very dark reddish-brown, mobile liquid with a characteristic aromatic, spicy odour, reminiscent of thymol, miscible with ethanol, with ether and with petroleum ether.

Identification: A. Examine by thin-layer chromatography, using silica gel as the coating substance.

Test solution: Dissolve 0.2 g of the substance to be examined in pentane and dilute to 10 ml with the same solvent. Reference solution: Dissolve 0.15 g of thymol, 25 μl of terpinen-4-ol and 40 μl of linalool in pentane R and dilute to 10 ml with the same solvent.

Apply separately to the plate as bands of 20 μl of each solution. Develop over a path of 15 cm using a mixture of 5 volumes of ethyl acetate and 95 volumes of toluene. Allow the plate to dry in air. Spray with anisaldehyde solution. Heat the plate at 100 °C to 105 °C for 5 min to 10 min while observing. Examine in daylight. The chromatogram obtained with the test solution shows three zones similar in position and colour to those in the chromatogram obtained with the reference solution: a violet zone corresponding to terpinen-4-ol, a violet zone corresponding to linalool and a brownish-pink zone corresponding to thymol and immediately below it a pale-violet zone corresponding to carvacrol. It also shows a large violet zone at the solvent front (hydrocarbons).

B. Examine the chromatograms obtained in the test for "chromatographic profile" (see below). The retention times of the principal peaks in the chromatogram obtained with the test solution are similar to those of the peaks in the chromatogram obtained with the reference solution.

Tests: Relative density: 0.915 to 0.935.

Refractive index: 1.490 to 1.505.

Chromatographic profile: examine by gas chromatography.

Test solution: The substance to be examined. Reference solution: Dissolve 0.15 g of β-myrcene, 0.1 g of γ-terpinene, 0.1 g of *p*-cymene, 0.1 g of linalool, 0.2 g of terpinen-4-ol, 0.2 g of thymol and 0.05 g of carvacrol in 5 ml of hexane.

The chromatographic procedures may be carried out using:

- a fused-silica column 25 m to 60 m long and about 0.3 mm in internal diameter coated with macrogol 20 000.
- helium for chromatography as the carrier gas (other gases can also be used, the author).
- a flame-ionisation detector,
- a split ratio of 1:100,

maintaining the temperature of the column at 60 °C for 15 min, then raising the temperature at a rate of 3 °C per min to 180 °C and maintaining at 180 °C; maintaining the temperature of the injection port at about 200 °C and that of the detector at 220 °C.

Inject about 0.2 μl of the reference solution. When the chromatograms are recorded in the prescribed conditions, the components elute in the order indicated in the composition of the reference solution. Record the retention times of these substances.

The test is not valid unless: the number of theoretical plates calculated from the *p*-cymene peak at 80 °C is at least 30 000; the resolution between the peaks corresponding to thymol and carvacrol is at least 1.5.

Inject about 0.2 μl of the test solution. Using the retention times determined from the chromatogram obtained with the reference solution, locate the components of the reference solution on the chromatogram (Figure 11.1) obtained with the test solution. Disregard the peak due to hexane.

Determine the percentage content of the components of the normalisation procedure.

The percentages range between the following values:

β-myrcene	1.0–3.0 per cent
γ-terpinene	5.0–10.0 per cent
p-cymene	15.0–28.0 per cent
linalool	4.0–6.5 per cent
terpinen-4-ol	0.2–2.5 per cent
thymol	36.0–55.0 per cent
carvacrol	1.0–4.0 per cent.

Storage: Store in a well-filled, air-tight container, protected from light and heat.

German Pharmacopoeia – DAB 2000

Serpylli herba – Quendelkraut

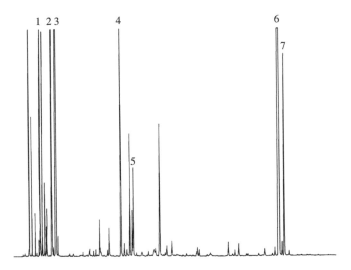

Figure 11.1 Gas chromatogram of thyme oil (*T. vulgaris*) according to the Pharmacopoeia Europaea, on Macrogol 20 000, 30 m.

Notes

1, β-myrcene.

2, γ-terpinene.

3, *p*-cymene.

4, linalool.

5, terpinen-4-ol.

6, thymol.

7, carvacrol.

Definition: It consists of the whole or cut dried aerial parts of *Thymus serpyllum* L. s.l. collected in blossom. It contains not less than 3 ml/kg essential oil and not less than 0.1 per cent phenols, expressed as thymol ($C_{10}H_{14}O$; M 150.2), calculated with reference to the anhydrous drug.

Characteristics: The herb has a characteristic, aromatic odour.

Identity: A. Morphological characteristics: Stem in the cross-section vaguely quadrangular to cylindrical, about 1.5 mm thick, weakly woody at base, side shoots thinner. Leaves decussate, usually 3 mm to 12 mm long and up to 7 mm wide, entire, lanceolate to ovate, wedge-shapedly ending in a short petiole, the edges are seldom rolled up towards the abaxial surface; petiole and leaf basis often ciliate; on both sides of the leaf glandular hairs (magnifying glass). Flowers crowded into a terminal capitate inflorescence; tubular calyx has two lips, the lower with two teeth, the upper one has three lobes; the calyx tube is closed by a crown of long, stiff hairs; the corolla is bilabiate, purple to light-red, wrinkled; four protruding stamens or only four rudiments.

B. Microscopical characteristics: see "Thyme herb" in the European Pharmacopoeia.

C. See "Thyme herb" in the European Pharmacopoeia; thin-layer chromatography of the essential oil on silica gel; reference: thymol. Mobile phase: toluene + ethyl acetate (93 + 7). Detection UV 254 nm and daylight after spraying with anisaldehyde-

H_2SO_4-reagent (a mixture of 0.5 ml anisaldehyde, 10 ml acetic acid, 85 ml methanol, and 5 ml H_2SO_4).

Purity: Foreign matters: Not more than three per cent

Water: Not more than 10.0 per cent after 2 h at 100 to 105 °C.

Total ash: Not more than 10.0 per cent.

Acid-insoluble ash: Not more than 2.0 per cent

Assays: Essential oil and phenols: see Thyme herb in the European Pharmacopoeia.

Storage: In air-tight and light-protected containers.

Thymi extractum fluidum – Liquid thyme extract – Thymianfluidextrakt

Definition: Thymi extractum fluidum contains not less than 0.03 per cent phenols, expressed as thymol (C10H14O; M 150.2).

Production: Thymi extractum fluidum is produced from 1 part freshly powdered thyme extracted with 2–3 parts of a mixture of 1 part ammonia 10 per cent (m/m), 20 parts glycerol 85 per cent, 70 parts ethanol 90 per cent (V/V) and 109 parts purified water by maceration.

Characteristics: Dark brown liquid with an odour of thymol and a spicy, weakly burning taste.

Identity: Thin-layer chromatography on silica gel of an extract from 5 ml liquid thyme extract with 3 ml methylene chloride (application 20 µl); reference: thymol (0.02 per cent in methylene chloride, application 20 µl). Mobile phase: methylene chloride. Detection UV 254 nm and daylight after spraying with anisaldehyde-H_2SO_4-reagent (see "Thyme herb").

Purity: Ethanol content (of the extract): 30–37 per cent (V/V)
Methanol, 2-propanol: content according to the permitted percentages in extracts [see Pharmacopoeia Europaea, the author].

Assay: 20.0 g liquid extract are mixed with 80 ml water and distilled until 85 ml have condensed in 10.0 ml ethanol 96 per cent prepared in a 100 ml volumetric flask. Top up with water to 100.0 ml. 5.0 ml of this solution are mixed with 45 ml water, 0.5 ml liquid ammonia, 1 ml aminopyrazolone (20 g/l). Mix and add 4 ml of a freshly prepared 20 g/l solution of potassium ferricyanide and mix again. Allow to stand for 5 min, add 25 ml of chloroform and shake. Separate the chloroform layer and filter through a plug of absorbent cotton moistened with chloroform into a 100 ml volumetric flask. Shake the aqueous layer with two quantities, each of 25 ml, and with 10 ml of chloroform, filter the chloroform layers through the plug of absorbent cotton. Rinse the plug with chloroform and top up to 100.0 ml with the same solvent. Measure the absorbance at 450 nm using chloroform as the compensation liquid.

The reference is prepared with 10.0 mg thymol in 25 ml Ethanol 96 per cent and solved in water to 100.0 ml. 5 ml of this solution is treated as described above. The calculation of the phenol content is performed according to the following formula:

$$x\% = \frac{100 \times A_1(m_2)}{A_1 \times m_1}$$

A_1 = Absorption of the test solution
A_2 = Absorption of the reference
m_1 = weight of liquid extract (g)
m_2 = weight of thymol (g)

(The blue colour develops according to the Emerson reaction, the author).

Storage: In tight vessels and light protected.

Swiss Pharmacopoeia – Pharmacopoea Helvetica (2000)

Thymi extractum liquidum normatum – preparative thyme extract – Eingestellter Thymianextrakt

Definition: Thymi extractum fluidum normatum contains not less than 0.025 and not more than 0.035 per cent phenols, expressed as thymol ($C10H14O$; M 150.2).

Production: Thyme 100 g are macerated with 300 g of a mixture of 1.5 parts of ammonia 10 per cent, 30 parts glycerol, 105 parts ethanol and 163.5 parts purified water over 5 days at room temperature. The liquid is pressed and filtered. In the filtrate the phenol content is quantified and adjusted with a mixture of 3.5 parts ethanol 96 per cent and 6.5 parts purified water at the required content.

Characteristics: Brown to dark brown liquid with a characteristic odour of thymol and a bitter taste. It can be mixed with water in a clear to opalescent liquid.

Identity: A. 1 ml of the extract is diluted with water to 50 ml. After the addition of 0.15 ml of a solution of potassium ferricyanide a green-brownish colour develops and later a flaky, brown precipitate.

B. Thin-layer chromatography: (see "Thymi extractum fluidum", the author).

Test: Content of ethanol: 35 to 45 per cent (V/V).

Methanol, 2-Propanol: Not more than 0.05 per cent (V/V).

Assay: Spectralphotometric evaluation of thymol after reaction with aminopyrazolone and potassium ferricyanide (see "Thymi extractum fluidum", the author).

Storage: In tight bottles, light protected.

Thymiansirup – Thyme syrup – Thymi sirupus

Definition: Thymi sirupus contains not less than 0.013 and not more than 0.017 per cent (m/m) phenols, expressed as thymol ($C10H14O$; M 150.2).

Production: 500 g saccharose is dissolved in 300 g purified water by warming on a water bath. After cooling the water loss must be replaced. Afterwards 150 g Thymi extractum liquidum normatum, and a solution of 0.10 g thymol in 40 g ethanol 96 per cent is added.

Characteristics: Clear, light-brown liquid with an odour of thyme; miscible with water, ethanol 70 per cent and ethanol 96 per cent.

Identity: A. Thin-layer chromatography: (see "Thymi extractum fluidum", the author).

B. 1 ml sirup is mixed with 10 ml water. An aliquote of 0.05 ml of this solution is warmed with 0.5 g resorcinol and 2.5 ml hydrochloric acid. After 10 min a dark red colour develops.

Test: Relative Density: 1.20 to 1.23

Refractive Index: 1.420 to 1.430

Preservative: not allowed.

Assay: Spectralphotometric evaluation of thymol after reaction with aminopyrazolone and potassium ferricyanide (see "Thymi extractum fluidum", the author).

Storage: In tight bottles, light protected.

WHO monographs

Herba (Part 1)

Plant material of interest: dried leaves and flowering tops of thyme.

General appearance: Same test as in the Pharmacopoeia Europae.

Organoleptic properties: Odour and taste aromatic (Pharmacopoeia Europaea 1995, Materia medika Indonesia 1980; British Herbal Pharmacopoeia 1979; Youngken, 1950).

Microscopic characteristics: In leaf upper epidermis, cells tangentially elongated in transverse section with a thick cuticle and few stomata, somewhat polygonal in surface section with beaded vertical walls and striated cuticle, the stoma being at a right angle to the two parallel neighbouring cells. Numerous unicellular, non-glandular hairs up to 30 μm in length with papillose wall and apical cell, straight, or pointed, curved, or hooked. Numerous glandular hairs of two kinds, one with a short stalk embedded in the epidermal layer and a unicellular head, the other with an 8- to 12-celled head and no stalk. Palisade parenchyma of two layers of columnar cells containing many chloroplastids; occasionally an interrupted third layer is present. Spongy parenchyma of about 6 layers of irregular-shaped chlorenchyma cells and intercellular air-spaces (Youngken, 1950).

Powdered plant material: Grey-green to greenish-brown powder; leaf fragments, epidermal cells prolonged into unicellular pointed, papillose trichomes, 60 μm long; trichomes of the lower surface uniseriate, two- to three-celled, sharp pointed, up to 300 μm in diameter, numerous labiate trichomes with 8 to 12 secretory cells up to 80 μm in diameter; broadly elliptical caryophyllaceous stomata. Six- to eight-celled uniseriate trichomes from the calyx up to 400 μm long; pollen grains spherical; pericyclic fibres of the stem (British Herbal Pharmacopoeia 1979; Materia medika Indonesia 1980; Pharmacopoeia Europaea 1995).

Geographical distribution: Indigenous to southern Europe. It is a pan-European species that is cultivated in Europe, the United States of America and other parts of the world (Youngken, 1950; British Herbal Pharmacopoeia 1979; Materia medika Indonesia 1980; Van den Broucke and Lemli, 1983).

General identity tests: Macroscopic and microscopic examinations (Youngken, 1950; Pharmacopoeia Europaea, 1995;), and chemical and thin-layer chromatography tests for the characteristic volatile oil constituents, thymol.

Purity tests: Microbiology: The test for *Salmonella* ssp. in Herba Thymi products should be negative. The maximum acceptable limits of other microorganisms are as follows (Deutsches Arzneibuch 1996; Pharmacopoeia Europaea 1997; WHO, 1998). For preparation of infusion: aerobic bacteria – not more than 107/g; fungi – not more than 105/g; *Escherichia coli* – not more than 102/g. Preparations for oral use: aerobic bacteria – not more than 105/ml; fungi – not more than 104/ml; enterobacteria and certain Gram-negative bacteria – not more than 103/ml; *Escherichia coli* – 0/ml.

Foreign organic matter: Not more than ten per cent of stem having a diameter up to 1 mm. Leaves with long trichomes at their base and with weakly pubescent other parts not allowed. The leaves and flowering tops of *Origanum creticum* or *O. dictamnus* are considered adulterants (British Herbal Pharmacopoeia 1979; Youngken, 1950). Other foreign organic matter, not more than two per cent (Materia medika Indonesia 1980).

Total ash: Not more than 15 per cent (Pharmacopoeia Europaea 1995).

Acid-insoluble ash: Not more than 2.0 per cent (Pharmacopoeia Europaea 1995).

Moisture: Not more than ten per cent (Pharmacopoeia Europaea 1995).

Pesticide residues: To be established in accordance with national requirements. Normally, the maximum residue limit of aldrin and dieldrin in Herba Thymi is not more than 0.05 mg/kg (Pharmacopoeia Europaea 1997). For other pesticides, see WHO guidelines on quality control methods for medicinal plants (WHO, 1998) and guidelines for predicting dietary intake of pesticide residues (WHO, 1997).

Heavy metals: Recommended lead and cadmium levels are not more than 10.0 and 0.3 mg/kg, respectively, in the final dosage form of the plant material.

Radioactive residues: For analysis of strontium-90, iodine-131, caesium-137, and plutonium-239, see WHO guidelines on quality control methods for medicinal plants (WHO, 1998).

Other purity tests: Chemical, alcohol-soluble extractive, and water-soluble extractive tests to be established in accordance with national requirements.

Chemical assays: Herba Thymi contains not less than 1.0 per cent volatile oil, and not less than 0.5 per cent phenols. Volatile oil is quantitatively determined by water/steam distillation, and the percentage content of phenols expressed as thymol is determined by spectrophotometric analysis. Thin-layer chromatographic analysis is used for thymol, carvacrol, and linalool (Pharmacopoeia Europaea 1995; Twetman *et al.*, 1995).

German Homoeopathic Pharmacopoeia – Homöopathisches Arzneibuch 2000

The monographs of the Homoeopathic Pharmacopoeia of Germany characterise the quality of ingredients of homoeopathic drugs prepared from plants, animals, minerals and also synthetic drugs. The test methods and quality requirements are identical with those of the DAB and the Pharmacopoeia Europaea, respectively. In homoeopathy, plants are mostly used in the form of tincture standards (German: Urtinktur) and liquid decimal dilutions (German: Decimalpotenz). Within the Homoeopathic Pharmacopoeia thyme is represented with 2 monographs, *T. vulgaris* and *T. serpyllum*. The tincture standards from both plants are prepared by a ten days maceration at room temperature of homogenised fresh plant material with ethanol 86 per cent (depending on the water content of the fresh plant from 1.0 to 1.3:1, ethanol content in the end product about 60 per cent). The D1 dilutions are prepared by mixing 3 parts of the tincture standards with 7 parts ethanol 62 per cent (D1). D2 dilutions are prepared by mixing 1 part D1 with 9 parts ethanol 62 per cent, D3 dilutions by mixing 1 part D2 with 9 parts ethanol 62 per cent, and so on. For the quality control of the plant material a description of the plants is given more or less identical to the descriptions in the DAB and the Pharmacopoeia Europaea, respectively.

The liquid standards of both plants are described as brown liquids with an aromatic smell and taste. Identity control of the liquid standards is performed by mixing them with water and iron(III) chloride yielding a characteristic green colour. A mixture of the liquid standards (0.5 ml) with water (10 ml) develops a characteristic blue colour when a ten per cent solution of sodium carbonate (0.1 ml) and a two per cent solution of dichlorochinone chlorimide in ethanol (0.1 ml) is added. Further evidence of the identity is given when a chromatographic separation of the apolar fraction on TLC according to the methods described in the DAB and the Pharmacopoeia Europaea, respectively, is performed. The assays concern the relative density of the liquid standards (for both herbs 0.900 to 0.920) and the residue on drying (1.4 per cent for *T. vulgaris*, 1.2 per cent for *T. serpyllum*). The liquid standards have to be stored light-protected.

CURRENT SAMPLE PRODUCT FORMULATIONS IN INDUSTRIES

The "Rote Liste 1999", a list of drugs sold on the German market, supplies about 100 "hits" of different drugs which contain thyme as an active principle. Such phytopreparations not only contain thyme as an herb but also in the form of dry extracts, liquid extracts (tincture, fluid-extract), semi-solid extracts, pressed juice from the fresh plant, and homoeopathic tinctures. Also the essential oil of thyme alone can be a constituent of those drugs. The uses of phytopreparations of thyme are mostly given with "for treatment of catarrhs of the upper respiratory tract, of bronchial catarrh, pertussis and hoarseness; they act as expectorants". Only four phytopreparations containing thyme show other applications: in mixtures with other herbal drugs as a roburant and for homoeopathic indications.

Internal application

For internal application liquids and syrups are the most usual formulations which are administered drop by drop or by the spoonful depending on the concentration of the active principle. For such formulations mostly fluid-extracts of thyme are processed with a herb to liquid ratio of 1:2.5 or 1:3 according to the pharmacopoeial instructions "Thymianfluidextrakt" in the German and Swiss pharmacopoeia (ethanol content 35–45 per cent). According to modern knowledge which prefers monopreparations to combinations, many of the liquid phytopreparations (18 of 100) contain thyme extracts as the only active compound. But there are also several combinations with other herbal extracts from plants which are known to be effective against catarrhs of the upper respiratory tract listed in Table 11.1. In this respect the most important herbs are *Primula* root, ivy leaf or *Drosera* herb. We also can find thyme tinctures (herb to extract ratio 1:5) as the active principle in liquids and syrups. One of these preparations

Table 11.1 Thyme liquid extracts in product formulations including combinations with other herbal products

Herbal products	Drops, syrups, and liquids														Bath additives		
Thyme fluid extract	x	x	x	x	x	x	x	x	x	x	x	x	x	x	x	x	x
Anise oil											x	x		x			
Camphor			x														x
Drosera					x		x										
Eucalyptus leaf extract												x					
Eucalyptus oil												x					x
Fennel oil												x		x			
Horsetail herb													x				
Ivy leaf									x	x					x		
Khella fruit												x					
Liquorice root										x				x			
Marshmallow root					x												
Peppermint oil												x					
Pine oil												x					
Plantain herb							x					x					
Primula root							x		x			x		x			
Snail extract																x	
Turpentine oil												x					
Verbascum flower												x	x				
Ephedrin												x	x				
Guaifenesin		x	x														
Gypsophila saponin				x													
Castanea leaves													x				
Number of product formulations	17	1	1	1	1	1	2	5	2	1	2	1	1	1	1	1	1

Table 11.2 Thyme dry extract in product formulations including combinations with other herbal products

Herbal products	Tablets		Teas	Capsules			Dragees	Suppositories
Thyme dry extract	x	x	x	x	x	x	x	x
Anise oil					x			
Drosera							x	
Eucalyptus oil				x				
Ivy leaf			x					
Primula root dry extract		x		x	x	x	x	
Number of product formulations	3	4	1	1	1	1	1	1

combines thyme tincture with tinctures prepared from *Grindelia* herb, *Quebracho* bark, *Saponaria* root and *Primula* root. One product on the market contains the pressed juice from the fresh thyme plant, won by a very special procedure.

Dry extracts contain the effective principle of thyme in higher concentrations than the liquid extracts, because the usual herb to extract ratio is 6 to 10:1 meaning a 6- to 10-fold accumulation. Water, ethanol 70 per cent, and ethanol 96 per cent are used as the solvents. Dry extracts represent the effective principle of capsules, tablets, dragees, and suppositoria or instant teas. The dry extracts are processed alone or together with other plant extracts (Table 11.2). Semi-solid extracts of thyme can be found in 7 preparations in the form of liquids, syrups, and pastilles (herb to extract ratio 4.5 to 8:1, mostly 5.5:1). Again combinations with other plant extracts, e.g. from *Primula* root, *Drosera*, or combinations with the essential oils from anise, fennel, and *Eucalyptus* are common.

Perhaps due to the bitter taste of thyme, infusions (a hot tea prepared with thyme) are not very common. In order to improve the taste teas are offered in mixtures with other herbal drugs used for the same purpose, e.g. lime tree flowers and elder flowers, fennel fruits, *Primula* flowers and plantain herb (Table 11.3). The most complex tea in the market represents a mixture of thyme with fennel fruits, Iceland moss, mullein

Table 11.3 Herbal teas of thyme in product formulations including combinations with other herbal teas

Herbal products	Herbal teas		
Thyme	x	x	x
Lime tree flowers	x		x
Elder flowers	x		
Primula flowers		x	x
Plantain herb		x	
Fennel fruit		x	x
Iceland lichen			x
Mullein flowers			x
White deadnettle flowers			x
Herb of black knotweed			x
Marigold flowers			x
Raspberry leaves			x
Number of product formulas	1	1	1

Table 11.4 Thyme oil in product formulations including combinations with other oils or chemicals

Herbal products	Bath additives					Balsams				Syrups	Ointments	Instant teas	Nasal Ointments
Thyme oil	x	x	x	x	x	x	x	x	x	x	x	x	x
Anise oil											x	x	
Camphor			x					x			x		
Coal tar				x									
Clove oil					x								
Conifer oil											x		
Dwarf pine oil										x			
Liquorice												x	
Marshmallow												x	
Peppermint oil	x												x
Pine oil		x				x							
Pine oil, Sibirian										x			
Rosemary oil									x				
Primula root extract												x	
Thymol											x		
Turpentine										x			
Levomenthol							x						
Dihydrocodein										x			
Number of product formulas	3	1	1	1	1	1	1	1	1	1	1	1	1

flowers, lime tree flowers, *Primula* flowers, white deadnettle herb, herb of black knot-weed, marigold flowers, and raspberry leaves.

External application

For external applications thyme oil is exclusively used against colds in bath additives, balsams, ointments and ointments for the nose. Especially the bath additives contain other essential oils, mostly *Eucalyptus* oil or camphor (Tables 11.1; 11.4). During application the oil is thought to reach the nose when warmed on the skin and to penetrate into the upper bronchial tract, and there it develops its disinfectant effect.

REFERENCES

Monographs

Commission E monographs: English version: Blumenthal, M. (1998) *The complete German Commission E Monographs – Therapeutic guide to herbal medicines.* American Botanical Council, Austin, Texas; Integrative Medicine Communications, Boston, Massachusetts.

ESCOP Monographs: Fascicules 1 and 2 (1996), Fascicules 3–5 (1997), Fascicule 6 (1999), can be obtained from ESCOP Secretariat, Argyle House, Gandy Street, Exeter EX4 3LS, United Kingdom.

Rote Liste 1999, Editio Cantor Verlag GmbH, Aulendorf.

WHO monographs on selected medicinal plants, Vol. 1, World Health Organization, Geneva, 1999.

Pharmacopoeias

Pharmacopoeia Europaea 3rd ed., Supplement 2001, Council of Europe, Strasbourg Cedex.

Deutsches Arzneibuch, Ausgabe 2000, Deutscher Apotheker Verlag, Stuttgart.

Homöopathisches Arzneibuch, Ausgabe 2000, Deutscher Apotheker Verlag, Stuttgart.

Pharmacopoea Helvetica 8, Suppl. 2000, Eidgenössische Drucksachen- und Materialzentrale, Bern.

References quoted in the ESCOP- and WHO-monographs

European Pharmacopoeia 2nd (ed.) 1995, Council of Europe, Strasbourg Cedex.

European Pharmacopoeia 3rd (ed.) 1997, Council of Europe, Strasbourg Cedex.

Materia medika Indonesia. IV Departemen Kesehatan, Jilid. Jakarta, Republik Indonesia,1980.

British Herbal Pharmacopoeia, Part 2. British Herbal Medicine Association, London, 1979.

Deutsches Arzneibuch 1996, Deutscher Apotheker Verlag, Stuttgart.

Adzet, T., Granger, R., Passet, J. and San Martín, R. (1977) Le polymorphisme chimique dans le genre *Thymus*: sa signification taxonomique. *Biochem. Syst. Ecol.*, 5, 269–272.

Adzet, T., Vila, R. and Cañigueral, S. (1988) Chromatographic analysis of polyphenols of some Iberian Thymus. *J. Ethnopharmacol.*, 24, 147–154.

Allegrini, J. and Simeon de Bouchberg, M. (1972) Une technique d'étude du pouvoir antibactérien des huiles essentielles. *Prod. Probl. Pharm.*, 27, 891–897.

Azizan, A. and Blevins, R.D. (1995) Mutagenicity and antimutagenicity testing of six chemicals associated with the pungent properties of specific spices as revealed by the Ames *Salmonella* microsomal assay. *Arch. Environ. Contam. Toxicol.*, 28, 248–258.

Bruneton, J. (1995) *Pharmacognosy, Phytochemistry, Medicinal Plants.* Lavoisier, Paris.

Chalchat, J.C. and Garry, R.Ph. (1991) Corrélation composition chimique/activité antimicrobienne: V – contribution à la comparaison de 2 méthodes de détermination des CMI. *Plant. Méd. Phytothér.*, 22, 195–202.

Czygan, F.C. (1989) Thymian. In M. Wichtl (ed.), *Teedrogen.* Wissenschaftliche Verlagsgesellschaft mbH, Stuttgart, pp. 498–500.

Deans, S.G. and Ritchie, G. (1987) Antibacterial properties of plant essential oils. *Int. J. Food Microbiol.*, 5, 165–180.

Dorsch, W., Loew, D., Meyer, E. and Schilcher, H. (eds) (1993) *Empfehlungen zu Kinderdosierungen von monographierten Arzneidrogen und ihren Zubereitungen.* Kooperation Phytopharmaka, Bonn, pp. 100–101.

Englberger, W., Hadding, U., Etschenberg, E., Graf, E., Leyck, S., Winkelmann, J. and Parnham, M.J. (1988) Rosmarinic acid: a new inhibitor of complement C3-convertase with antiinflammatory activity. *Int. J. Immunopharmac.*, 10, 729–737.

Farag, R.S., Salem, H., Badei, A.Z.M.A. and Hassanein, D.E. (1986) Biochemical studies on the essential oils of some medicinal plants. *Fette Seifen Anstrichm.*, 88, 69–72.

Farnsworth, N. R. (1995) NAPRALERT database. University of Illinois at Chicago, IL, Chicago. An on-line database available directly through the University of Illinois at Chicago or through the Scientific and Technical Network (STN) of Chemical Abstract Services.

Freytag, A. (1933) Über den Einfluß von Thymianöl, Thymol und Carvacrol auf die Flimmerbewegung. *Pflügers Archiv; Europ. J. Physiol.*, 232, 346–350.

García Marquina, J.M. and Gallardo Villa, M. (1949) Saponinas del *Thymus vulgaris* L. *Farmacognosia (Madrid)*, 9, 261–276.

Ghazanfar, S.A. (1994) *Handbook of the Arabian Medicinal Plants.* CRC Press, Boca baton, Fl, p. 128

Gordonoff, T. and Merz, H. (1931) Über den Nachweis der Wirkung der Expektorantien. *Klin. Wochenschr.*, 10, 928–932.

Gracza, L., Koch, H. and Löffler, E. (1985) Über biochemisch-pharmakologische Untersuchungen pflanzlicher Arzneistoffe, 1.Mitt.: Isolierung von Rosmarinsäure aus *Symphytum officinale* und ihre anti-inflammatorische Wirksamkeit in einem *In-vitro*-Modell. *Arch. Pharm.*, 317, 222–228.

Hegnauer, R. (1966) *Chemotaxonomie der Pflanzen.* Vol. IV, Birkhäuser Verlag, Basel, p. 321 and 328.

Hochsinger, K. (1931) Die Therapie des Krampf- und Reizhustens. *Wiener Med. Wschr.*, 13, 447–448.

Janssen, A.M. (1989) *Antimicrobial activities of essential oils.* Thesis, Leiden: Rijksuniversiteit te Leiden, University of Leiden, 91–108.

Janssen, A.M., Chin, N.L.J., Scheffer, J.J.C. and Baerheim Svendsen, A. (1986) Screening for antimicrobial activity of some oils by the agar overlay technique. *Pharm. Weekbl., Scientific Edition*, 8, 286–292.

Janssen, A.M., Scheffer, J.J.C. and Baerheim Svendsen, A. (1987) Antimicrobial activity of essential oils: a 1976–1986 literature review. Aspects of the test methods. *Planta Med.*, 53, 395–398.

Juven, B.J., Kanner, J., Schved, F. and Weisslovicz, H. (1994) Factors that can interact with the antibacterial action of thyme essential oil and its active constituents. *J. Appl. Bacteriol.*, 76, 626–631.

Lamaison, J.L., Petitjean-Freytet, C. and Carnat, A. (1990) Teneurs en acide rosmarinique, en dérivés hydroxycinnamiques totaux et activité antioxydante chez les Apiacées, les Boriginacées et les Lamiacées médicinales. *Ann. Pharm. Fr.*, 48, 103–108.

Lens-Lisbonne, C., Cremieux, A., Maillard, C. and Balansard, G. (1987) Methods for evaluation of the antibacterial activity of essential oils: application to essences of thyme and cinnamon. *J. Pharm. Belg.*, 42, 297–302.

Litvinenko, V.I., Popova, T.P., Simonjan, V., Zoz, I.G. and Sokolov, V.S. (1975) "Gerbstoffe" und Oxyzimtsäureabkömmlinge in Labiaten. *Planta Med.*, 27, 372–380.

Menghini, A., Savino, A., Lollini, M.N. and Caprio, A. (1987) Activité antimicrobienne en contact direct et en micro-atmosphère de certaines huiles essentielles. *Plant. Méd. Phytothér.*, 21, 36–42.

Mossa, J.S., Al-Yahya, M.A. and Al-Meshal, I.A. (1987) *Medicinal Plants of Saudi Arabia*. Vol. 4, King Saud University Libraries, Riyadh, Saudi Arabia.

Nakatani, N., Miura, K. and Inagaki, T. (1989) Structure of new deodorant biphenyl compounds from Thyme (*Thymus vulgaris* L.) and their activity against methyl mercaptan. *Agric. Biol. Chem.*, 53, 1375–1381.

Natake, M., Kanazawa, K., Mizuno, M., Ueno, N., Kobayashi, T., Danno, G. and Minamoto, S. (1989) Herb water-extracts markedly suppress the mutagenicity of Trp-P-2. *Agric. Biol. Chem.*, 53, 1423–1425.

Paster, N., Menasherow, M., Ravid, U. and Juven, B.(1995) Antifungal activity of oregano and thyme essential oil applied as fumigants against fungi attacking stored grain. *J. Food Prot.*, 58, 81–85.

Patáková, D. and Chládek, M. (1974) über die antibakterielle Aktivität von Thymian- und Quendelölen. *Pharmazie*, 29, 140–143.

Perrucci, S., Cecchini, S., Pretti, C., Varriale Cognetti, A.M., Macchioni, G., Flamini, G. and Cioni, P.L. (1995) *In vitro* antimycotic activity of some natural products against *Saprolegnia ferax*. *Phytotherapy Res.*, 9, 147–149.

Petersson, L.G., Edwardsson, S. and Arends, J. (1992) Antimicrobial effect of a dental varnish, *in vitro*. *Swed. Dent. J.*, 16, 183–189.

Qureshi, S., Shah, A.H., Al-Yahya, M.A. and Ageel, A.M. (1991) Toxicity of *Achillea fragrantis-sima* and *Thymus vulgaris* in mice. *Fitoterapia*, 62, 319–323.

Reiter, M. and Brandt, W. (1985) Relaxant effects on tracheal and ileal smooth muscle of the guinea pig. *Arzneim. Forsch.*, 35, 408–414.

Samejima, K., Kanazawa, K., Ashida, H. and Danno, G. (1995) Luteolin, a strong antimutagen against dietary carcinogen, Trp-P-2, in peppermint, sage, and thyme. *J. Agric. Food Chem.*, 43, 410–414.

Schilf, F. (1932) Einfluß von Acetylcholin, Adrenalin, Histamin und Thymianextrakt auf die Bronchialschleimhautsekretion; zugleich ein Beitrag zur Messung der Bronchialschleimhaut-sekretion. *Naunyn-Schmiedeberg's Arch. Pharmacol.*, 166, 22–25.

Simeon de Bouchberg, M., Allegrini, J., Bessiere, C., Attisso, M., Passet, J. and Granger, R. (1976) Propriétés microbiologiques des huiles essentielles de chimiotype de *Thymus vulgaris* L. *Riv. ital. EPPOS*, 58, 527–536.

Skopp, K. and Hörster, H. (1976) An Zucker gebundene reguläre Monoterpene Teil I. Thymol- und Carvacrolglykoside in *Thymus vulgaris*. *Planta Med.*, 29, 208–215.

Stahl-Biskup, E. (1991) The chemical composition of *Thymus* oils: a review of the literature 1960–1989. *J. Essent. Oil Res.*, 3, 61–82.

Tantaoui-Elaraki, A. and Errifi, A. (1994) Antifungal activity of essential oils when associated with sodium chloride of fatty acids. *Grasas y aceites*, 45, 363–369.

Twetman, S., Hallgren, A. and Petersson, L.G. (1995) Effect of antibacterial varnish on mutant *Streptococci* in plaque from enamel adjacent to orthodontic applances. *Caries Res.*, 29, 188–191.

Vampa, G., Albasini, A., Provvisionato, A., Bianchi, A. and Melegari, M. (1988) Études chimiques et microbiologues sur les huiles essentielles de *Thymus*. *Plant. Méd. Phytothér.*, 22, 195–202.

Van den Broucke, C.O., Lemli, J.A. and Lamy, J. (1983) Spasmolytic action of the flavones of different species of *Thymus*. *Plant. Méd. Phytothér.*, 16, 310–317.

Van den Broucke, C.O. (1980) Chemical and pharmacological investigation on thymi herba and its liquid extracts. *Planta Med.*, 39, 253–254.

Van den Broucke, C.O., Dommisse, R.A., Esmans, E.L. and Lemli, J.A. (1982) Three methylated flavones from *Thymus vulgaris*. *Phytochemistry*, 21, 2581–2583.

Van den Broucke, C.O. and Lemli, J.A. (1981) Pharmacological and chemical investigation of Thyme extracts. *Planta Med.*, 41,129–135.

Van den Broucke, C.O. and Lemli, J.A. (1983) Spasmolytic activity of the flavonoids from *Thymus vulgaris*. *Pharm. Weekbl. Sci.*, 5, 9–14.

Van den Dries, J.M.A. and Baerheim Svendsen, A. (1989) A simple method for detection of glycosidic bound monoterpenes and other volatile compounds occurring in fresh plant material. *Flavour Fragr. J.*, 4, 59–61.

Van Hellemont, J. (1988). In J. Van Hellemont (Ed.), *Fytotherapeutisch compendium*. 2nd (ed)., Bohn, Scheltema & Holkema, Utrecht, pp. 599–605.

Vollmer, H. (1932) Untersuchungen über Expektorantien und den Mechanismus ihrer Wirkung. *Klin. Wochenschr.*, 11, 590–595.

Vollon, C. and Chaumont, J.P. (1994) Antifungal properties of essential oils and their main components upon *Cryptococcus neoformans*. *Mycopathology*, 128, 151–153.

Von Skramlik, E. (1959) über die Giftigkeit und Verträglichkeit von ätherischen Ölen. *Pharmazie*, 14, 435–445.

Wagner, H., Wierer, M. and Bauer, R. (1986) *In vitro*-Hemmung der Prostaglandin-Biosynthese durch ätherische Öle und phenolische Verbindungen. *Planta Med.*, 52, 184–187.

Weiss, B. and Flück, H. (1970) Untersuchungen über die Variabilität von Gehalt und Zusammensetzung des ätherischen Öles in Blatt- und Krautdrogen von *Thymus vulgaris* L. *Pharm. Acta Helv.*, 45, 169–183.

World Health Organization (1997) Guidelines for predicting dietary intake of pesticide residues. (unpublished document WHO/FSF/FOS/97.7; available from Food Safety WHO, 1211 Geneva 27, Switzerland).

World Health Organization (1998) Quality control methods for medicinal plant materials. WHO, Geneva.

Wüthrich, B., Stäger, P. and Johannson, S.G.O. (1992) Rast-specific IGE against spices in patients sensitized against birch pollen, mugwort pollen and celery. *Allergologie*, 15, 380–383.

Youngken H.W. (1950) *Textbook of pharmacognosy*. 6th (ed). Blakiston, Philadelphia.

Zani, F., Massimo, G., Benvenuti, S., Bianchi, A., Albasini, A., Melegari, M., Vampa, G., Bellotti, A. and Mazza, P. (1991) Studies on the genotoxic properties of essential oils with *Bacillus subtilis* rec-assay and *Salmonella* microsome reversion assay. *Planta Med.*, 57, 237–241.

Index

(Continued)